# Scurvy

# SCURVY

## THE DISEASE OF DISCOVERY

*Jonathan Lamb*

WITH A CODA WRITTEN BY
JAMES MAY AND FIONA HARRISON

PRINCETON UNIVERSITY PRESS   PRINCETON & OXFORD

press.princeton.edu

Jacket art by Port Jackson Painter, c. 1788–98. Detail from *Aborigines Attacking a Sailor*.
Courtesy of the Natural History Museum (London).
Graphic courtesy of Shutterstock
Jacket design by Pamela Lewis Schnitter

Library of Congress Cataloging-in-Publication Data

Names: Lamb, Jonathan, 1945– , author.
Title: Scurvy : the disease of discovery / Jonathan Lamb ; with a coda written by James May and
Fiona Harrison.
Description: Princeton : Princeton University Press, [2017] | Includes bibliographical references and
index.
Identifiers: LCCN 2016010625 | ISBN 9780691147826 (hardcover : alk. paper)
Subjects: | MESH: Scurvy—history | Scurvy—complications | Expeditions—history | Confusion |
History, Modern 1601–
Classification: LCC RC627.S36 | NLM WD 140 | DDC 616.3/94–dc23 LC record available at
http://lccn.loc.gov/2016010625

British Library Cataloging-in-Publication Data is available

This book has been composed in Garamond Premier Pro

Printed on acid-free paper. ∞

Printed in the United States of America

1  3  5  7  9  10  8  6  4  2

To the memory of

Sir James Watt, KBE FRCS (1914–2009),

physician, sailor, and scholar

The number of people who perish annually at sea, by famine [and] the scurvy … affords matter for another shocking calculation.

—J. J. Rousseau, Appendix to *A Discourse on the Origin of Inequality* (1755)

She soon grew pregnant & brought forth
Scurvy and spott'd fever.
The father grinn'd & skipt about,
And said, "I'm made for ever!

For now I have procur'd these imps
I'll try experiments."
With that he tied poor scurvy down
& stopt up all its vents.

And when the child began to swell,
He shouted out aloud,
"I've found the dropsy out, & soon
Shall do the world more good."

—William Blake, *An Island in the Moon* (1787)

# CONTENTS

# ILLUSTRATIONS

*Color plates of figures 2, 6, 13, 17, 21, 22, and 26 follow page 150*

# ACKNOWLEDGMENTS

There are many people and institutions deserving thanks for the help they have given me during this enterprise for, had I gone without their support and guidance, the world would have missed another book. First of all, Vanderbilt University has been generous as always with leave and research financing. I am grateful to the National Maritime Museum for a Caird Fellowship, which launched my research and writing in 2011, and the John Simon Guggenheim Foundation for funding another year of leave that proved crucial to the momentum of the exercise. A Research Scholar's Grant from Vanderbilt was an invaluable resource, funding two trips to Australia and two more to Britain. I shared the grant with Fiona Harrison who, with her colleague Jim May, has been steadfast in support of the book since the very start. I have never worked with neuroscientists before, and this was an education for me not only in the biochemical mysteries of scurvy but also in the marriage of two widely differing disciplines. Thanks to Fiona and Jim, I now am familiar with the function of the choroid plexus and of many other moving parts of the body, especially the nerves, of which I had hitherto been entirely ignorant. Their contribution is evident throughout and, of course, clearly manifest in the last chapter. While I was on the Caird Fellowship, the National Maritime Museum very generously hosted a conference on scurvy. I thank Nigel Rigby for acting as coordinator of the event, Lisa de Jaeger for administering it, Robert Blyth for editing the proceedings, and Kevin Fewster and Margarette Lincoln for putting the Queen's House at our disposal. This came in the early days of the project and was a very great boost to it. Mark Harrison and Erica Charters, as well as participating in the conference, invited me to Oxford for two colloquia when I first arrived in Britain on the Caird, and their advice, as well as their work, has been extremely useful.

For over twenty years, all the work I've done on the sea has been heavily indebted to the knowledge and advice of Glyn Williams, an historian wonderfully patient with the more intuitive approaches to truth often taken by students of literature—some of them not strictly consistent with good historical practice. He read part of the manuscript when it was in the making and has offered invaluable help in its revision. His colleague and close friend, James Watt, was the first to investigate the possibility of significant psychological effects arising from nutritional deficits. It was he who suggested that

Captain James Cook's extravagant reactions to theft and disobedience during the third voyage might have been owing to pellagra. When I started work on scurvy, he was a generous and courteous source of encouragement: really a mainstay for what was still largely a conjectural project. Anthony Ossa-Richardson of Queen Mary University of London translated Walter Charleton's difficult medical Latin, a terrific help. The chapter on nostalgia has benefitted from more revision than any other, and this is in large part owing to the very careful reading and advice given by Kevis Goodman and Jonathan Schroeder. I am greateful for all they have given me in advice and bibliographical direction. On the topic of photic damage, a conversation I had with John Mollon of Caius College Cambridge was of immense value. While I was working on the heroic age of Antarctic exploration, Karen May helped me a great deal. Michael McKeon, Simon Schaffer, Peter de Bolla, Noelle Gallagher, Margaret Cohen, and latterly Alistair Sponsel have been great supporters. Two former graduate students, Adam Miller and Killian Quigley, have participated so much in colloquia and in the field generally that it would be hard to specify how much they have helped me develop ideas, especially about prosthesis and aesthetics. My research assistant Katie Miller has been patience itself, smiling at chaos. Never in my experience getting books ready for the press has the awkward business of copyediting been more graciously and promptly handled than by Kelly Clody, to whom my sincerest thanks are due.

The Australian section of the book I have found a very exciting topic. I owe a special thank you to Iain McCalman. At so many stages in my career, he has been an inspiration and an active supporter, no less in this. Alison Bashford assisted me greatly and continues to do so. Paul Pickering and Alex Cook have looked after me in Canberra, and Simon During and Lisa O'Connell in Brisbane. It was a great discovery to find that Michael Rosenthal, whom I used to meet frequently at the National Humanities Centre at the Australian National University, was working on a history of early Australian painting. He has been very generous with his material, particularly concerning the vexed identity of the Port Jackson Painter. I am indebted also to Jeanette Hoorn for information about this mysterious figure. On a brief trip to Sydney and Tasmania, I was treated so cordially by Jane Harrington and Susan Hood at the Port Arthur Historic Site, by Mary Casey, who took me on an archeological adventure to Parramatta, and by Kieran Hosty at the Sydney NMM, that I find it difficult to make an adequate acknowledgment of their kindness. It is such a pleasant memory; I am eager to renew these links.

To Bridget Orr, my wife, and Charlie Lamb, my son, I offer apologies for a scorbutic ellipsis that rendered me absent so often from their company and sometimes absent-minded when I was in it. I hope I am cured.

# ABBREVIATIONS

| | |
|---|---|
| HAJ | House of Assembly Journals, Hobart, Tasmania |
| HRA | Historical Records of Australia |
| HRNSW | Historical Records of New South Wales |
| LC | Christopher Lloyd and Jack Coulter |
| RCA | Report of the Committee of the Admiralty |
| RCPE | Library of the Royal College of Physicians at Edinburgh |
| RSCHC | Report of the Select Committee of the House of Commons |
| TNA | The National Archives, Kew |
| TSA | Tasmanian State Archives, Hobart |

# SCURVY

# Prolegomena

I saw those things, which the rude Mariner
(Who hath no Mistress but Experience)
Doth for unquestionable Truths aver,
Guided belike by his external sence:
But Academicks (who can never err,
Who by pure Wit and Learning's quintessence,
Into all Nature's secrets dive and pry)
Count either Lyes, or coznings of the Eye.
—Luiz Vaz de Camoens, *The Lusiad*, Richard Fanshawe's
translation, (1655)

Death is the penaltie impos'd, beware,
And govern well thy appetite, least sin
Surprise thee, and her black attendant Death.
—John Milton, *Paradise Lost* (1674)

T his book grew out of a chapter in an earlier book on the South Seas,
where it was placed as a kind of pendant to another on leprosy. Scurvy
being a disease afflicting voyagers in the Pacific, and leprosy an imported one
that devastated indigenous populations, they combined to show how vulnerable humans were in the eighteenth century to the effects of imperial expansion, whether they were its agents or its victims. I thought of writing a book
on scurvy in the early 2000s and signed a contract with a British press spe-
cializing in history titles, but procrastination put it out of reach. When I was
asked at the end of the decade to sign a contract for a book to be entitled
"Scurvy, the Disease of Discovery," it seemed like a happy rejuvenation of an

old idea, and I agreed. At first I thought the title laid down a pretty straightforward track through the history of scurvy—an illness that accompanied the extreme conditions incident to navigating the ends of the known world and that was gradually banished as nutrition and hygiene on ships grew more sophisticated. But it didn't take long to understand what many people had understood before, namely, that the course of scurvy, like that of true love, was nowhere near so smooth.

While writing the book, I have been teased by the double meaning lurking in the word "discovery," especially with regard to James Cook's first voyage into the Pacific on the *Endeavour* (1768–71). This was the first time a British naval ship had sailed there with a set of instructions supplied not just by the Admiralty but also with a set of hints drawn up by the Royal Society. Between them, they set out specific goals, cartographical and scientific, including mapping the coastline of New Zealand, finding out whether there was a Great Southern Continent, measuring the transit of Venus, and studying the customs of the Tahitians, whose islands had been discovered the previous year by Samuel Wallis in the *Dolphin*. The vessel was also stocked with a variety of foods supplied by the Victualling Board, all thought to have some antiscorbutic value—citrus juice, dried beans, portable soup, sauerkraut, vitriol, and malt wort—along with a direction to pay particular attention to the malt. On the one hand, Cook was charged with a set of objective tasks of discovery, some definite and some speculative, and on the other, a set of scientific experiments and ethnographic observations, among which was a nutritional trial that was to take place on the ship's company, a self-experiment as it were. Discovery was divided then between the external world and the space of the vessel itself, but there was no friction between the two enterprises, for Cook surpassed all expectations. His discoveries of new lands and archipelagos was a matchless feat of navigation, the measurement of the transit of Venus was reckoned a success, and on the basis of William Perry's clinical trial during the passage, malt wort was recommended as the leading antiscorbutic, for no one during this arduous voyage had died of scurvy.

But in his next voyage (1772–75), Cook succumbed to an illness whose symptoms, if not scorbutic, almost certainly were owing to vitamin deficiency; many of his crew and all of his scientists exhibited signs of scurvy; it was particularly bad on his consort, the *Adventure*, commanded by Tobias Furneaux. This was not surprising, for the search for a Great Southern Continent had taken the two ships on great loops into Antarctic waters, where for months on end no fresh food was to be obtained. Johann Reinhold Forster's journal exhibits with great vividness a rapidly widening gap between sentiments appropriate to the public mission of discovery and the reactions of an experimental gentleman to the failure of the nutritional experiment conducted on his own body. They are shrill, voluble, personal, and urgent. In these circumstances, scurvy acquires an importance of its own. It is not a discovery according to an established plan of public policy but rather an unfortunate event taking place in the organs, bones, and flesh of the voyager's own body, often privately lamented and never easily communicable. The *Resolution* voyage was the one Samuel Coleridge relied on most heavily for *The Rime of the Ancient Mariner* (1798), the very same on which his schoolmaster William Wales discovered to his chagrin that he had scurvy. The poem is exclusively concerned with this latter kind of discovery, delivered as a strangled confession in which scorbutic symptoms function not as a frame but as the very substance and burden of the narrative. The purpose of the voyage is never so much as mentioned.

If there were two sorts of discovery going on—on the one hand, unrolling the map of the world and, on the other, revealing what had lain hidden in the sensations, passions, sickness, and contingent circumstances of the voyager—of what exactly did the second discovery consist? Well, scurvy was an inevitable accompaniment of long periods at sea. Although it was capricious with regard to the speed of its onset and the signs of its presence, no one could avoid it who lived for more than three months on preserved food; yet despite what many surgeons and physicians believed, it wasn't a disease a person caught, like yellow fever or typhus, it came about when the source of some vital principle in food ceased to be supplied to the body. That is to say, it was owing to an innate infirmity in the human constitution that the mere mass of food could not remedy. It was a sort of essential starvation. Specialists in gases like Thomas Beddoes even referred to it as suffocation, or dry drowning, believing that victims were suffering from a huge deficit of oxygen. The symptoms plied between horrid dissolution of the body and an extraordinary susceptibility of mind; they were evinced by magnified sensations and tearful melancholy, seasoned every now and then with visions of the right kind of food and sometimes the ravishing experience of tasting and swallowing it. So if scorbutic patients were to open their mouths, they would *discover* (in the sense of *reveal* as well as *find*) blackened gums and very loose teeth; or if they bared their legs, they would *discover* blood spots and ulcers. If they talked (and they were prone to be very circumstantial), they would spend many words trying unsuccessfully to provide a full tale of the spectrum of their miseries, exhibiting a weakness of mind and fraught emotions typical of the disease. Was this why scurvy could lay claim to being a disease of discovery, because in the effort of finding exotic lands and peoples, it came upon its victims like a surprise, manifest in their own bodies and feelings?

Scurvy's first attested appearance in literature is to be found in Luis Vaz de Camoens's *The Lusiads* (1572), a poem celebrating Vasco da Gama's expedition into the Indian Ocean, where the word "discover" is handled in the two different ways I have outlined. It bears the primary meaning of finding out by toil the remotest boundaries of the sea, but it has another deriving from the fact that the places da Gama discovered were already thriving ports, not at all receptive to overtures from a tattered mariner regarding trading treaties with Portugal. In rejecting da Gama's embassy, the king of Calicut demanded (according to the journal of the first voyage) why they came so far without gifts: "To this the captain rejoined that he had brought nothing because the object of his voyages was merely to make discoveries.... The king then asked what it was he [had] come to discover: stones or men?" (Ravenstein 1898: 62). Timothy Hampton draws this interesting conclusion: "What da Gama 'discovers' is, in effect, his own journey" (Hampton 2009: 110). What might have been sustained as a fiction of embassy is transformed into an exercise of self-authentication, da Gama's own greater truth concerning the accidents and perils of traversing an unknown sea: his imagined role as representative of the Portuguese Crown supplanted, like Othello's, by a tale of moving accidents (108). A significant element of that ordeal was the arrival of scurvy off the coast of Mozambique, which is introduced like this in Richard F. Burton's translation: "Who but eye-witness e'er my words could trust? / Of such disform and dreadful manner swole / The mouth and gums, that grew proud flesh in foyson / Till gangrene seemed the blood to poison, / Gangrene that carried foul and fulsome taint, / Spreading infection through the neighbouring air" (Camoens 1881: 5.81 [2.620]). At a critical moment in the unfolding of this voyage, the truth is available only to someone who was there, present at the emergency.

This embarrassment of the voyaging first-person narrative is so common in seagoing journals as to seem threadbare, yet in the company of scurvy, it deserves closer attention than it generally gets, for it brings the problem of discovery directly into the process of communication itself and sets a challenge for narrator and audience alike. For instance, when Captain Matthew Mitchell of the *Gloucester* describes men dying and dead of scurvy as his vessel vainly tries to make land in the South Seas, the rhetoric is familiar: "So miserable was the Scene, that words cannot Express the Misery that some of the Men dyed in" (Mitchell 1740–43: n.p.). What he discovers in his own experience, yet cannot discover as public testimony, is utterly unfamiliar, another kind of world of which we have no idea. In the fictions derived from voyages like da Gama's, such as Francis Bacon's *New Atlantis* (1627; from Ferdinand Magellan's [1525]) or Thomas More's *Utopia* (1516; from Amerigo Vespucci's [1505]), it is a licence for sheer imagination and used confidently by narrators such as Ralph Hythloday: "Yf you had bene with me in Utopia, and had presentelye sene their fashions and laws, as I dyd ... then doubtles you wolde graunt, that you never sawe people ordered, but onlye there" (More 1898: 53). But you weren't, and you can't. Usually the narrator is in a much more forlorn position, wanting sympathy that is impossible to procure. Here is Philip Carteret, a forerunner of Cook, at his wits end in the Pacific, the fabric of his ship disintegrating, and he and his men down with scurvy: "The ravages of the scurvy were now universal, there not being one individual among us that was free.... The mind participated in the sufferings of the body, and a universal despondency was reflected from one countenance to another.... And it is not perhaps very easy for the most fertile imagination to conceive by what our danger and distress could be increased; yet debilitated, sick, and dying as we were, in sight of land we could not reach, and exposed to tempests we could not resist, we had the additional misfortune to be attacked by a pirate" (Carteret in Hawkesworth 1773: 1.405–6). Whether a futile appeal for sympathy is made or transcended, readers find themselves on the near side of the boundary of discovery, usually content to leave it at that. An extraordinary example of what it is like to straddle that limit—to discover, if you like, its mutuality—is provided by Lemuel Gulliver, William Dampier's cousin, in the fourth book of his travels as he tries to mediate between the incredulity of the Houyhnhnms and the disbelief of his own people. Having given an account of how matters stand in Europe, particularly with regard to horses, he tells his host, who plainly thinks he is saying the thing that is not, that the story of this very encounter, if told in Britain, would be rejected out of hand as a lie. It is so singularly and privately true that neither the equine public nor the European is able to accept it as anything but a fiction. The circumstances of discovering that a truth for him is necessarily a barefaced lie among strangers and compatriots alike, leave Gulliver with no market for his greater truth. He discovers from his voyage only how absolute is his isolation from the two communities he tried to represent in the course of it.

The reason that scurvy is so often part of an incommunicable story has to do with the nature of the disease. Its victims are often described as urgently bent on communicating the least change in their feelings to anyone who will listen, betraying extreme impatience if no one obliges. Their surest audience, it might be supposed, would be found among their fellow sufferers, but Carteret shows why that is not so. On his crazy ship, so John Hawkesworth's official account goes, "despondency was reflected from one countenance to another." What Carteret actually wrote in his journal omitted the scene of a shared misery: "We were now in a sad deplorable situation[,] the Sickness was become general.... This occasioned a universal dejection onboard[,] particularly the Sick that were past coming upon deck" (Carteret 1965: 1.212). In a lightless forecastle, universal dejection doesn't forge a commonalty of woe by the exchange of glances. It consists of a crowd of individuals in the dark, simultaneously yet singly believing that there is no grief like their own. So the challenge I am willing to meet in writing a book about scurvy as a disease of discovery involves pushing at that door until I can discover what lies behind it.

Some time during the course of evolution, a liver enzyme necessary for the synthesis of vitamin C ceased to be produced in humans and anthropoid apes. Although they still possessed the gene that did this job for other animals, it had mutated to the point that all vitamin C had to be incorporated. This did not result in adverse selection, presumably because their diet was rich in the vitamin they needed. In parallel evolutionary events, the same genetic mutation occurred in guinea pigs, some species of birds, and fruit bats. Fish, on the other hand, have never needed the extra enzyme, deriving all the vitamin C they need from their environment. Humans, therefore, depend on food to sustain an adequate level of ascorbate (roughly 1500 mg in the body of a healthy adult). If it falls below 300 mg, signs of scurvy will appear.

Except in nonsense verse, apes and guinea pigs do not put to sea, but developments in the technologies of shipbuilding and navigation made it possible for humans to cross large expanses of ocean and finally to circumnavigate the globe. Generally it takes three months without fresh food to reduce the body's ascorbic acid to the level at which scurvy will supervene, roughly the time it took to get from the Atlantic into the Indian and Pacific Oceans. Sailors became aware how cruelly the failure of that mutant gene was to affect them when, like da Gama and Magellan, they made the breakthrough via the Cape of Good Hope or Cape Horn and found their limbs growing stiff and their skin bruised and ulcerous. Their gums grew black with corrupted blood and swelled so much the mortified flesh hid their teeth, which finally fell out when there was no longer anything solid in which to anchor them. As well as their bodies, their moods and minds were affected, alternating between blank lassitude and eager longings for the food they needed, or between fits of tears and pointless exuberance. Kenneth Carpenter estimates that scurvy caused two million deaths in navies and merchant marines between 1500 and 1800, ranking as the premier occupational disease of the great maritime era (Carpenter 1986: 253).

Land scurvy had been reported for centuries, occurring during famines, epidemics, and sieges, as well as in prisons and army camps. Lucretius handles its symptoms very accurately in the sixth book of *De rerum natura* when describing the plague at Athens. They are found again in the histories of the early settlement of Australia. But never had there been a shock to the public mind resembling the one caused by the return of the sole surviving ship of George Anson's squadron in 1744 after a four years' cruise in the South Seas. Although the expedition seized a fabulous amount of treasure, four ships and thirteen hundred sailors and marines had been lost, two-thirds of the total complement, the bulk of them to scurvy. When he became First Lord of the Admiralty, Anson was determined to have no repetition of this disaster. So for as long as was possible, he maintained a policy of supplying British fleets in home waters with fresh food; but he was dealing with an insoluble problem, not only logistically, for it was a hugely expensive task to send out tenders filled with livestock and greens to ships on station, but also medically, for no one knew for certain the cause of scurvy until two hundred years later. Moreover, its cure at sea was a matter of dispute right up until the discovery of vitamin C. Unless the nutritional conditions apt for living on land could be reproduced at sea, sailors on long voyages eating preserved food were bound to contract it.

The epidemiological impasse—whether scurvy was owing to the absence of something necessary for health, or to the presence in the air or diet of some malign gas or effluvium—induced an inevitable partiality in all concerned when it came to defending a favorite theory or practice. Meanwhile, the victims felt anxious and neglected, obsessing over their symptoms. Commanders were ashamed to have the disease on their ships and often took steps to have scurvy called by another name. Physicians and surgeons were roughly divided between those with a theory and those with a practical regimen, so they quarreled over rival treatments, sometimes claiming cures with medicines that could not possibly have worked, and sometimes recommending diets difficult to obtain on long voyages. Maritime historians, finding themselves unable to offer the reader a coherent account of scurvy's *longue duree*, have often settled for local examples of success. For instance, John Woodall, author of *The Surgeon's Mate* (1617), is singled out as the first naval surgeon fully to confront the puzzles of the disease while recognizing the unequivocal value of fresh fruit and vegetables in its prevention and cure. George Anson and Edward Hawke are praised as pioneers of the provisioning of naval vessels on blockade duty with fresh food, an expensive but reliable method of dealing with scurvy. Among the physicians most celebrated now, but neglected in his day, stands James Lind, whose clinical trials on scorbutic patients aboard the *Salisbury* and at the Haslar Naval Hospital proved beyond any doubt that the juice of oranges and lemons cured scurvy. Captain Cook, whose regimen of warm clothing, sufficient sleep, scrupulous attention to hygiene, and frequent landfalls combined to keep scurvy at bay during his first navigation of the South Seas, is generally praised most of all. Sir Gilbert Blane, who persuaded the Sick and Hurt Board of the British Admiralty to stock naval ships with lime juice, did much to save sailors in the Royal Navy from scurvy. Yet all of those remedies—fresh oranges and lemons, lemon juice preserved in bottles, dry clothes, regular cleaning and purging of the ship, personal hygiene, and frequent landfalls—had been known and recommended two hundred years earlier, for example, in the journals of James Lancaster and Richard Hawkins (Purchas 1906: 2.396; 17.77–78), meaning that discoveries of cures for scurvy were simply being lost or discredited and then made again. At the end of the nineteenth century, Sir Almroth Wright stood supreme among germ theorists, all of whom were convinced that scurvy was the result of eating tainted food. None of these interventions added up to a coherent account of the cure of the disease for reasons that will become apparent. Efforts to integrate the fragments of this history into a triumphal narrative make for rousing claims but partial arguments, such as those gathered under the subtitle of Stephen Bown's book, *How a Surgeon, a Mariner and a Gentleman Solved the Greatest Medical Mystery of the Age of Sail* (2003), where the fifty years dividing Lind from Blane are neatly edited to explain how the problem of scurvy was solved just in time for the Napoleonic Wars.

More to the point, however, are the reminders of the uneven and awkward rhythm of this tale of a disease not easily identified or readily curable. Thomas Trotter's astonishment at Lind's failure to draw the conclusion in favor of citrus juice as a preventive as well as a cure, which the Haslar Naval Hospital's trials had indicated; or Christopher Lloyd's and Jack Coulter's amazement that the delay in making it a standard preventive should be owing to Cook's unequivocal support for malt wort, a remedy of no use to a scorbutic patient; or Janet Macdonald's discovery that many British warships after 1795 were not supplied with enough antiscorbutic juice for a regular issue—these moments of incredulous recognition of the nondefinitive and nonheroic phases of the narrative are useful correctives to its triumphalist inflections. Readers who believe that the curse of scurvy at sea was indeed lifted by regular doses of preserved lime juice might ponder the following irony, fully worthy of the malady: namely, that the distribution of lime juice in all vessels flying the British flag was achieved in the very same decade (1870s) that lime juice was widely discredited and supplanted by a theory of toxic food that prevailed as medical orthodoxy until the discovery of vitamin C in the 1930s. At such time, another irony surfaces, for one of its discoverers, Tadeusz Reichstein, synthesized the vitamin in 1933 while analyzing nitrogenous cyclic compounds responsible for the flavors in coffee and for the toxic dust on the wings of monarch butterflies. He was not intent at all on discovering the chemical that would enfranchise humans to an unlimited stint upon the bosom of the ocean.

That is to say, a narrative of scurvy, whether found in a log or journal, a medical treatise or a general history, pretends to offer a coherent, faithful, and factual report in proportion as it elides the exceptions, contradictions, dissents, and interruptions that are inescapable even in a description of the disease. Trotter himself, surely one of its most acute observers, was driven to confess, "The Scurvy is attended by a train of symptoms peculiar to itself, and which the genius of the distemper has rendered extremely difficult to explain" (Trotter 1792: 149). The malign genius of scurvy, elsewhere called its je ne sais quoi, was sufficiently active to embarrass the descriptive resources of its victims. How frequently one is going to meet phrases such as this, which imprisons the narrator in a memory with which no reader will be capable of sympathizing: "Our mental sufferings were such as defy description, and nothing but being placed within the same situation could convince those who have not the power to imagine its monotonous dreariness" (Beale 1839: 310). The same scorbutic genius is likely to irritate any historian wishing to record a gradual accumulation of practical knowledge: "One of the most bewildering aspects of the history of scurvy," writes J. J. Keevil, "is the manner in which a cure was repeatedly found, only to be lost again" (Keevil 1957: 1.102).

In the West, we are used to approaching medical and technical breakthroughs by a series of promising steps that steadily advance toward a full disclosure of the useful truth. Heart disease and cancer have yielded a good deal of their malignity to the pressure of medical research. This is exactly what did not happen in the case of scurvy, although considerable investments were made to find a cure. When the biochemical function of ascorbate was finally revealed in the twentieth century, the prevailing view among doctors was hostile to any notion of nutritional deficit. Robert Falcon Scott's senior surgeon had laid it down for an axiom that there was no antiscorbutic property in any food or drug: the disease was introduced into the body on the vehicle of ptomaine poisoning. A medical officer of the Booth Shipping Line thought it arrived in the food chain from bacteria generated by a parasite living in the guts of cockroaches (Lloyd and Coulter 1963: 4.120). The only constant feature of this amazing conflict between prevailing orthodoxies and the facts was the long-running division between those who were attached to theories of deficiency and those who stood up for toxicity. One party was sure that scurvy occurred for the same reason as the suffocation of a living creature in a vacuum, namely, the loss of an invisible and unknown element critical to life that had to be restored if the organism was not to perish. The other was equally certain that it was a corruption of the body owing to inimical particles caught from mephitic air or rotten provisions. The reason that Lind was unable to draw the conclusion his clinical trials had justified was that he did not believe there was anything in fresh fruit that prevented scurvy, citric acid acting only as a remedy for the bodily effects of tainted food. Trotter, on the other hand, believed there was an important but unknown constituent in food that was missing in a sailor's diet, which explains perhaps why he was seduced by the plausible analogy with oxygen. Even experts closely allied in the politics of naval nutrition, such as Trotter and Blane, found themselves on opposite sides in this debate. Many surgeons recommended lime juice as a mouthwash, not as a supplement. Many, including Trotter, warned of its corrosive effects, particularly upon the stomach.

It might be objected that lots of diseases were like scurvy insofar as deep disagreements existed about their causes, leading to an uncertain methodology for treating them. However, it is generally the case that periodic epidemics of illnesses that are not, or were not, fully understood—such as the bubonic plague formerly and now ebola, zika, swine flu, and bird flu—are intermittent, unlike diseases that constantly affected populations, such as yellow fever, typhus, cholera, dysentery, and scurvy. In the case of yellow fever, as Mark Harrison has shown recently, careful empirical monitoring of its symptoms and circumstances, together with efficient nursing, had very good results even though the etiology of the affliction was still a mystery. The same empirical attention to the details of ailments such as typhus and cholera, particularly with respect to hygiene, had similar valuable results. It is commonly assumed that captains such as Cook and physicians such as Trotter exerted the same pressure on scurvy with lenitives such as clean berths, warmth, fresh air, and dry clothes. But of the many remedies available to Cook on board the *Endeavour*, only two were antiscorbutic, and neither significantly so. These were properly fermented sauerkraut and freshly made spruce beer, neither popular with his crews. Otherwise the well-known resource of fresh fruit and vegetables required either an itinerary interspersed with easy landfalls or a well-supplied market at the home port. Apart from those resources, there was nothing to rely on when scurvy struck. Cook had no means at hand to stem the proliferation of scorbutic symptoms on the *Resolution* during its cruise along the margin of the ice in the Southern Ocean; nor did Trotter when scurvy attacked the Channel Fleet in the winter of 1794 while it was based in Torbay. "Lemons indeed were now so scarce, and the consumption of juice had been so great, that little was left in the kingdom" (Trotter 1804: 1.417).

Empirically, he concluded that "recent vegetable matter imparts a something to the body, which fortifies it against the disease," but not only did he not know what this something was, there were no recent vegetables available (1.424). To make matters worse, both Cook and Trotter entertained theories at odds with their empirical practices, for Cook believed concentrated malt to be the best antiscorbutic available, even though it contained no vitamin C (any virtue ascribed to malt must have been owing to its vitamin B, which may have eased symptoms of pellagra and beriberi, diseases often coincident with scurvy). Trotter, as we have noticed, was convinced that scurvy was owing to depleted supplies of oxygen, which he thought could only be restored by fresh food, a pneumatic conjecture his colleague Thomas Beddoes had persuaded him to embrace. This explains why his endorsement of oranges and lemons was not total, for he believed that too much citric acid might damage the digestion and hinder the conversion of food into oxygen.

Although Camoens had included scurvy in *The Lusiads*, the epic of Portuguese navigation, there was no epic written about scurvy itself. Its genius resisted progress by bringing its students back to where they set out, offering no option but a renewal of a quest that was going nowhere and, rather like the tale told by the Ancient Mariner, a narrative without an ending. Or perhaps it is fairer to say that it was progress in the shape of voyages of discovery and the plantation of colonies that awakened scurvy from its earthbound dream, giving it life and scope in the oceans surrounding the rapidly shrinking Terra Incognita. Every enterprise of voyaging intended to vindicate the resourcefulness of human beings exposed a weakness in their constitutions that was inevitably to become hideously visible, yet nobody knew why it should appear, or what it was. Nor was it only epidemiological disagreements that made scurvy's story difficult to tell; its effects on those trying to tell it firsthand were fatal to the continuity as well as the probability of any tale of maritime prowess. When he arrives at the black miracle of scurvy off the coast of Mozambique, Camoens, speaking for da Gama, lists the symptoms of "a disease more cruel and loathesome than I ever before witnessed," accompanied by the usual gesture of exclusion: "Would any credit without seeing it?" Despite copious invocations of Virgilian epic, Camoens had already disarmed his narrative of discovery of a common truth when he begins the fifth canto by acknowledging that da Gama's singular observations, especially of this disease, were dismissed by sailors sharing the same standards of empirical observation as "false or feebly understood" (Camoens 1997: 114, 5.81; 101, 5.17).

So it is worth emphasizing that the fundamental difference between scurvy and all other epidemic maladies during the age of discovery does indeed revolve around the mutant gene that fails to prompt the synthesis of ascorbate in the human body. Scurvy is not owing to the entry of bacteria or viruses into the organism, nor to parasites nestling in a human host; scurvy occurs because the body has not absorbed in sufficient quantity a chemical habitually ingested and crucial to the functioning of the nerves, blood, bone, cartilage, and tissue. In this respect, it resembles other illnesses that arise because of the deficiency of a necessary biochemical agent, such as caries, rickets, or goiter. These afflictions have had to await recent developments in organic chemistry in order that they be understood and treated. So although Thomas Trotter, and before him Walter Charleton, had a strong intuition that there was some sap or essence in certain kinds of food that was needed if humans were not to fall sick, they had no means of isolating it.

It is worth asking why this intuition was not more widely shared, considering the empirical evidence in its favor, and the shamefulness of scurvy must have some bearing on the answer—its association with dirt, depression, caprice, and laziness; its resistance to any adequate explanation or description; the stench of its revolting lesions—so much so that it was often deliberately misdiagnosed, or misreported, or referred to tactfully as "sickness." It is hard not to think that the nature of voyaging to the ends of the earth in search of sights no one else had seen and territories no one else (apparently) owned—adventures amidst exceptional circumstances prolific in singular surprises—also had its effect. So there is a great emptiness in which scurvy makes its appearance, and a corresponding vacuum where explanations and cures of scurvy ought to have been found. Without an epic vehicle, then, what generic options remained? I shall be discussing romance at some length in chapter 5, but at the outset, it is worth mentioning the resources of revealed religion and myth.

When Adam is treated to the extensive view of the discoverable world by the archangel Michael in *Paradise Lost*, it is not to introduce him to the wealth of his patrimony but to the consequence of having eaten something that made him mortal: "Adam, now ope thine eyes and first behold / Th'effects which they original crime hath wrought" (Milton 1958: 252; 11.423–24). Adam's eyes are hard to open because his body is already degraded by sin. Raphael has already warned him where this will end: "Death is the penaltie impos'd, beware, / And govern well thy appetite, least sin / Surprize thee, and her black attendant Death" (7.545–47). The prime reason he and Eve are to be expelled from the garden is because the seed of corruption is already evident to God, and two bodies that had been perfect are now disgusting. God informs Michael, "Those pure immortal Elements that know / No gross, no unharmonious mixture foule, / Eject him tainted now, and purge him off / As a distemper gross to aire as gross, / And mortal food, as may dispose him best / For dissolution wrought by Sin, that first / Distemper'd all things, and of incorrupt / Corrupted" (11.50–57). The large promise made by science in the age of exploration and empire, "to represent to mankind, the whole Fabrick, the parts, the causes, the effects of Nature ... to have [human] Eyes in all parts, and to receive information from every quarter of the earth," has its price in the unavoidable exposure of the innate infirmity of the agents of the enterprise—starting with Adam's eyes, so blurred by sin that he needs an angelic medicine made of euphrasy and rue in order to see not only the world he must discover but also the misery its settlement will breed (Sprat 1667: 20; Warren 2013: 569). Like Adam, the voyager in unknown seas finds things out in an adventure determined from start to finish by nutrition, succumbing to a hideous foulness in his fallen human constitution because he has eaten the wrong food.

A myth apt for scurvy is to be found in the story of Glaucus told by Ovid in his *Metamorphoses*; for Glaucus stands either side of a division between the known world of the land and the unknown one of the undersea, analogous to that da Gama thought he was piercing when he sailed into the Indian Ocean, or Magellan when he entered the straits named after him. Observing how the fish he has caught have begun to eat the grass upon which he laid them and, how, recruited by this new food, they find the strength to leap back into their proper element, Glaucus eats some too and directly dives into the ocean, determined to forsake the land forever. He becomes a sea god, with dark green hair, blue skin, and a fish's tail, immortal and no longer human. Sometimes in pictures he brandishes the vegetable trophy from which his divinity derives (Fig. 1). The fish, beguiled into taking a bait that proves them mortal, mend themselves with the same vegetable and reenter a world where scurvy never comes. In following their example, Glaucus eradicates the seed of mortality lurking in himself. It is a tale of plants with magical properties prompting fish and humans to cross the boundary between mortality and eternal life. Scorbutic sin makes way for recovered innocence. Dampier reports that near the Celebes his companions who were troubled with scorbutic ulcers on their legs found great relief from a poultice made of the leaves of a local vine pounded with pig fat. Everywhere in his voyages, Dampier is assiduous in testing the benefits of plants and fruits, like Glaucus on the lookout for a leaf or berry to purge corruption.

Here it is necessary to specify what areas of the body and the brain are influenced by low levels of ascorbate, and how that influence is manifested and felt. Vitamin C is an antioxidant. Oxidation occurs in cells because of metabolic activity or inflammation, generating free radicals, the waste matter of sentience. Ascorbate scavenges these radicals, clearing up the detritus of a

Fig. 1. Cornelis Bloemaert, *Glaucus als Meermann*. Public domain.

working brain. A second important task of ascorbate is to serve as cofactor for the enzymes responsible for the final step in the production of collagen, the glue of the cell and the scaffold of the body. Without collagen, cartilaginous material would disappear from our joints, bones would become fragile, and scar tissue would begin to unknit. People in the last stages of scurvy were said to rattle and creak when they were moved. A third important function of ascorbate is to recycle and regenerate tetrahydrobiopterin, a requisite cofactor for several key enzymes. One of these generates nitric oxide, actually a gas that diffuses from the cells lining the blood vessels and relaxes the surrounding muscle, thus allowing the arteries and capillaries to dilate when needed. Related to this, the fourth function of ascorbate is to prevent the leakage that occurs when loss of collagen renders the wall of a blood vessel permeable. Weakness in this seal is evident in the blood blisters (petechiae) that form around hair follicles in the first stage of scurvy, in the ulcerous ecchymoses that affect the legs and in the periosteal bleeding from the muscles that blackens the bones (Fig. 2, Fig. 3, and Plate 1). In *The Rime of the Ancient Mariner*, Death's bones are recognizably those of a scorbutic cadaver, exhibiting not only the hue but also the striations in the bone typical of a collagen deficit: "His bones were black with many a crack, / All black and bare I ween, / Jet-black and bare" (Coleridge 2004: 152, ll. 181–83). At its worst, scurvy weakens arterial walls so badly that the slightest movement, smell, or sound provokes the seizure or aneurysm that brings scurvy to its fatal conclusion: Dampier's Captain Cook, who had been ailing for some time in the South Seas, "died of a sudden, though he seemed that Morning as likely to live, as he had been some Weeks before" (Dampier 1999: 212, 61).

A fifth function of ascorbate is to serve as a cofactor for the synthesis of neurotransmitters such as serotonin and dopamine, crucial for the efficient operation of the synapses, the junctions of the neuronal pathways. When they are disturbed, not only is the mood of the scorbutic patient liable to be affected, but also the perceptual apparatus is compromised (May 2013: 95–105). This is when differences between joy and sorrow, surfaces and depths, fancy and fact all grow uncertain: when feelings are intense but rapidly supplanted by emotions quite opposite, and when impressions are so impossibly vivid they cannot be transmitted, belonging exclusively to the individual who enjoys or endures them.

It was often reported that scorbutic sailors suffered personality changes. For example, rugged sea dogs became prone to tears, describing their symptoms with dreadful volubility, as if no amount of particulars was adequate to convey the nature of their condition or allay their anxieties about it. Their sense of community was so far diminished they would fall prey to childish egoism and sometimes passionate and violent jealousy. Hallucinations were common, especially with regard to the color and consistency of the sea. Although the most common symptom in the later stages was severe debility and lassitude, it was remarkable how often sensory impressions were enlarged in scurvy, as if the neuromodulators had been put out of action, and there were no longer any inhibitions to sensory excitement. Smells became overwhelming, often disgusting yet sometimes exquisite; sights were dazzling; sounds fascinating or intolerable; the sense of touch morbidly acute; and taste fatally

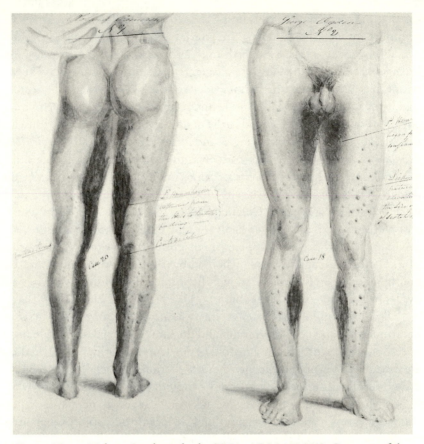

Fig. 2. Henry Mahon, Scorbutic limbs. TNA, ADM 101/7/8. Courtesy of the National Archives of the United Kingdom. See also Plate 1.

voluptuous. Sailors often died from pleasure in the moment they ate the fruit and drank the sweet water for which they had been yearning. As Dampier says bluntly of Cook's sudden death, "But it is usual with sick Men coming from the Sea, where they have nothing but Sea-Air, to die off as soon as ever they come within view of the Land" (ibid.). Many are the accounts of scurvied seamen dying while being taken ashore, either because they succumbed to the shock of physical movement or were swept away by the joyful sight of vegetation.

To the extent that reports of scurvy are autobiographical, they exhibit a typical attention to the last detail of a phenomenon, often associated with a powerful ambiguity in their responses to it, as if it were at once the most attractive and the most sinister thing. We find James Cook on the Antarctic

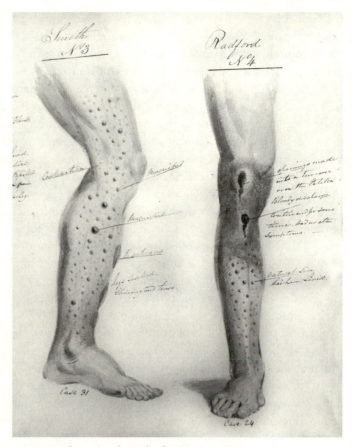

Fig. 3. Henry Mahon, Scorbutic limbs. TNA, ADM 101/7/8. Courtesy of the National Archives of the United Kingdom.

sweep of his second Pacific voyage transfixed by the moonlit beauty of the ice on the rigging, until suddenly he recollects what the extra weight might do to the balance of the ship in a swell. In *Omoo* (1847), Herman Melville mentions a scorbutic whaleman who shrieked in agony at the smell of flowers. In *The Rime of the Ancient Mariner*, the sea snakes and even the albatross go through a similar rotation, for the water creatures strike the Mariner first as vile then as beautiful, while the bird is seen first as a good omen, then as a evil one, and finally as a kind of messenger of the Antarctic spirit world. In the first-person singular, then, it is hard to establish a point of interchange with an audience when talking of the experience of scurvy or of sensations received while in a scorbutic state because there is no correspondence between the egotism and passion of the sufferer, on the one hand, and the consensual measure of value subscribed to, on the other: "Yes, so you shot a bird?" Time and again, this rupture is made plain not only in the scepticism of the audience (so many journals from remotest parts were disparaged as fictions), but equally in the challenges issued by eyewitnesses to the limited experience and imagination of those who were not present at the emergency. When Richard Walter describes the arrival of the *Centurion* off the shore of Juan Fernandez, he makes the usual exception: "It is scarce credible with what eagerness and transport we viewed the shore.... Those only who have endured a long series of thirst, and who readily recall the desire which the ideas of springs and brooks have at that time raised in them, can judge of the emotion with which we eyed a large cascade of water" (Walter 1838: 111). Of course the exclusively private rapture claimed here does not define a community of sufferers all suddenly enjoying together the sight of relief: "we" stands for a discrete collocation of individuals each possessed by an unparalleled emotion, analogous to no other. "We" in a state of scorbutic transport is really "I" multiplied: a crowd, as Thomas Hobbes would say, but not a community. "O Wedding-Guest! this soul hath been / Alone on a wide wide sea: / So lonely 'twas, that God himself / Scarce seemed there to be" (Coleridge 2004: ll. 597–600).

The foreignness traditionally associated with loathsome and dangerous diseases, plunging those who succumb to them into the loneliness of a castaway or the solitude of an unwelcome stranger (Bewell 1999: 6–7), is especially evident in scurvy. Community is suspended while the travails of the lonely individual are perfected. The Ancient Mariner complains most of all about his loneliness and alienation from everything familiar, ringing the changes on the word "alone," which like a bell tolls him back to his sole self. In chapters 2 and 3, this foreignness is examined first under the heading of scientific experiment and then of nostalgia in order to show how scurvy's estrangement of domestic and familiar things travestied, and perhaps shadowed, the course of empirical science itself, until in 1800 the two trajectories came to a strange juncture. Yet home and all its customary pleasures are inseparable from the scenes of eating and drinking that populate the dream-life of scorbutics. Trotter coined the term "scorbutic nostalgia" to signalize the vividness of the reveries of his patients and the extravagant grief they indulged when they awoke to the dismal and exiguous realities of their illness. Is it the actuality of home and its comforts they miss, or life on land, or has a malfunction in the brain coined a phantom that has no equivalent in the real world? Is it possible to recover from pathological foreignness and return to the familiar?

Scurvy stood in an odd set of relations to science, at once its *cause celebre*, its antithesis, and its double. A cure for scurvy was the chief object of naval medical science in the eighteenth century, so when Cook's vessels set sail, well stocked with scientific instruments for measuring the heavens and the earth along with a large variety of supplemental foods, the experimental nature of the voyage extended, as I have already suggested, outward to the globe and the cosmos and inward to the digestive tracts of the crew, who were asked to consume a variety of specialized foods then report the results. On his return, Cook addressed the Royal Society solely on this topic of nutrition, not on his measurements of the transit of Venus or the spellbinding discoveries he had made of lands and cultures unknown to Europeans. However, this journey and many others to the most foreign of the world's exotic regions left scientists themselves troubled, embattled, and even nostalgic, as if the actual pursuit of knowledge had thrown up awkward facts that did not fit their protocols of investigation or expectations of consensus. Although historians of science such as Steven Shapin generally assume that the culture of the Royal Society depended on a company of civil eyewitnesses achieving agreement about the nature of an experimental event, nothing could have been more exorbitant to gentlemanly discourse than the intemperate exchanges between William Wales, the astronomer on Cook's second voyage, and Johann Reinhold Forster, the naturalist, after they returned, so vituperative as to indicate a level of mutual exasperation and ill-will intense enough to bring the authenticity of their testimony into doubt. Both of them admitted to having contracted scurvy, suggesting that the failure of the internal experiments of the voyage were having a deleterious effect on the external ones, too.

Nevertheless, a direction had been taken in scientific practice during the seventeenth century that did not always require consensual confirmation of a fact, and it was more apt for the stress, surprise, and loneliness experienced by scientists in distant places. First of all, the improvement of specialized scientific equipment such as Robert Boyle's air pump and Robert Hooke's microscope brought facts to light that could not possibly have been noticed without the action of a machine. A great debate was carried on between Hooke, an enthusiast for prosthetic assistance to the sense organs, and Margaret Cavendish, who despised it. John Locke put his objection in a manner that situated the operator of a mechanical eye or ear and the specialist aboard an experimental vessel in parallel predicaments when he said that if sight or audition were increased ten thousand fold, the reporter would find himself "in a quite different World from other People: Nothing would appear the same to him, and others" (Locke 1979: 303 [II, xxiii, 12]). In this respect, improvements in navigation ran directly parallel, for Cook went off in search of the Great Southern Continent, the Terra Incognita that formed the southern border of all maps up until this time; and what the microscope disclosed, according to the eager Hooke, was its miniature counterpart, "new Worlds and Terra-Incognita's to our view" (Hooke 2003: xvi). Even if the experiment came closer to home, the trick was to give it distance and strangeness: "An Observer should endeavour to look upon such Experiments and Observations that are more common, and to which he has been more accustom'd, as if they were the greatest Rarity, and to imagine himself a Person of some other Country or Calling, that he had never heard of, or seen the like before" (idem 1969: 61–62; cited in Daston and Park 1998: 315). Boyle thought that uncommon experiments often disclosed facts whose relation to other phenomena was not ascertainable, inducing a state of mind he compared to reading fiction: "The full discovery of Natures Mysteries, is so unlikely to fall to any mans share in this Life, that the case of the Pursuers of them is at best like theirs that light upon some excellent Romance, of which they shall never read the latter parts" (Boyle 1671: 118–19). Bacon and Cavendish were not the only scientists to locate their fictions on the far side the known world, where the boundaries of the known and the foreign blur: Boyle wrote a romance of his own, as did Sir Kenelm Digby.

The state of mind Hooke and Boyle were alluding to, the feeling of stemming toward a new world with eyes wide open in wonder, lacking any analogy or comparison that might communicate the pressure of these foreign novelties, takes center stage in one of the more remarkable episodes of scurvy's discontinuous history. In the last decade of the eighteenth century, Samuel Mitchill, an American chemist, took a hint from Joseph Priestley and made experiments on two gases with similar molecular structures, nitric oxide and nitrous oxide. From the results obtained from his work on the latter, he concluded that it was the gas responsible for all contagious diseases, including scurvy, and he named it "septon." When he tested nitrous oxide at the Pneumatic Institute in Bristol, Humphrey Davy found no evidence for what Mitchill alleged, but he did fall into ecstasies that made him feel very strange indeed: "I lost all connection with external things.... I existed in a world of newly connected and newly modified ideas" (Davy 1800: 487). Besides this peculiar rearrangement of his mind, which he could neither directly express nor analogize, he found himself overloaded with sensations that brought him, via septon, into the same realm of inarticulate and isolated hypersensitivity that distinguished the state of the nerves in full-blown scurvy: "I imagined that I had increased sensibility of touch: my fingers were pained by anything rough.... I was certainly more irritable, and felt more acutely from trifling circumstances.... My visible impressions were dazzling and apparently magnified.... When I have breathed it amidst noise, the sense of hearing has been painfully affected even by moderate intensity of sound" (464, 487, 491).

It will be evident throughout this book how important to its conclusions is the history of Epicurean materialism, not only as it affects natural scientists such as Robert Boyle, Margaret Cavendish, Lucy Hutchinson, Sir Kenelm Digby, and John Evelyn, but also the physicians who first began to discriminate the symptoms and possible causes of scurvy, in particular Thomas Willis and Walter Charleton. The appeal exerted on these thinkers by Lucretius's interpretation of Epicurus lay in its emphasis on atomic motion. For an age that was about to penetrate the secrets of calculus, or "fluxions" (as Newton called motion along the curvature of parabola), by tracing minute deviations from a straight line to its limit—a limit that for Lucretius met in vast spirals and vortices, like the clouds, lightning, and thunder of the turbulent last book of *De rerum natura*—the promise of a fluid mechanics seemed at once more extensive and more disturbing than the physics of solids. Boyle conceived of the body as a pump, along the lines of his favorite machine except that it sucked in all sensations as a flow of perceptions and all matter as a stream of nutrition, like a "hydraulo-pneumatical Engine." Digby had an intuition of the anarchic energy of this process when he studied the bean plant, where he beheld a willful monad following "its own swing" and inclination for dominion, hoisting itself up over rival growth. So the idea of aliment as simply adding to the mass of a plant or a body had to be adjusted in favor of a perpetual tide of energies, appetent and resistant, and never perfectly in equilibrium (Digby 1669: 211). To account for the flow of energy, Digby supposed a crucial additive whose presence caused the bean to flourish out of all proportion to the size of its seed, and whose absence caused it to sicken, shrink, and die. Digby thought of this as a saline juice or nitrous salt. Willis and Charleton entered eagerly into this debate about nutrition as a swirl of fluid events uniting the body-engine, and the nerves in particular, to the world beyond it. The mysterious element propelled into the body by eating and drinking they called a "sap" or "latex," and it set the terms for a flux and reflux of matter that could no longer be accounted for by considering nutrition simply as the addition of solids to body mass, or by viewing health merely as the maintenance of an internal balance according to the humoral pathology of Galenic medicine.

The doctrine of the four humors derived from Hippocrates, who had assigned their equilibrium not only to food but also to airs, waters, and places, offered a wider and more dynamic circuit that was to provide the basis of environmental accounts of health and indeed of speciation among natural historians such as Georges-Louis Leclerc, Comte de Buffon; Jean-Baptiste Lamarck; and Charles Darwin. However, even an environmental humoral doctrine supposed that a healthy life was spent in an even relationship to airs,

...s, and places rather than in a flow that might at any moment become turbulent and cause explosive alterations not only to the organs but also the nerves. In concentrating on the animal spirits and the structure as well as the distribution and activity of the nerves, Willis and Charleton were anticipating developments in medicine that took place later in the eighteenth century, propelled by the researches of John Brown and Robert Whytt into the principles of nervous irritation, which were findings that shaped the investigations of Trotter and Beddoes first into scurvy and latterly into nervous diseases. Here we arrive at another important moment in the disagreements between the school of deficient food as opposed to the school of defective food and toxic air. The latter worked in a framework of solid mechanics, believing that body mass was corrupted by foreign material coming into the organism and disturbing the relation of parts; while the former operated in a system of fluxions where everything was in motion and nothing was inherently foreign, consequently the relation of matter, feeling, and thought was a perpetual interchange. Disease was an impediment to flow, not to mass. Scurvy was caused by the absence of the fluid particle responsible for human "swing," the expanding curve of vitality and motion, without which the body sustained blockage, stagnation, and putrefaction.

In any discussion of homesickness, the foreign has to have some sort of relation to the familiar and domestic. If the Epicurean account of health, sensation, and perception, not to mention imagination, is the fruit of a perpetual commerce between atoms moving on the inside and those patrolling at large, then to be at home is to be always in passage to somewhere else, and to be lost or forlorn is to be stuck, like the Ancient Mariner, in a calm or a trance. In chapter 3, the word "situation," which has already acquired some thicker definition in chapter 2, is developed a little further in order to show how, in the exigencies of scorbutic experience, being alone and far from home do not in themselves excite homesickness; but to be stuck inside a dream of home that corresponds to none that exists, or to go back home under the influence of such an illusion and to have it shattered, these conform to a definition of homesickness first offered by Johannes Hofer and recently refurbished by Helmut Illbruck to the effect that it arises from a lesion of the imagination resulting in a reverie entirely without a referent. So I agree with Kevis Goodman, who has argued that nostalgia, like scurvy, is a disease of motion and displacement, the pathology of imperial expansion, in which home stands as the norm from which ambitious nations cause their servants to deviate. Her emphasis on volition as an oscillation between free will and unintended momentum is a valuable model for what I call, with respect specifically to scorbutic nostalgia, an ellipse of loss and satisfaction. If a degree of fluidity is to survive the experience of being immobilized in a predicament or a fantasy, some kind of liaison has to be sustained between imagination and the objects of the senses. So when scorbutic voyagers find the familiar things they have been dreaming of, that connection is briefly confirmed, or corroborated, but never finally consummated, otherwise all movement would cease. There is no report of scorbutic landfall where pleasure is not rendered less than sheer by its opposite—confusion, apprehension, fear, discomfort, or terror. Take, for example, Anders Sparrman's experience of shooting ducks in Dusky Bay in New Zealand, where the *Resolution* landed with almost everyone exhibiting signs of seriously low levels of ascorbate or of manifest scurvy: "The blood from these warm birds which were dying in my hands, running over my fingers, excited me to a degree I had never previously experienced.... This filled me with amazement, but the next moment I felt frightened" (Sparrman 1944: 49). The same oscillation between pleasure and horror is to be found in Cook's reaction to ice on the rigging; or in J. R. Forster's belief that the exquisite manifestations of phosphorescence were owing to putrescence; or in Robert Louis Stevenson's coral islands, which he found were inhabited by fat, fleshy worms of a most unseemly pink hue. The most celebrated instance is provided by the Ancient Mariner's sea snakes, which appear either as slimy monsters, offences to creation, or as elfin tracks of iridescence, irresistibly beautiful. Specifically, what is breaking down in cases such as these is any sense of situation as a definite relation between time and space, leaving room for an intuition of flows between opposite positions—whether these are of the subject and the object, disgust and delight, the familiar and the exotic, the real and the fictional, or the dangerous and the vulnerable—prompting what might best be termed a condition of sensory dazzlement, where two contrary impulses are acting in relation to each other, neither predominating.

The settlement of Australia is the subject of chapter 4, a notable example of a landfall that provided none of voluptuous satisfactions generally associated with the Indian Ocean or the South Seas, although many of those coming ashore were scurvied, especially after the arrival of the second fleet. As a joint experiment in penology and colonization, it was utterly novel, being as far from England as it was possible to get and consisting of unreconnoitred territory where no previous European settlement had been made. It was on all fronts a monstrous gamble that easily could have failed because scurvy had not been factored into any of the arrangements for making the trip. The first fleet was lucky in this respect with a smooth passage and good supplies of fresh food at the Cape, but subsequent transports arrived in Australia with many dead or dying from scurvy, owing partly to the poor condition of many of the convicts when they were first shipped, partly to overcrowding in wet conditions, and partly, of course, to their diet, which was standard naval fare containing a reasonable amount of carbohydrate, protein, and fat, but no vitamins. This was a burden the first fleet also had to bear because, owing to the peculiarities of Port Jackson and its environs, it was impossible adequately to supplement their stocks of food with fresh greens. There were a few antiscorbutic herbs in the bush, but nowhere near enough to supply more than a thousand people, soon to be joined by hundreds more sick with scurvy. Besides, even these scant resources were hard to access because of difficult terrain and hostile Aborigines. Starving prisoners desperate to get hold of the first sign of a green leaf thwarted attempts to grow vegetables. I am not altogether convinced, therefore, by Joyce Chaplin's claim for scorbutic nostalgia as global "earthsickness," an ecumenical desire for land of any sort as long as it is not sea, since the example of Botany Bay runs so counter to the assumption that land, any kind of land, provides a cure.

Those whom scurvy did not ruin were imperiled by a regime that became proportionately savage as food stocks were threatened, a situation the failure of the first storeship, the *Guardian*, which was badly damaged by an iceberg and turned back to the Cape, did not ease. The redemptive relation of scurvy to sin and expulsion explored by Milton was exemplified not only in the utopian pretensions of the enterprise and the reformative aims of its more idealist advocates, but also in the despair it provoked. Despatched to an antipodean Eden where a combination of exile and slave labor, lasting anywhere from seven to fourteen years and beyond, was meant to secure emancipation, scurvy acted as a constant temptation to theft and the repetition of the crime that was already being punished. For a while David Collins believed the devastation of market gardens and plantations of maize imperiled the viability of the colony. He thought this was owing to sheer anarchy and a proclivity to crime, but the thieves were people frantic for food with vitamins in it. Whether this induced a culture of recidivism, or whether the intolerable anguish of the subsequent punishments inflicted on these involuntary emigrants drove them beyond a limit at which social bonds made sense, some years later a government enquiry concluded that this great penological experiment had been a total failure and was now reduced to a scene of disorder and depravity without parallel in the civilized world.

Life in early Australia was like being stuck on a vast dismasted ship in the middle of the ocean. As the largest and best-documented example of so-called "land scurvy," it provides a unique glimpse into what the culture of scurvy might amount to. Using the accounts of David Collins and Watkin Tench as mainstays, along with the 1837 Parliamentary Report on the colony and the work of the Port Jackson Painter as visual evidence, I try to make the case for a culture of dazzle. That is to say, the absence of any settled cognitive relation of sensations to ideas extended the confusions arising from a place as hostile as it was strange until the natural history, the topography, the indigenes, the parlous physical circumstances of the settlement, and the inhuman laws that governed it drew a tight band of impermeable misery around its European inhabitants. Never had a civil society so inauspicious a beginning. One of the members of the Parliamentary Select Committee suggested that the most promising future for New South Wales was as a pirate colony like Tunis. Another, adverting to the extraordinary number of lashes inflicted for petty thefts, thought it simpler to execute all convicts on arrival as a more humane and swifter route to death than the reigning circuit of flaying and disease. Eventually, horror and a kind of desperate exuberance seemed to make friends, especially in the theater, a resource from the earliest days of the settlement. It is evident also in the best work of the Port Jackson Painter, whether he is turning the sandstone cliffs of the harbor into dizzying baroque scallops and striations or representing in the faces of his best portraits a rivetting blend of apprehension and aggression.

Dazzle and romance have much in common. Describing the situation of the first European inhabitants of Australia, Matthew Flinders explained the confusions of the landscape in negatives: "A history of this establishment at the extremity of the globe in a country where the astonished settler sees nothing, not even the grass under his feet, which is not different to whatever had before met his eye, could not but present objects of great interest to the European reader" (Flinders 1814: 1.xcv). What he supposes to be of interest is what one cannot possibly know as itself or compare with anything else. I have already suggested the scorbutic ingredient in voyages of discovery is inimical to the epic because it emphasizes the internal and external confusions that multiply when the boundary between the knowable and the unknown is crossed. Here is Henry James on romance contrasted with the real: "The real represents . . . the things we cannot possibly not know, sooner or later, in one way or another. . . . The romantic stands, on the other hand, for the things that . . . we never can directly know; the things that can reach us only through the beautiful circuit and subterfuge or our thought and desire." Romance, he goes on, is experience liberated from "the inconvenience of a *related*, a measurable state, a state subject to all our vulgar communities" (James 1947: 31, 33). The loneliness of voyaging, the onset of scurvy, and the cognitive disturbance arising from the morbid irritability of the nerves and the senses, the failure of the tongue to express what it is like to be so proximate to things vibrating with sensational stimuli, and the strange pathological coalition of the beautiful and the horrible induced by such a chaos of impressions—dazzle in short—are responsible for the translation of ships' journals into fictions, such as Pedro Fernandez de Quiros's account of the New Jerusalem he founded in Vanuatu, or "Hildebrand Bowman's" sequel to the massacre of the crew of the *Adventure's* cutter in New Zealand in 1773, or Jacques-Henri Bernardin de St. Pierre's conversion of his scorbutic arrival at Mauritius into the love story of *Paul et Virginie* ([1788] 1907). Likewise, it is responsible for the customary opening of utopias, where voyagers make scorbutic landfalls at places unknown, such as in Bacon's *New Atlantis* (1627), Henry Neville's *The Isle of Pines* (1668), Margaret Cavendish's *Description of a New World, called the Blazing World* (1666), and Jonathan Swift's island of the horses (1726). At the furthest extreme of the real and known world, another is found to be adhering where values are reversed, history doesn't work, things are so fertile in contradictions that they need never change, and the autonomy of the imagination is held forth as a kind of noble bastardy. This is the subject of the fifth and last chapter, where I try to strap the culture and aesthetics of scurvy to its poetics.

There is a final note (the Coda) in which my colleagues James May and Fiona Harrison, having assisted me so generously throughout this project, unite in spelling out the neurological basis of the claims I have been making. I hope the whole book will resituate the discussion of scurvy as something more serious in a maritime context of discovery than the astonishing revelation of the "real" and "true" heroic history of its cure.

# CHAPTER ONE

# Enigma

Throughout the symptoms of this disease, there is something so
peculiar to itself, that no description, however accurate it may be,
can convey to the reader a proper idea of its nature.
—Thomas Trotter, *Observations on the Scurvy* (1792)

In Daniel Defoe's *Journal of the Plague Year* (1722), the calamity of the
arrival of bubonic plague in London in 1665 is sharpened considerably by
the ignorance of everyone in the city concerning the cause of the disease, its
mode of transmission, the fatal tendency of its symptoms, and the method of
its cure. While generally it is assumed that it arrived in a cargo from the East,
nobody knows where it was lodged, whether in fabric, ointments, or food,
nor how it was at first communicated, whether by touch, effluvium, or the
hand of God. The cures attempted range from the probable (fumigation) to
the superstitious (amulets). The symptoms vary, and the progress of the disease is uncertain. People who have been in close touch with victims survive,
the afflicted who have been isolated succumb; some die as soon as the marks
are on them, others recover if the swellings burst. Although the narrator
seems powerfully of the opinion that shutting up stricken houses with the
survivors still inside is the only way to contain the infection, he is full of
praise for those who evade municipal controls by getting out of the window,
or blowing up the watchmen, or escaping to the country, or by supporting
their diseased families with the income from ferrying passengers, freight, and
possibly the plague itself up and down the river. Every attempt made by the
narrator to nail an effect to a cause or a problem to a solution is inevitably
contradicted by events or his own judgments, leaving him confronting unrelated events that coalesce into a kind of miracle: "A dreadful plague in London
was / In the year of sixty-five, / Which swept an hundred thousand souls /
Away, yet I alive!" (Defoe 1966: 256).

His narrative explodes into a montage of astonishing scenes—a naked man
walking down the street with a pan of burning coals on his head; another

mistaken for a corpse and making a joke of it as he climbs out of a death pit at Mount Mill; grass growing in the streets; eldritch shrieks of the dying and bereaved. He is unable to trace a sequence leading to any predictable outcome. The plague haunts everything, because although everyone knows what it does, nobody knows what it is; and the result is a confusion that involves both the event and every effort to make sense of it. "Here was no difference made," says the narrator of the Mount Mill pit, where corpses are tumbled in regardless of age, sex, and nakedness, whether they are quick or dead, advertising a chaos in which all distinctions are lost—between the living and the dead, the poor and the rich, the decent and the shameful, the effectual and the ineffectual, belief and horror (80). Doubtless Defoe's decision to pass off his fiction as history was influenced by the consideration that amidst such sheer confusion the difference between what is imagined and what is true is immaterial and that fidelity to such a prodigious interruption of civil life requires more than facts to do it justice.

The modern reader of H. F.'s narrative is placed in a position rather like that of the female spectator who shouted out to Othello that he was a fool for not believing Desdemona innocent. For the reader is perfectly acquainted with the truth; namely that bubonic plague is incubated and transmitted by the fleas that live on rats; or, in the case of the pneumonic variant, by the breath of the victims. But the fascination and tension of this historical moment arises from the fact that those in peril did not know this and had no hope of knowing it. We observe how epidemiology is transformed into something like fate, and acts of self-preservation into dramatic irony. We assume for the purpose of intelligent inspection of any epidemic that there is a methodical relation between the empirical facts of the disease and the biochemical breakthrough that will put an end to its dominion. We assume further that the lapse of time is a series that can be understood and that the structure of a historical situation depends on an orderly adjustment of the inside of private experience to the outside of public knowledge. But those in the midst of the infection in 1665 reacted to it as an apocalypse from which the differences between past and future, and between an interior space and an external point of observation, had been obliterated. And it is to that dreadful intuition of lost coordinates that H. F. is loyal, rather than to the reader's medical hindsight or faith in providence. He calls the plague a "speaking sight" because it communicates only with those implicated in it, not to readers of subsequent accounts of it. "It is impossible," he says, "to say anything that is able to give a true idea of it to those who did not see it, other than this, that it was indeed very, very, very dreadful, and such as no tongue can express" (80, 79). Time and space puzzle him. He has no explanation for how the affliction began or how it ended; why it traveled where it did and why it stopped when it did. He is equally confused about social policy and personal resourcefulness in this kind of emergency, whether quarantine or escape is the best plan, or whether resignation is not, in the end, the only feasible response.

Over a much longer span of time and on a much broader canvas, scurvy exhibited the same dramatic overthrow of the structure of meaning on which its students and victims wanted to rely. Even though it had been known from the first that food played a part in the onset and the cure of this disease, the identification of a preventive in fresh vegetables, meat, and fruit, and of a remedy in concentrated citrus juice (called "rob"), was never universally embraced or systematically pursued. Although victims were regularly restored to health, provided they could make land and start eating the herbs and salads they found growing there, disagreements among medical experts about the causes and treatment of the malady, combined with clinically uncertain results from remedies used at sea, inclined them to rely on favorite theories rather than experiments because facts simply compounded the confusion. Close observation of symptoms yielded "a strange jumble," Sir Gilbert Blane complained, where no one datum was consistent with another (Lloyd and Coulter 1961: 3.298). Thomas Willis spoke for many subsequent physicians when he called it, "an Iliad of Diseases": "It extends itself into so various and manifold Symptoms, that it can neither be comprehended by one definition, or scarce by a singular description" (Willis 1684: 170, 169). "The Scurvy," wrote Sir Everard Maynwaringe, "owns not one univocal cause, but is the Bastard of many Parents" (Maynwaringe 1666: 5). Thomas Trotter was by no means the only one to notice that scorbutic symptoms were as likely to multiply in the head as the body, with a repertoire of temperamental peculiarities that added to the difficulty of making an adequate diagnosis. Thomas Beale, a victim of the disease, announced the difficulties of comprehending an illness that overflowed all points of observation and resistance: "Our mental sufferings were such as defy description, and nothing but being placed within the same situation could convince those who have not the power to imagine its monotonous dreariness" (Beale 1839: 310). Like Defoe's narration of the plague and like Shakespeare's dramatization of the extravagances of jealousy, a gulf divides those absorbed by pain and passion, and those who are not. If the latter were able to apprehend the full extent of scorbutic dreariness, then they would know the eyewitness speaks truly of it; but as this is impossible, they cannot even form a judgment of what it was like.

The history of scurvy is heavily marked by this conflict of interests between those who dwell inside the drama and those who don't. The biochemical breakthrough that isolated ascorbic acid and explained its physiological and neurological importance to the human organism was never aligned with the history of empirical knowledge of the disease. That is to say, biochemistry and naval medicine never shared an inevitable and common destination, although there were many occasions when a coalition of the two was accidentally and briefly achieved. But it will be evident in the course of this chapter that it is difficult for anyone outside the disorganized and passionate situation of scurvy itself not to consider its history as anything but periodic fits of willful ignorance that blinded the world to a necessary truth and an obvious cure: a dismal record then of lost opportunities and culpable amnesias. And in proportion, as the desire to understand the disease from the perspective of its allegedly final elucidation inflects the account of it, another angle of it is ignored. This is the history of immethodical passion, variously productive of misery, exhilaration, mistakes, madness, heroism, and death—the sort of history H. F. writes of the plague and that the lady in the audience thought Othello might coolly have sidestepped, had he only known the truth.

Perhaps in an effort to redress the balance, it has been common in recent years to sound a triumphal note when mentioning the clinical trials of James Lind aboard the *Salisbury* in 1747 and James Cook's conquest of scurvy during his Pacific voyages. The bearing of Lind's work on the practical successes of Cook in avoiding fatal outbreaks during his decade of navigating the South Seas (1768–79), and the navy's later decision to distribute rations of lemon juice to all sailors, are greeted as phases of a real and solid progress in eradicating the affliction, thus the title of Francis E. Cuppage's book, *James Cook and the Conquest of Scurvy* (1994). The subtitle of a history of the disease, *How a Surgeon, a Mariner and a Gentleman Solved the Greatest Medical Mystery of the Age of Sail* (Bown 2003), is intended as triple homage to the experimental breakthrough of Lind, the prophylactic successes of Cook, and the administrative genius of Sir Gilbert Blane, each praised for discovering and applying the virtues of antiscorbutic remedies and regimes broadly enough to keep the British Navy in fighting trim throughout the Napoleonic Wars.

The truth is that these men achieved much in their respective fields, but none provided an unequivocal solution to the problem of scurvy. Lind had empirical evidence to show that citrus juice removed its effects, but he was much more hesitant in adducing causes, suggesting that indigestion, blocked pores, and the weather might be factors as malign as the shortage of good food. His specialty was cure, not prevention, and even then he was cautious, warning toward the end of his career that health at sea could not demonstrably "be ascribed to any diet, medicine, or regimen whatever" (quoted in Carpenter 1986: 70). In their survey of medicine and the navy in the eighteenth century, Christopher Lloyd and Jack Coulter have suggested that Cook's neglect of lemon juice in favor of malt wort, which then gained his powerful endorsement in a paper he gave before the Royal Society in 1776, set the cause of citrus back for a generation (Lloyd and Coulter 1961: 3.302). For reasons which must have had more to do with naval economies and medical politics than clinical accuracy, malt (valueless as an antiscorbutic) became the chief official remedy for thirty years, relied on by many surgeons afterward, thanks largely to Cook, whose banishment of scurvy from his ships (particularly the *Endeavour*) lent practical authority to his judgment in its favor. As for Gilbert Blane, his espousal of the cause of preserved lemon juice was to result in something like its general distribution as a preventive in the navy from the mid-1790s; but he was candid enough to admit that the efficacy of the concentrated juice was not guaranteed, being significantly less than that of fresh lemons, probably owing to the method of preparation (3.132, 319–21). Although Blane arranged a repeat of the success of Lind's clinical trial aboard the *Suffolk* in 1793, which sailed to Madras without a single case of scurvy, his preference was always for the fresh juice rather than the rob, a position consistent with his insistence on nutrition as the major factor in the cause and cure of scurvy, one that was shared by the other chief proponents of citrus, such as Thomas Trotter and Nathanael Hulme, although with rather different emphases. On the night before Trafalgar, Admiral Nelson penned a note on the importance of the supplies of onions held on his ships, which, after three months at sea, formed his most reliable stock of antiscorbutic food (*Times*, 3 May 2013, 3).

If we are looking for a hero in the history of scurvy, Lind would certainly be a leading candidate had the Admiralty not entirely neglected his achievements for forty years, and had his own eclectic approach not misled him. In the context of a disease so manifold and inconstant, perhaps it is inevitable that the person who comes closest to the heart of its mystery is ignored, especially if his discoveries are empirically and not epidemiologically produced. Any empirical cure has to compete with all the other practical and theoretical remedies in which so many people, some of them powerful, have an interest. Not until Albrecht Szent-Gyorgyi isolated the antioxidant molecule he called hexuronic acid, which he found in high concentrations in the adrenal cortex, and which Charles Glen King and A. W. Waugh subsequently found in lemon juice, was the indisputable cause of scurvy known to lie in the depletion of vitamin C, or ascorbic acid (Carpenter 1986: 186–90). Previously, it was referred to variously as a charm, a je ne scais quoi, a sap, a miasma, or a taint in food that entered the body and corrupted it. And because of the obscurity of the cause, Lind was seduced into theories that lessened the value of the results of his clinical experiment on the *Salisbury*. Another ignored hero in the history of scurvy is Hugh Platt, a contemporary of Francis Drake and Richard Hawkins, who seemed to have understood a lot earlier than most that there is a virtue in citrus that is at its maximum when fresh and requires careful processing and storage if it is to be preserved in any useful quantity. His "philosophical fire" seems to have been a method of pasteurizing juice to prevent fermentation while preserving its curative content. Furthermore, he had devised a method of airtight bottling that likewise kept the juice free from contamination and decay. But as no one was willing to pay for this secret, the secret was kept until Nathaniel Hulme borrowed it for Sir Joseph Banks (Keevil 1957: 1.107–9). Platt knew that it was one thing to find out that fresh oranges, lemons, and limes would cure the scurvy, but quite another to ensure that their hidden virtue lasted over distance and time.

The handkerchief in the *Othello* tragedy of scurvy is citrus juice. In the play, the embroidered fabric circulates as an equivocal sign of innocence, and at sea, the juice acts fitfully as a lesson in antiscorbutic medicine that is no sooner learned than it is misused, mistaken, or forgotten. Neither ever coheres into a satisfying and final proof of worth and virtue. Favored by the British and eventually the Americans, but never much fancied by the Dutch, French, Spanish, or Russians, citrus juice makes many appearances that are decisive and elusive by turns (Keevil 1957: 1.223). From the earliest days of long-distance navigation the medicinal value of oranges and lemons was known and exploited. Richard Hawkins and James Lancaster had used them and vigorously recommended them. John Woodall endorsed the antiscorbutic benefits of citrus in his influential book on maritime health, *The Surgeon's Mate* (1617). Long before Cook, Drake managed a voyage of a thousand days with only a single victim of scurvy. The regimes pursued by these mariners were very like Cook's, for besides the use of fruit and fresh greens, they all appreciated the importance of having a clean and dry ship, a well-clothed crew, and frequent stops for refreshment. One of the reasons for the Admirality's neglect of Lind was the embrace it extended to broad-spectrum prevention (of which he himself was a proponent) evident in the wide variety of dietary supplements supplied by the Sick and Hurt Board to John Byron, Samuel Wallis, Philip Carteret, and Cook in their explorations of the Pacific during the 1760s.

Before David MacBride published a theoretical account of scurvy that claimed for malt the curative potency Lind had found in oranges and lemons, it was commonly understood that a variety of precautions and remedies would defeat scurvy, not a single wonder cure. This was basically Cook's approach, until he was edged into a public recommendation of malt after he returned from his second Pacific voyage. Lind himself had never proposed the juice of oranges and lemons as a superlative remedy, only rather more effective than other medicines, such as electuaries, malt, vinegar, and vitriol. This was a fact he had experimentally disclosed, not a program of general health he espoused. Therefore it was not he, but physicians such as Gilbert Blane, Nathaniel Hulme, and Thomas Trotter who recommended lemon or lime juice as an invaluable resource for the navy. And even they had reservations—Blane about the quality of rob at sea, Trotter about the effect of frequent doses of acidic liquid on the stomachs of seamen, while Hulme was aware that the preparation of reliable rob was elaborate and expensive.

A point upon which these defenders of citrus juice were technically unanimous, however, was their contempt for theorists. Their empiricism was joined to a common and powerful belief in the value of clinical trials, not in conjectures concerning air, moisture, salt, acidosis, putrefaction, or bad food. Unlike those who argued that scurvy was an imbalance of the humors, they agreed that a diet of fresh food would infallibly prevent it, and cure it once it had appeared, owing to an ingredient in fresh fruit, vegetables, and citrus juice that was lacking in preserved stores such as salt meat, flour, biscuit, or dried peas. While he was blockading Brest at the start of the Seven Years' War, George Anson—a commander who had more cause than most to fear a disease that cost him two-thirds of the men who sailed with him in 1740–44—pioneered the first real success in keeping British crews healthy for long periods of time at sea . His was a method that could successfully be practised only near home waters or friendly ports, for it involved supplying ships on blockade duty with fresh meat, vegetables, and fruit. It was prohibitively expensive and physically arduous, but utterly reliable. Sir Edward Hawke, who was serviced with fresh fruit by tenders while blockading the French fleet off Ushant in 1759, and Trotter, who, as Physician of the Fleet in 1794, purchased vegetables and lemons for crews trying to subsist on salt provisions in the hard winter of 1794, provide two notable examples (Rodger 2004: 281, 484–85; Trotter 1804: 1.405; Macdonald 2004: 13). Trotter concluded after his successful intervention, "The late occurrences have sufficiently established the fact, that scurvy can always be prevented by fresh vegetables, and cured effectually by the lemon" (Trotter 1804: 423). A variation of this regime was the plantation of large botanic gardens at strategic distances from home ports, such as that of the Dutch at the Cape, the French in Mauritius, and the British in Calcutta, St. Vincent, and St. Helena. Trotter urged the plantation of a similar garden in the grounds of the Haslar Naval Hospital at Portsmouth (Lloyd and Coulter 1961: 3.241).

The unanimity among proponents of fresh food extended to the battery of measures involving hygiene, clothing, warmth, rest, and proper storage of supplies, but was rather less evident in the matter of citrus. Like Lind, Trotter goes no further than the fact of juice's usefulness as a cure, for causes are beyond him. He knows that fresh food puts an end to scurvy, but not why. Lemon juice, therefore, does not strike him any more than it strikes Lind, Blane, or Cook, as an Ariadne's thread leading to a plenary solution of a medical mystery. Enthusiasts like Leonard Gillespie, who were much more exclusive in their devotion to juice, were destined to be disappointed, for on many occasions it was tried without success. It is we, from our postvitamin perspective, who understand the clinical and neurological importance of oranges and lemons and play the part of the lady shouting to Othello, exasperated that those with a vital interest in the matter are blind to what is so obvious or, having seen it, ignore it. It is as if Defoe's story of the plague experimentally demonstrated a connection between fleas, rats, and the incidence of the disease simply as a naked fact, and left it at that. Hindsight plays leapfrog with that kind of factuality, trying to force the pace toward the truth we know, but of which the experimentalist is ignorant or only imperfectly aware. The history of scurvy is especially tormenting in this respect, being strewn with red herrings, false starts, and mistaken conjectures that mock all teleological symmetries.

When the surgeon William Clowes examined the case of two scorbutic patients and suggested that lack of exercise, neglect of keeping clean and dry, and want of good air "were the only means they fell into the Scurvy," J. J. Keevil drily remarks, "Unfortunately, Clowes's 'only means' were so numerous that they completely obscured the primary cause" (Keevil 1957: 1.99). Again, when citing Hawkins's prophetic advocacy of sour citrus, he observes, "On of the most bewildering aspects of the history of scurvy is the manner in which a cure was repeatedly found, only to be lost again" (1.102). Lloyd and Coulter feel the same impatience with William Cockburn, a man so infatuated by his love of theory "he could prescribe lemon juice in a case of scurvy and yet fail to draw any conclusions from its success" (Lloyd and Coulter 1961: 3.40). They are incredulous at official obtuseness in the face of Lind's work on scurvy ("Neither the Admiralty, the Navy Board, nor the Sick and Hurt Board made any energetic attempt to implement his suggestions"), and they are frankly suspicious of the circumstances in which Cook threw his weight behind malt, postponing for a generation the naval adoption of lemon juice ("In our opinion the causes for the delay are ... devious") (3.43, 303). Yet there is nobody involved in this history who possesses sufficient facts to identify Keevil's "primary cause," and in the absence of such facts, conjecture was to some degree relied upon by everyone, even hard-bitten observers such as Lind and Trotter.

Undoubtedly, the most far-reaching event in the sorry tale of antiscorbutic discoveries forgotten or ignored was the failure of Sir George Nares's expedition to the Arctic in the *Alert* and the *Discovery* in 1875–76. After only a year on the ice, there was a serious outbreak of scurvy (sixty cases, three fatal) that the naval issue of lime juice was not alleviating. In one case, the symptoms, far from being eradicated, actually returned after liberal doses of the juice (Lloyd and Coulter 1963: 4.119; Williams 2009: 362). In an interesting reversal of the official position a century before, an Admiralty enquiry (the "Scurvy Committee") tried to vindicate the limes, but influential voices were arguing that lime juice had never worked, and they prevailed. "By the end of the century," Lloyd and Coulter point out, "all faith had been lost in the efficacy of lime juice," a shift in public opinion that was to have a malign influence on the victualing of Scott's expedition in the *Discovery* to the Antarctic in 1901–4, where a bad attack of scurvy put an end to Shackleton's participation in it. Problems of nutrition in British attempts on the South Pole were aggravated by the current theory in favor of ptomaine poisoning as the chief cause of scurvy, a theory harking back to Lind's ideas about bad food and strengthened by the partial but shocking evidence of the destruction of Sir John Franklin's expedition to the Arctic, which appeared to have been precipitated by infected canned meat. By the time of Scott's attempt on the pole in the *Terra Nova* (1910–13), when Edward Evans succumbed to scurvy and Edgar Evans died of it, ptomaine poisoning was still the official explanation, although many people (including Roald Amundsen) were aware of the importance of fresh seal and penguin flesh. Reginald Koettlitz, Scott's senior surgeon on the *Discovery*, stated baldly however, "There is no antiscorbutic property in any food or drug" (Lloyd and Coulter 1963: 4.120; Chick 1953: 216–17; Guly 2012: 1–7). Scott himself referred to citrus juice as an "antidote," as if its function were solely to neutralize a poison, and he thought its justification as either a preventive or a remedy was "beset with immense difficulties" (Scott 1907: 1.405). No citrus juice other than in small pharmaceutical preparations was prescribed on either expedition to the Pole (Anon 1962: 32). The range of vitamins from A to C were entirely inadequately represented in their diet of pemmican, chocolate, tea, tinned meat, pea flour, biscuits, sugar, tinned fruit, dried vegetables, and condensed milk (Watt, Freeman, and Bynum 1981: 168). To emphasize how far back into the past this turn of events had thrust medical knowledge, Lloyd and Coulter quote Woodall's confession of ignorance from a book now three and a half centuries old and as relevant as ever: "The causes of this disease are … infinite and unsearchable" (Lloyd and Coulter 1963: 4.119). The enigma of scurvy was fully restored.

The reason for this disastrous loss of faith in the standard antiscorbutic remedy, by then prescribed for sailors worldwide, lay in the growing preference of the navy for West Indian limes instead of Sicilian lemons—the very limes whose overwhelming success as an antiscorbutic had been saluted by Lind himself (Lind 1753: 199). By 1860, the West Indian fruit had entirely replaced Mediterranean citrus, and in 1867, the Admiralty decided to make all purchases of juice from the firm of Sturge, who had extensive plantations in Montserrat. Not until 1917, with an analysis conducted by the Lister Institute, was it known for certain that West Indian limes contain a negligible amount of ascorbate. Sir Clements Markham was right to assert that this sort of juice had never worked, although he could not explain why. It was not intense cold sapping its virtue, or too much boiling in its preparation that destroyed its efficacy, for it had never possessed much to begin with (Lloyd and Coulter 1963: 4.115–16). This left the field free for eminent bacteriologists such as Sir Almroth Wright, who not only identified the cause of scurvy in tinned meat (rendering suspect the chief supply of protein in the Antarctic and causing much of it be rejected) but also shifted the whole debate from nutrition, where it belonged, to bacteriology, where it didn't.

Casting back through the tangled story of lime and lemon juice, it is possible to see in the enigma of scurvy something like a symmetry between the imperfections of human judgment and the defects of the remedy. Basically there were two methods of preparing rob, either by preserving freshly squeezed juice in casks with 10 percent brandy or rum, or by boiling or reducing it and bottling the concentrate with a seal of oil (Hulme 1768: 57–59; Lloyd and Coulter 1963: 4.113). Lind had correctly rated the ascorbic properties of oranges as superior to lemons. For reasons of portability, he recommended inspissation, using a bain-marie to reduce the liquid to a third of its volume (Lind 1753: 207; Lloyd and Coulter 1961: 3.299). This was a technique widely in use, but it was tempting to short-circuit the process with some brisk boiling, which was known to make the product less effective, as Blane pointed out. However, Nathaniel Hulme concocted a small amount of rob by Lind's method for Sir Joseph Banks, who used it on the *Endeavour* when suffering early signs of scurvy, apparently successfully. Somewhere between the confidence of Hulme and Banks and the blanket rejection of rob by the Sick and Hurt Board in 1781 as "of no service" (Lloyd and Coulter 1961: 3.317), we locate the hesitations of Cook, Blane, and Trotter. These may well have had the same foundation as more decisive estimates of the failure of juice such as that of William Perry, the man who replaced William Monkhouse as Cook's surgeon on the *Endeavour* voyage, James Patten's, surgeon on his second, or William Anderson's, surgeon on the third, upon all of whom Sir John

Pringle relied for his rejection of further trials of juice (3.312, 314, 303; Pringle RCPE PRJ/1–9, 8.449).

In any trial of antiscorbutics in which citrus juice was involved, there must have been credible doubts of its efficacy, especially if it had been reduced by boiling rather than in a bain-marie, then stored for long periods and exposed to extreme heat or cold. On the other hand, how long the potency of fresh juice might last if it were merely preserved with an addition of spirits was impossible to determine, but the accounts of its success between the 1790s and 1850s would suggest that it must have been more reliable than the rob were it not for the fact recently discovered by Janet Macdonald that many naval ships' logs of that period have no record of receipt of any kind of lemon or lime juice on a scale sufficient for a regular issue (Macdonald 2004: 164). To make matters more complicated, Lloyd and Coulter point out that the words "lime" and "lemon" were used interchangeably, so there is no certainty (at least from the Seven Years' War onward) that the liquid being used was not obtained from the West Indian lime, *citrus medica acida*, the fruit whose properties Leonard Gillespie had with unconscious irony deemed "incalculable" and which Lind himself had singled out for praise (Lloyd and Coulter 1961: 3.322). With regard to the juice generally, we can only assume that some stores of it were inert, but that others (perhaps the greater amount) retained variable degrees of ascorbate and worked as a cure. As for the trials of malt in its various forms, whether as wort or spruce beer, we must ask how it was possible that a remedy less valuable as an antiscorbutic than even the West Indian lime could garner any measure of enthusiastic support. Well, if it were used in spruce beer, the leaves and sap of pine were known since the days of Jacques Cartier's exploration of the New World to have mildly curative powers in cases of scurvy (Keevil 1957: 1.86). Sauerkraut, if properly fermented, also had a small amount, and if it were consumed along with wort, as it was on Cook's ships, then malt would have shared the success of cabbage as it did of spruce. However, it is evident in some voyages that scurvy was confounded with beriberi and pellagra, each likely to erupt from situations of nutritional deficit (Watt 1979: 68–69). In cases involving a loss of thiamin or niacin, malt might even have contributed to recovery. Of course, this would have been claimed as malt's success against scurvy and, given the state of medical knowledge at the time, one not easily disputed.

This is to assume that all trials of antiscorbutics were made in good faith, which manifestly was not the case. Cook, prudent as well as judicious, was keenly aware of the cost of inspissated juice, writing to Pringle that, "the dearness of rob of lemon and oranges will hinder them from being furnished in large quantities" (Lloyd and Coulter 1961: 3.316). Even after it authorized the use of lemon and lime juice, the Admiralty was always in pursuit of economies, hence its preference for West Indian limes over more expensive kinds of Mediterranean citrus. The cost of the rob put on board the *Resolution* for Cook's second voyage in 1772 was 7 shillings and sevenpence per pint. In 1793, two years before the decision to make lemon juice available to any naval ship that wanted it, there were 45,000 active sailors on the navy's books. At a rate of 28,000 gallons for 10,000 men for a year, which was the rate determined in 1795, the cost would have amounted to 250,000 pounds sterling for the full 45,000 men. By the end of the wars with France, the strength of the navy stood at 140,000. Nelson said he could get lemon juice at Messina for a shilling a gallon, and with the navy at full strength, that would have represented a relatively modest expenditure, but there was no guarantee that prices would remain at that level. In 1701, lemons were £4 a chest, and when Trotter was searching for fresh food for the Channel Fleet in 1794, he said there was scarcely a lemon left in the kingdom (Trotter 1804: 1.417; Lloyd and Coulter 1961: 3.319). No doubt the price factor influenced the reception of reports comparing malt favorably with lemon juice and directly or indirectly qualified Cook's judgments in the paper he delivered to the Royal Society. Certainly the high price of rob helped MacBride gain influential support for his campaign on behalf of malt, and of this influence he was aware, writing, "As the Material for making this Liquor may be so cheaply purchased, and is capable of being much longer preserved than Oranges, Lemons, or indeed any other Kind of Fruit, there can be no Hesitation in giving it the Preference for general Use" (MacBride 1767: 60). The cost of the citrus cure explains the delay in its adoption by the Merchant Marine. Obliged by the Shipping Act of 1854 to make the same distribution of lime juice as the navy, owners of merchant vessels were so miserly that they purchased cheap supplies of adulterated juice or simply neglected to buy any at all. Merchant seamen (especially whalemen) continued to suffer and die from scurvy. Chapter 4 shows that few vessels on the Australian run, especially convict transports, were able to avoid it. It was reported in 1866 that "in almost every case of the manifestation of scurvy [in ordinary merchant ships] the lemon juice was proved to have been originally spurious or to have become deteriorated," so the Act was strengthened in 1867 (Dickson 1866: 5; Cook 2004: 224–29). It is ironic that the law gained potency just in time for the remedy itself to lose all credit.

If we return to the parallel between the history of scurvy and a performance of *Othello*, it becomes clear that the bewilderment and frustration of those at the front end of the struggle to find a cure for scurvy are transmitted to those with a more managerial interest in the same outcome. Although the wisest heads in maritime medicine were aware of the need for diverse approaches when confronting a disease whose symptoms were so confusing and contradictory, it is equally clear that the disease presented obstacles to voyages of discovery and naval strategy serious enough to generate official impatience with moderate and limited solutions. As naval blockades became the strategy of choice for the British after 1757, it was necessary to find a way of keeping crews healthy, for as Trotter pointed out, of all naval duties, blockading was the one where the absence of fresh supplies was most likely to provoke scurvy. Although Nicholas Rodger disputes some of the wilder estimates of death rates for scurvy, such as Francis Cuppage's (Rodger 2004: 308, n. 53), it is certain that it was treated as a clear and present danger by the time the navy was engaged in a transatlantic war. For every Admiralty Othello loudly calling for a handkerchief for his headache, there was an expert such as David MacBride ready to play the part of Iago, with an allegedly sure remedy that was likely to make things worse.

The peripeteia of a tragedy provoked by the very thing (handkerchief or lime juice) that might have averted it, involves everyone in a new obscurity of which Scott's two expeditions are unhappily exemplary. Naval medicine ends up where it began: in ignorance and with nothing more substantial to rely on than conjecture. In their irritation with what they take to be the carelessness causing this terrible step backward, its premier historians begin to share a little of the darkness and the passion they lament. I think this leakage from one category of knowledge, or frame of representation, to another—let us say from the inside of confusion and emotion to the outside of observational exactness or theoretical symmetry—is typical of any encounter with scurvy no matter how refined or academic. It is a disease that finds all boundaries porous and (like the plague) abolishes all differences, looping cause into effect, the future into the past, and tempting the audience out of their seats and onto the stage; and it is to this leakage I want now to turn, beginning with my own specimen of it.

In any narrative of scurvy, there are two relatively secure points from which triangulations may be made, and these are the rival positions maintained between those who thought scurvy was owing to a deficiency in diet and those who believed that it was a toxic state brought on by a miasma in the atmosphere or a poison in the food. Insofar as this debate concerned nutrition, it divided over a diet that was either deficient or defective: either there was something lacking in what was eaten, or something present in it that was inimical to health. In a recent essay, Mark Harrison has shown how neatly these opposing views were represented by Blane and Trotter, even though they were proponents of the identical remedy. Blane did not regard citrus as a food but as a medicine capable of correcting a debility arising from "a vitiated diet, consisting in salted animal food" (Blane 1799: 479). It was a remedy, not a supplement. Trotter, on the other hand, was convinced that a diet without fresh fruit, vegetables, or well-prepared lemon juice was failing to supply a catalyst for the oxygen the body needed (Harrison 2013: 7–25). Lind, like Blane, believed that toxicity in food was a powerful factor in scorbutic cases, arguing that any aliment producing a bilious reaction would undermine a remedy based on citrus or fresh food. "I am fully convinced," he wrote in his Postscript to the 1757 edition of his *Treatise*, "that whatever weakens the constitution, and especially the organs of digestion, may serve without any other cause to introduce this disease ... even among such as live on fresh vegetables, greens, or the most wholesome diet, and the purest air" (quoted in Trotter 1795: 78). At the time Lind wrote that, the consensus among naval administrators and seamen was about to fall behind malt wort and the theory that it supplied the body with the fixed air (carbon dioxide) that was lost in long voyages. But by the time Trotter was writing, there was an equally strong party in favor of the supplemental virtue of citrus juice, a view that was to last until the Admiralty report on the calamity of the Nares expedition. This was the last official endorsement of lime juice before the biochemical discoveries of the 1930s, but it was not unanimous: Dr. Rae stood up for deficiency of diet along with the bulk of the committee, but Dr. Leach believed the problem lay in the bad quality of the food, largely preserved in tins, a suspicion that had been getting stronger since the loss of Franklin's expedition (RCA 1877: vii).

By the time Scott embarked on his two expeditions, the dominant theory was that bacteria growing in food, especially imperfectly canned meat, were responsible for scurvy. This was espoused by the influential physician Sir Almroth Wright and by the surgeons on the two expeditions: Reginald Koettlitz, Edward Wilson, and E. L. Atkinson. When Atkinson gave a lecture on scurvy in 1911 to the people of the *Terra Nova*, he stated that the first cause of scurvy was tainted food, producing acid intoxication of the blood followed by the myriad symptoms of the disease (Scott 1947: 300–301). It is evident that the same germ theory governed the practice of Koettlitz and Wilson, for when scurvy struck the *Discovery*, the latter wrote, "The outbreak is unaccountable, for nothing radically wrong with the hygiene has been discovered" (Wilson 1967: 201). But by the time he sailed again with Scott, Wilson was convinced, for reasons that shortly will become apparent, that seal meat would stop scurvy; but Atkinson was skeptical and thought lime juice useful only if taken regularly, which evidently it was not. He said fresh vegetables would soon restore health, but they were simply not to be had in the Antarctic. Although proponents of citrus were not a powerful caucus by this time, seal meat had plenty of adherents. The Admiralty Committee of 1877 recommended crews in the polar regions consume as large a proportion of fresh meat as it was possible to obtain (RCA 1877: xxii). A Royal Society report on scurvy of 1900 likewise emphasized the value of fresh meat (Carpenter 1986: 149). The Norwegians Nansen and Amundsen were enthusiasts for this solution to the problem of scurvy at the poles: "Everyone, without exception, preferred seal meat to tinned foods" (Amundsen [1912] 1976: 51). Shackleton, who had suffered badly from scurvy on the *Discovery* expedition, believed in the defective food theory, but on the *Nimrod* expedition proved himself a great eater of seal and penguin, and did not become scorbutic.

Scott's experience on the *Discovery* expedition was not as alarming as Shackleton's, but he got scurvy along with Wilson and described very well the terrible lassitude that comes with it (Scott 1907: 2.121). There was another symptom that had been discussed in the Admiralty Report which he experienced but did not recognize: the fantasy or dream of food. Scott had begun by finding seal meat unpalatable and blubber so disgusting he thought it tainted the meat. Yet once scurvy made itself felt, he caught himself indulging a special relish for blubber and recalling with hopeless longing feasts of seal: seal soup made of seal blood and porridge strengthened with seal liver (2.79, 51). In the course of the Admiralty enquiry of 1877, Dr. Buzzard was asked about dreams of food: "Would you consider these phenomena as a mental expression, urging the dreamer to the choice of a food best suited to his condition?" "Yes," he replied, "I should certainly" (RCA 1877: 198). This is a conclusion Scott never allows himself to reach because the flesh and organs of seals are antiscorbutic only in the negative sense that they allow his men to consume protein uncontaminated by ptomaine or other poisons. When hunting was good, "I was able to give the satisfactory order that no tinned meat of any description should be issued" (Scott 1907: 1.401).

Apart from the suspicion that their canned meat was tainted, there was no explanation for the arrival of scurvy or for its eventual disappearance. Seal is an "antidote" or "antiscorbutic" not because it contains something the other food lacks, but because it allows the poisonous constituent in their diet to be jettisoned: "Many cures might have been attributed to the virtues of a supposed antidote which were really due to a discontinuance of the article of food that caused the disease.... The evidence shows [scurvy] was caused by the food the discontinuance of which led to recovery" (1.405, 409). The truth was that, besides canned meat, their supplies contained under half the necessary amounts of niacin, thiamine, riboflavin, and vitamin D, and nothing at all of ascorbic acid (Watt, Freeman, and Bynum 1981: 167). Even the sledded rations contained insufficient calories for the work these men were doing. Nevertheless, Scott declares, "We are still unconscious of any element in our surroundings which might have fostered the disease, or of the neglect of any precaution which modern medical science suggests for its prevention" (Scott 1907: 1.403). He is driven to treat seal meat like pemmican, cooking it "again and again" in an effort to reduce its mass and make it easier to carry. But, of course, this does to seal meat what others had done to orange and lemon juice when they reduced it and boiled away its virtue: "We are rather afraid that our seal-meat has suffered from the heat, and that it is not as antiscorbutic as it was" (2.84). When an opportunity arises experimentally to assess the virtue of fresh meat, it isn't taken. They have a Christmas feast with biscuit and seal liver, and everyone with scorbutic symptoms feels better; but Scott puts it down to the food in general—the flour, biscuit, sugar, butter, pea flour, tea, chocolate, jam, and marmalade—not to the liver, brains, blood, and flesh of seals that supplies the deficiency of the other. This is where one feels like shouting out down the years at him, "Can't you see that there is something in seal meat that is not in anything else you are eating, and that you are the better for it?"

\* \* \*

Scurvy was singular, like no other disease, and yet it was like them all. It was unique and multiple at the same time, unparalleled and common. Trotter wrote in his *Observations on the Scurvy* (1792), "Throughout the whole symptoms of this disease, there is something so peculiar to itself, that no description, however accurate it may be, can convey to the reader a proper idea of its nature" (Trotter 1792: 71). He went on, "The debility of Scurvy is of so singular a nature, that nothing seems analogous to it: certain it is, that no disease is related to it.... [It] is attended by a train of symptoms peculiar to itself, and which the genius of the distemper has rendered extremely difficult to explain" (106). He was enlarging on Lind, who was adamant that "the scurvy is not a hodge-podge or combination of various difference diseases, but is itself a simple identical malady," but given to protean manifestations (Lind 1757: 351). The exhibition of scorbutic "genius" was variously esteemed extraordinary (Lind), wonderful (Walter), miraculous (Benjamin Morrell), scarcely credible (Philip Saumarez), and (as we have heard) "infinite and unsearchable" (Woodall). Jacques Cartier called it an "unknowen sicknes ... after the strangest sort that ever was eyther heard of or seene" (Keevil 1957: 1.86). On the other hand, the "Iliad of Diseases," identified by Thomas Willis, "extends itself into so various and manifold Symptoms, that it can neither be comprehended by one definition, or scarce by a singular description" (Willis

1684: 170, 169). Blane called it a jumble, and many people (Thomas Sydenham among them) viewed it as a crowd or multitude of diseases masquerading as one: its name legion, for it was many (Carpenter 1986: 41). The poet
Abraham Cowley exclaimed, "A Thousand horrid Shapes the Monster wears, /
And in as many Hands fierce Arms it bears" (Cowley 1881: 2.260). At its
various stages, it would exhibit the signs of phthisis, asthma, sclerosis, tym-
pany, leprosy, gangrene, dysentery, rheumatism, sea-fever (calenture), venereal
disease, dropsy, and mania.

Here is Cowley's survey of its variety:

When Men for want of Breath can hardly blow
Nor Purple Streams in azure Channels flow,
Then the bold Enemy shews he's too nigh,
One so mischievous can not hidden lie.
The Teeth drop out, and noisome grows the Breath,
The Man not only smells, but looks like Death.
Qualms, Vomiting, and torturing Gripes within,
Besides unseemly Spots upon the Skin,
His other Symptoms are; and with Clouds the Mind
To Life itself makes Living an Offence. (Cowley 1881: 2.260).

To Cowley's list of shortness of breath, problems with the flow of blood, hem-
orrhaging under the skin, biliousness, constipation, and severe depression,
Camoens contributed the lurid addition reported by da Gama, namely that
the blackened gums of the victims began to rot and had to be pared away with
knives. As time went on, others were added, such as thickening and coarsen-
ing of the skin, contraction of the tendons of the legs, severe pains and swelling in the joints, and bloody stools. When opened, scorbutic corpses were
found to have lost cartilage from the knees, ankles, and ribs, and their bones
were black with blood that had leaked from the muscles. In his contribution
to the Royal Society's *Philosophical Transactions* (1721) concerning the outbreak of scurvy in Paris in 1699, Francois Poupart reported that the bones
of the victims clattered as they were carried away, and that "their Ribs [were]
separated from their Gristles ... the Ligaments corroded ... the Bones separated from each other ... black, worm-eaten and rotten" (Poupart in Jones
1721: 5.360). Lind could hear scorbutic ribs creaking against the thorax, and
Harvey detected "a Cracking and Noise of the Joynts" (Harvey 1675: 27;
Lind 1753: 315). Henry Ettrick, surgeon of Anson's *Centurion*, noted the
same blackening of the bones among the dead of Anson's voyage, "Their
Bones, after the Flesh was scraped off, appeared quite black" (Williams 1999:

Fig. 4. David Jones, *Death and Life in Death*. © Trustees of the Estate of
David Jones.

60). This macabre detail was reported of Crusaders dying during the siege of
Damietta in 1218, "Their hips and shin bones turn[ed] black and putrefied"
(Bollet 2004: 174). Those in the last stages of decay were regarded as living
dead, rotting before they were in the grave. James Rymer thought, "A person
so affected is really in a state of actual dissolution and decomposition"
(Rymer 1793: 36), an idea and an image imported by Coleridge into *The
Rime of the Ancient Mariner* for the personifications of Death and his mate,
the one exhibiting the authentic color and texture of scorbutic bone—"His
bones were black with many a crack, / All black and bare, I ween; / Jet-black
and bare" (Coleridge 2004: 152, ll. 181–85)—and the other the unnatural
hue of roughened skin that made it look like leprosy, ornamented in the accompanying illustration with a subcutaneous hemorrhage masquerading as a
beauty spot on her thigh (Fig. 4).

The reason for the variety of physical symptoms of scurvy is to be found in
the crucial part played by the antioxidant molecule of vitamin C in the syn-

thesis of collagen. Collagen knits the cells together, and when it fails, the cells dissolve. The disintegration of tissue is indicated in the collapse of the walls of blood vessels, beginning with the smallest. The tiny hemorrhages at the seat of hair follicles, called petechiae, are usually among the earliest physical signs of scurvy, along with the sponginess of the gums, also owing to the collapse of the capillaries (Harrison and May 2009: 719–30). Petechiae are followed by ecchymoses, large areas of discolored skin that soon ulcerate, leaking an evil-smelling fluid. The foul breath of the victims comes from the extravasated tissue that begins to blacken and bulge from the mouth (called "bullocks' liver" by sailors), hiding the teeth that are left. Sir Richard Burton in his translation of Camoens names it gangrene, "Gangrene that carried foul and fulsome taint, / Spreading infection through the neighb'ring air" (Burton 1881: 2.620). The arteries themselves are then affected, becoming increasingly fragile, while the blood thickens, often ending in an aneurysm, seizure, or infarction when the patient attempts to move, breathes fresh air, drinks cold water, or hears a sudden noise. Difficulty in breathing would be compounded in these cases by the degeneration of the lung tissue, so that coughing blood was occasionally an accompaniment, often mistaken for consumption. Ernest Shackleton was badly afflicted with this symptom of the disease. Thomas Willis pointed to a "dyscrasie" and "foeculence" of the blood as lying at the heart of the thirteen symptoms he lists of scurvy (Willis 1684: 171, 173, 179). The walls of the stomach, intestines, and bowel were known to leak blood, triggering the bilious symptoms often associated with scurvy, along with constipation or diarrhea, until, in the final stages, bleeding from the anus would exhibit the same symptom as dysentery. Subperiosteal bleeding from the muscles accounts for the blackness of bones, and it is also lack of collagen that explains the disappearance of cartilage, leaving he joints uncushioned and, in the end, unhinged from the frame of the skeleton. The destruction of cells causes the old wounds of scorbutic victims to reopen and old fractures to break apart again. It is mentioned as one of the chief causes of Captain Lawrence Oates's difficulties shortly before his death on Scott's second Antarctic expedition that a broken femur from the Boer War became unknit owing to scurvy and began to crepitate, causing him atrocious pain (Cherry-Garrard 1994: xxxiii). All organic functions are compromised in this state of affairs: the weakening of the surface, circulation, digestion, sinews, and frame of the body makes it impossible for it to fuel and animate itself, introducing the languor, loss of blood, and deterioration of tissue that dramatically signposts the progress of the disease toward putrefaction and death. Scott remembered, "It was an effort to move.... I had an almost unconquerable inclination to sit down wherever a seat could be found.... To write, or even to think had become totally distasteful, and sometimes quite impossible" (Scott 1907: 2.121–22).

However, there is no symptom of scurvy that may not fail to manifest when expected or to remit when least likely. There is no telling why Columbus should have crossed the Atlantic without a single case while Cartier's expedition forty years later was severely stricken; or why Cook in the *Endeavour* should have escaped its visitations while Louis-Antoine de Bougainville, pursuing a similar route through the Pacific, did not. Some experts said scurvy was most likely to attack the young and vigorous, while others said that the most vulnerable were the weak and dejected. Once established, scurvy did not always start with petechiae and swollen gums; it could begin with swollen joints and aching limbs, or with a bilious gut. "The Captain has a return of swelled feet & Obstructions," Forster reported on the second voyage of Cook, "my son likewise has swelled Legs, all owing to the Scurvy" (Forster 1982: 4.689). On the basis of his observations on the *Centurion*, Richard Walter reported the inconstancy of symptoms together with the irregularity of their appearance, "For scarcely any two persons have complaints exactly resembling each other; and where there has been found some conformity in the symptoms, the order of their appearance has been totally different" (Walter 1838: 102).

Victims usually succumbed to dejection and homesickness; on the other hand, some were reported in good spirits, even ebullient, until the very end. "Some Hours before they die, they are taken light-headed, and fall a joking and laughing; and in this Humour they expire" (Bulkeley and Cummins 1927: 97). Some recovered with miraculous speed as soon as they reached shore, while others were affected for the rest of their lives (Lloyd and Coulter 1961: 3.125). Many examples have already been cited of cures achieved with remedies whose success could only have been as placebo, a fact not lost on certain astute commanders such as the Prince of Orange, who gave pills of chalk and sugar to his scurvied troops at the siege of Breda, whereupon "Cheerfulness appears in every countenance, and an universal faith prevails in the sovereign virtues of the remedies" (Falconer 1791: 127). William Falconer gets the story from Lind, who said it was an anecdote "in every respect consonant to the most accurate observations, and best attested descriptions of the disease" (Lind 1753: 128). "It would look like reviving the age of miracles," says Trotter, taking the opposite view of this trick at Breda, "to say, that the bloated face assumed its ruddy hue, the contracted ham became straight … in proportion to the glad tidings" (Trotter 1792: 122). Nowhere is the sudden shift from prostration to joy more common than when a ship with scurvy comes in sight of the shore, and men haul themselves on deck to behold the vegetable world that will save them. Walter assures the reader that even the worst affected on the *Centurion* drew strength and spirits merely from viewing the prospect of Juan Fernandez. Trotter confirms that "the patient in the inveterate stage of the disease seems to gather strength even from the sight of the fruit" (141). Yet there are stories of sailors getting worse after eating it and becoming scorbutic after drinking lime juice (Keevil 1957: 1.116). The genius of scurvy was never likely to fail before the force of a straightforward remedy. A fresh sweet-smelling breeze will revive the languid mariner, Richard Mead assured his readers (Mead 1794: 112); yet Saumarez observed that his invalids no sooner crawled on deck to breathe the uncontaminated air than instantly they perished, as if a fetid atmosphere and a filthy berth had become their natural element (Saumarez 1739–43, 23 April and 30 May 1741).

I have suggested that Perry's and MacBride's use of malt as a cure of scurvy may have seemed to succeed because without their knowing, it was curing pellagra, caused by a shortage of vitamin B that malt restored. If scurvy comprehends a host of other diseases, it sometimes enlists under the flag of another malady. Illnesses arising from nutritional deficits were bound occasionally to accompany scurvy, confounding symptoms that were already confounded enough, so it is worthwhile briefly pointing out what these are. The three confederates of scurvy are night blindness (nyctalopia), pellagra, and beriberi, owing respectively to deficiencies of riboflavin, niacin, and thiamin. The first induces a retinopathic condition that makes it impossible to see in the dark, or in poor light. Pellagra is defined by the four Ds—dermatitis, diarrhea, dementia, and death—and is most widely known for the personality changes associated with alcoholism, which generally provokes a rapid depletion of niacin. It is noteworthy that improvements in the milling of corn-meal created an epidemic of pellagra that affected three million people in the American South between 1906 and 1940, causing a hundred thousand deaths (Bollet 2004: 160). It has been blamed for Cook's odd behaviour before his death. Sir James Watt supposed that Cook's bad attack of colic on the second voyage was caused by a roundworm infection caught in Tahiti that left his intestines unable to absorb vitamin B, leaving him prey to the ungovernable swings of mood on the third voyage that are typical of the disease (Watt 1979: 155). Pellagra is also accompanied by eruptions on the exposed skin of the arms, neck, and face, and by a staggering walk and a swaying habit of the upper body. Mania and convulsive tears are typical too (Carpenter 1981: 9–11). Kerguelen, the French navigator, complained of erysipelas on his second voyage, so it is possible that his erratic and violent actions during that expedition to the Indian Ocean were, like Cook's during his third voyage, owing at least partly to a lack of niacin.

Apsley Cherry-Garrard remembered that Captain Scott, who did not believe that any nutritional factor was involved in scurvy, was a victim of moods and depressions lasting for weeks, and that "he cried more easily than any man I have ever met" (Cherry-Garrard 1994: 206). The famine aboard Jean de Lery's ship while making its way from Brazil to France resulted in symptoms that sound like the dementia of pellagra: "The senses are alienated and the wits dispersed; all this makes one ferocious, and engenders a wrath that can truly be called a kind of madness" (de Lery 1992: 212). Captain Cheap lost his wits after the wreck of the *Wager*, shooting one of his midshipman fatally in the face and then not long afterward forgetting his own name (Byron 1768: 173). Bligh said that some of the men in the Bounty's launch had "apparent debility of understanding" by the time they got to Timor (Bligh 1937: 2.219). The Muselmaenner of Auschwitz, identified by Primo Levi and so called because the victims would sway like students in a madrassa reading the Q'ran and stagger when they walked, were exhibiting classic symptoms of pellagra, anciently known as "the dance of St. Guy" (Agamben 1999: 45; Carpenter 1981: 9–11). Ptyalism, or the overproduction of saliva, is a symptom of pellagra, and so is black tongue—both listed by Willis as symptoms of scurvy (Carpenter 1981: 214; Willis 1684: 63, 179).

As for beriberi, John Davis's description of his terrible journey from the South Atlantic to Ireland includes the swollen legs and genitals characteristic of the disease. Foster identified similar swellings as a recurrence of scurvy in the lower limbs of Cook and his own son, but it may have been beriberi (Forster 1982: 4.689). Certainly the dropsical symptoms associated with scurvy, the flesh pitting when pressed, is typical of beriberi, as is nerve damage. Cook's prohibition against the eating of "slush," the fatty residue left on the sides of the copper boilers after cooking salt meat, was intended to prevent a set of reactions he called scorbutic, but that may very well have belonged to beriberi or to tropical sprue. Trotter describes its effect on some Chinese cooks who were especially fond of slush: "They became extremely corpulent, and were shortly overrun with scurvy.... [Their] skin appeared as if it would burst ... [yet] their cunning to procure the slush of the meat exceeded all description" (Trotter 1795: 54). Byron remembered Cheap at his worst with "his legs ... as big as mill posts, though his body appeared to be nothing but skin and bone" (Byron 1778: 173), suggesting a combination of all three diseases: scurvy, pellagra, and beriberi.

The genius of scurvy overwhelms the distinctions between all other diseases, shuffling their symptoms as it pleases and dealing them out in macabre permutations. This genius extends to the environment, too, defined not simply by the body of the victim, but reaching to the timbers of the ship, the air through which it moves and the ocean on which it floats. Seldom was a serious outbreak of scurvy unattended by horrid transformations of the ship, for the sick often outnumbered the healthy and, finding it difficult to move about and with no one to help them, they were left unclean. While that would produce a noisome state of affairs under any circumstances, it was exacerbated by the putrescence that characterized the disease: the breath and stools of the victims were widely reported to be exceptionally malodorous, like the lesions on their skin. Witnesses describe a circuit of infection, with men falling sick—or sicker—from the filth sick men produce. Alain Corbin reports that in the catalogue of bad smells scurvy has always held first place (Corbin 1986: 41). In the course of Alvaro de Mendana's navigation of the Pacific, Pedro Fernandez de Quiros found himself with only six men capable of work, the rest below deck groaning and succumbing to the miasma they were making: "The sick became rabid from the effluvia of mud and filth that was in the ship" (de Quiros 1904: 1.105). A surgeon on the prison ships at Brest described a scene of multiplying miseries: "As the heat and stench increased, many fainted, some broke out into raving madness, uttering the most incoherent and horrid imprecations" (Crosfield 1797: 5). Philip Saumarez on the *Centurion*, in Anson's squadron, reported, "Our sick ... caus'd such an unwholesome Stench as contributed to Infect the rest" (Saumarez 1739–43, 23 April 1741): putrefying bodies "spreading infection through the neighb'ring air," as Richard Burton puts it in his translation of Camoens. One of the most powerful theories concerning the origin of scurvy traced it from the air, asserting that nothing was more corrupting than the foul air emanating from scorbutic flesh (Addington 1753: 3–4).

Although contagion is often discounted in discussions of scurvy, Lind was interested in the idea, while the *Encyclopedie* (1751–77) identifies it directly as a contagious disease, a position later adopted by the American chemist Samuel Mitchill (Lind 1753: 2; "Scorbut" 1765; Mitchill in Trotter 1804: 2.406). The circular contagion obscenely evident in de Quiros's and Saumarez's accounts provides a notable example of scurvy's genius insofar as it is seen causing itself. Like all personifications, and like the gods, it abolishes the difference between cause and effect: it is what it does (Foxhall 2012: 131). Scott seldom refers to it as anything else but a formidable intelligence, inimical and cunning: "The evil having come, the great thing now is to banish it.... [Our] determination [is] not to give our dread enemy another chance" (Scott 1907: 1.399–400). A. H. Markham recalled what it was like to fall under its influence as he and the crew of the *Alert* sleighed northward toward the Pole: "The terrible idea forced itself on us that this fearful disease was slowly but surely laying its hands on us" (Markham 1880: 275). It was not difficult to imagine a scorbutic hold filled with rotting bodies as the center of evil. Having visited Paradise, Louis-Antoine de Bougainville was compelled to report as he approached Indonesia, "There have been many arguments about where Hell is situated, truly we have found it" (Bougainville 2002: 135).

The effluvia of scorbutic bodies were believed to rot the very fabric of the vessel. Thomas Trotter gained his first knowledge of scurvy aboard slave ships, where it was rife. These vessels were commonly supposed not to last more than ten voyages, for "the heat and stench arising from diseased bodies rot the very planks" (Coleridge [1796] 1970: 138). In the poem which owes so much to Coleridge's interest in the effects of scurvy, the decayed state of Death and Life-in-Death is answerable to the skeletal and disarticulated state of the spectre-bark on which they sail, "a plankless Thing / A rare Anatomy!" (Coleridge 1993: 31). In scorbutic voyages, this diseased equivalence between the ship and its company is often noticed. At the furthest reach of Mendana's expedition, the corruption of bodies ("ulcers coming out on feet and legs") was matched by the parlous state of the vessel: "Of the hull of the ship, it may be said with truth that only the beams kept the people above water, for they were of that excellent wood of Guayaquil called *quatchapeli* .... The ship was so open in the dead wood that the water ran in and out ... when we sailed on a bowline" (de Quiros 1904: 1.105–7). Similarly, when he was sailing in bad weather off the northeastern coast of Australia, Matthew Flinders reported many men down with scurvy, himself covered with scorbutic sores, and the ship in a crazy condition, its bottom worm-eaten and leaky, the stemson decayed, with "rotten ... timbers, planking, bends, tree-nails &c" (Flinders 1814: 2.135; Flinders 2015: 2.179). Carteret found himself in the same condition off Pitcairn Island, the ship taking water, sails splitting, the rigging worn out, and scurvy making great progress among the crew (Carteret 1965: 1.151–54). On several occasions, British crews taking French prizes were convinced they had been tainted by the enemy craft (Lloyd and Coulter 1961: 3.133, 160). Perhaps the most lurid example of a ship disintegrating alongside its crew, prey to its own unique scorbutic taint, is provided by John Davis in the journal of his voyage from Brazil, their only food a freight of dried penguins.

> Our dried Penguins began to corrupt, and there bred in them a most lothsome and ugly worme of an inch long.... There was nothing that they did not devoure, only yron excepted: our clothes, boots, shooes, hats, shirts, stockings: and, for the ship, they did so eat the timbers as that we greatly feared they would undoe us by gnawing through the ships side ... so that at last we could not sleep for them, for they would

eate our flesh and bite like Mosquitos. In this woful case ... our men began to fall sick of such a monstrous disease, as I thinke the like was never heard of: for in their ankles it began to swell; from thence in two days it would be in their breasts, so that they could not draw their breath. (Davis 1880: 125)

It is a short step from the twin corruption of a ship and its crew to the ocean itself. The Ancient Mariner's memorable exclamation, "The very deep did rot: O Christ! / That ever this should be! / Yea, slimy things did crawl with legs / Upon the slimy sea" (Coleridge 2004: 171, ll. 123–26; Fig. 5), is indebted to Sir Richard Hawkins's description of a rotting sea during a calm in the South Atlantic when his crew were already scorbutic, infected by the "loathsome sloathfulnesse" of the ocean itself. "The Sea became so replenished with severall sorts of gellys, and formes of Serpents, Adders, and Snakes, as seemed wonderfull: some greene, some blacke, some yellow, some white, some of divers colours, and many of them had life." It was Hawkins's opinion that the sickness was latent in the sea, "And were it not for the moving of the Sea by the force of windes, tydes, and currants, it would corrupt all the world" (Hawkins 1906: 17.75–76). His experience of the sinister extent of the wonders of scurvy was not unique. On his voyage to the Pacific, John Byron found himself in the South Atlantic with all his senior officers incapable of duty, surrounded by a sea "as red as blood" (Byron in Hawkesworth 1773: 1.12). On the same ship, John Bulkeley remembered the dying captain prophesying with terrifying exactness the sufferings in store for the living just before the ship sailed through a bloody sea. Johann Reinhold Forster made experiments on phosphorescence in the Indian and Pacific Oceans, believing that the beauty of the phenomenon was owing to the putrefaction of tiny mollusks and fish, rottenness at the heart of "that luminous appearance which we so much admire"—a mixture that alternately attracted and repelled Stevenson when he sailed past the "precarious annular gangways" of the Southern Ocean (Forster 1996: 27; Stevenson 1922: 16.212).

From the first, reports of scurvy had been considered an unhappy notice of a disgraceful affliction. It was foulness itself, having its source in foulness, and capable of making all things foul. Its presence betokened dirtiness, idleness, and loss of morale. After Cook's success with hygiene, diet, and medication on the *Endeavour*, there was apparently no reason why any well-stored and well-run ship should suffer from it, no matter how far it had to sail. So when George Vancouver, who had been a midshipman on the *Resolution* during the second voyage, and on her consort the *Discovery* during the third, found scurvy on his own ship when he was off the northwest coast of North

Fig. 5. David Jones, *Death Fires*. © Trustees of the Estate of David Jones.

America, he was devastated. It caused him "utter astonishment and surprise." All the measures taken by him in imitation of Cook's model "on this occasion seemed to have lost their effect," and his disappointment was "inexpressible" (Vancouver 1984: 4.1471). Bligh had been master of the *Resolution* and was under no illusion of what was expected of a naval commander on voyages to the South Seas. "The scurvy is really a disgrace to a ship where it is at all common, provided they have it in their power to be supplied with dried malt, sour Krout, and portable soup" (Bligh 1937: 219). When scurvy made its appearance on the *Bounty*, during a voyage when food and its distribution had been the subject of furious disputes, Bligh was having none of it. He had been punctilious in following Cook's rules, which he had learned at firsthand; so when his surgeon, who was an alcoholic, reported what were fairly reliable symptoms of scurvy, although they did not include ulcerated gums or loose teeth, he put them down at first to prickly heat, and finally to rheumatism (1.359–63). James Morrison was in no doubt it was scurvy, reporting it in his journal, adding that the worst affected were given malt (Morrison 1935: 29). When he read this, Bligh made a marginal comment in the third person: "Captain Bligh never had a symptom of Scurvy in any Ship he commanded" (Bligh 1937: 153). A blend of the same surprise and denial is found in Scott's journal when scurvy attacks the *Discovery* expedition. Unable to explain the presence in his company of a disease he cannot but acknowledge as a reproach, Scott finds himself caught between two stools. His candid resolution to "have everything above suspicion" sits uneasily next to his prudential reservation: "The great thing is to pretend that there is nothing to be alarmed at" (Scott 1907: 400, 399).

He took up a position on scurvy that has always helped confound the facts of it. It is possible that Cook's proud boast of never losing a man to scurvy in the Pacific did not extend to Tupaia, the Ra'iatean priest who perished in Batavia, probably of a mixture of typhoid fever and the scurvy he contracted off the Australian coast four months before. Cook's cold summary of the case indicates problems with food and temperament, none of which he cared to blame directly on scurvy: "Tupia['s] death indeed cannot be said to be owing wholly to the unwholsom air of Batavia, the long want of a Vegetable diat [*sic*] which he had all his life before been use'd to had brought upon him all the disorders attending a sea life. He was a Shrewd, Sensible, Ingenious Man, but proud and obstinate which often made his situation on board both disagreeable to himself and those about him, and tended much to promote the deceases which put a period to his life" (Cook 1955: 442; Druett 2011: 209). Cook and Bligh perhaps did not want their records blemished by any unambiguous trace of scurvy, but there were other reasons for loathing and disowning it. Lassitude was the most predictable symptom of this infirmity, and the least likely to get a tolerant reception on a ship, where heavy labor was the priority. "The truth is," wrote Melville in *Omoo*, "that among sailors as a class, sickness at sea is so heartily detested, and the sick so little cared for, that the greatest invalid generally strives to mask his sufferings" (Melville 1847: 48). By "sickness," Melville almost always means scurvy.

This kind of discretion has often acted as a filter when reporting nutritional diseases. E. L. Atkinson is supposed (I believe erroneously) to have sunk the evidence of scurvy on the four corpses in Scott's tent, believing such an affliction to be incompatible with tragic heroism. Wilson did not want to tell Scott when he diagnosed it. He wrote: "Royds was talking to Armitage in his cabin, so I reported it to them, and at present we are the only three who know about it" (Wilson 1967: 192). When the bad news was communicated to Scott the thought crossed his mind that he might keep it a secret, either by pretending to the men it was nothing serious, or by not including it in his official report: "I had been tempted to omit this matter as calculated to cause unnecessary anxiety, but, reflecting the rumour might spread from some other source and become greatly exaggerated, I had finally decided to state the facts exactly as they were" (Scott 1907: 240). However, he was very relieved to see the scorbutic Shackleton carried off to New Zealand. His Jonah had been satisfactorily ejected, along with the disease he carried: "The word 'scurvy' has not been heard this year, and the doctor tells me there is not a sign of it in the ship" (150).

Commercial self-interest prevented the full disclosure of the causes of infant scurvy in the United States in the 1930s; and when in 1942 in Britain, a Manchester doctor reported to the Ministry of Health that he had diagnosed eight cases of scurvy in the previous four months, the official response entirely neglected serious food shortages as a threat to public health, and read (much like Cook's judgment of Tupaia) as follows: "It is quite clear that such persons as do suffer from scurvy do so of their own free will" (Carpenter 1986: 157, 164; *The Observer*, 20 September 2000, 9). Even after the importance of vitamins to human health was well established in the twentieth century, a powerful preference for germ theories over nutritional regimes, combined with corporate cunning and political opportunism, led to the disgrace of figures such as Thomas Babcock who led the fight against pellagra in the southern states between 1908 and 1914. His epidemiological work was condemned as injurious to the state of South Carolina (Bollet 2004: 161).

Among the reasons that prompted the denial of scurvy on a ship, shame is certainly a powerful one, and not simply because scurvy is associated with dirt, bad smells, and loss of esprit de corps. In a world where order was maintained by means of hierarchies of command and an elaborate division of labor, any interference with the steady achievement of outcomes was dangerous. Greg Dening's analysis of Bligh's bad language showed that vituperative and abusive speech was not only impolite and demeaning but destructive of the grammar of command on a ship, allowing passion to usurp meaning and insults to supplant and cheapen the words which, according to the customary confirmation of any order ("Make it so!"), rendered things as they ought to be. In the history of whaling, where skippers deliberately stayed at sea in order not to lose their crews, their own tempers quickly became ungovernable. Melville and Beale suffered from this, Beale remembering how his captain became intolerably abusive, but notwithstanding that he had ordered the second mate to overload a cannon, cried like a child when it took off the man's arm (Beale 1839: 343–44). Ahab is the fictional synthesis of the infatuation of scorbutic commanders. Inexplicable, terrible, and capricious, they exhibited the properties of scurvy itself, lieutenants of a dangerous and unpredictable force menacing the system and even the physical structure of a ship, just as it menaced the taxonomy of knowledge and the propriety of language. The leakage of the sickness from one part of the body to another, from that body to other bodies, from bodies to minds and back again, from bodies to the frame of the vessel itself, and finally from there to the ocean and the air, proved how immune scurvy was to methods of definition and control. It could overrun every boundary on which physical coherence depended and every distinction on which language and specialist knowledge relied.

To some extent, any disease on a ship (especially an infectious one such as typhus or yellow fever) created disorder, but with scurvy, the leakage and loss of distinction was not limited to physical factors. In one way or another, it entered into the temperaments and minds of those involved and set loose those passions that people of a scientific or professional bent would rather not own they feel. We have heard Cook, Bligh, Vancouver, and Scott reacting to scurvy with painful astonishment and degrees of denial, finding it an impediment to the business in hand but also evidently disturbed by an ill-defined guilt they would rather disavow. With those unfortunate enough to succumb to the scorbutic temperament, it is clear that a plain statement of the case comes hard, evident from this querulous entry in William Wales's journal: "Brewed Wort for some of the People who began to have symptoms of the Scurvy. I suppose I shall be believed when I say that I am unhappy in being one of them" (Cook 1961: 64, n. 3). Why should he not be believed, and what was the reason for his defensiveness? Presumably Wales was aware of Cook's formidable reputation as a commander superior to the vagaries of this disease and knew how heretical his confession would sound. Possibly he went further, as if it were more than mere rhetorical or formal unhappiness that was his burden, but real misery at being part of a disgraceful community of suffering. Perhaps Wales was peering over the gap that separates the inside of unreflective passion, where he sits, from the outside of cool observation, where he used to belong, and it is a position he does not enjoy. Nevertheless, he recognizes in his skeptical audience a group of people with whom (for the moment at least) he has not much in common, especially of mutual confidence, whereas with his scorbutic companions, he is affected by what is affecting them, relying on them for company of a sort. Beginning his eyewitness account of the wreck of the *Wager*, John Byron feels obliged to excuse his position in the narrative as if starvation and scurvy had already excluded him from the company of his readers: "The greatest pain I feel in committing the following sheets to the press, arises from an apprehension that many of may readers will accuse me of egotism" (Byron 1778: v). William Funnell frankly offers his own miseries as entertainment for the public, convinced that pity is not a reaction he can expect: "The great Variety of Accidents we met with, and the ... particular Accounts of the manner how our Attempts miscarried, I hope cannot but be very acceptable to the inquisitive Reader" (Funnell 1707: [ii]). In scurvy's last great exhibition of its genius, the position of the sufferer was almost formalized. One of Shackleton's team wrote, "If we had everything we wanted we should have no privations to write about and that would be a serious loss to the 'book.' Privations make a book sell like anything" (Alexander 1919: 195). The estimate of the reader's attitude is not mistaken, for in Nancy Mitford's *The Pursuit of Love* (1947) there is a character called Davey: "Fascinated by Polar expeditions, he liked to observe, from a safe distance, how far the body can go when driven upon thoroughly undigestible foodstuffs deficient in vitamins. 'Pemmican,' he would say gleefully falling upon the delicious food for which Aunt Emily's cook was renowned, 'must have been so bad for them!'" (Mitford 1947: 191–92). Whatever makes Davey's victims take the risk of showing their scars in the public, it is evidently as imperative as the impulse that drives the Ancient Mariner periodically to give his story to a listener not apparently selected for his compassionate nature. That is to say, it is not an appeal for sympathy, nor is it a plea for absolution; and though it may be defensive, it is far from diffident. It is as if the person on the inside is saying to the one on the outside, "Maybe I cannot say anything you will credit or tolerate, and maybe my misery amuses you, but I have seen what you will never see, and felt what you will never feel"—the candor of passion showing off its impenetrability to prudential selfishness.

First of all, Thomas Willis and subsequently Thomas Trotter took up the challenge of scorbutic passion, showing how the gruesome alterations of the body were accompanied by alterations sometimes quite as shocking in the mind. Willis thought that the scorbutic taint generally weakened the "moisturizing juice" and "nervous stock" of the animal spirits, leaving them thin and languid (Willis 1684: 179). This sour liquid occasioned a shrinkage of the self he calls "a dejection and as it were a falling down of the whole Soul" (ibid.). But typically, this is a scorbutic symptom that can make way for one quite opposite in its energy and drift. "For indeed, the Liquor watering both the nerves, and the nervous parts, sometimes disceding from its naturall disposition, is so much stuff'd with heterogeneous and explosive particles, that the animal spirits ... are irritated into continuall, as it were cracklings, or convulsive explosions" (61). There are some memorable accounts of these cracklings and irritations in the literature of scurvy, two of which I shall briefly mention.

In a strange aside in the journal of his first voyage, Cook confesses that if it were not for the pleasure of being the first discoverer even of trifling hydrographic details such as sands and shoals, his work would be insupportable. By the third voyage, his addiction to the private pleasure of seeing things for himself became distinctly stronger and more exotic when, for instance, his rapidly growing interest in local ceremonies required he take his clothes off to witness one in Tonga. Along with this active interest in anthropology, Cook became increasingly passionate if his will were thwarted, indulging the kind of bad language that lost Bligh his ship. He would stamp on the quarterdeck in a rage, "paroxysms of passion, in which he often threw himself upon the slightest occasion" (Cook 1967: 1.cliii). Even more sinister, he began to order punishments for local thieves that were torturous (slicing to the bone, cutting off their ears) and well outside the remit of his instructions and commission. At least two of his officers thought he was infatuated in the weeks before his death, so prone was he to intense tides of feeling and irrational actions, including fits of foot-stamping so extravagant they compared them to the Maori rhythmic dance, the *heiva*. If Cook were suffering from severe pellagra at this stage, as Watt suggests, it is not entirely idle to suppose his dancing the heiva was symptomatic of the spasmodic beat of the ancient dance of St. Guy, one of its notable symptoms. In any event, Watt's interpretation of this phase of his life as consistent with the nervous or irritated manifestations of a deficit of niacin seems entirely plausible as an explanation for behavior so odd that it still troubles his biographers to explain it.

Ives-Joseph Kerguelen-Tremarec was a Breton officer of the French Navy who made two voyages of discovery to the island in the southern Indian Ocean that bears his name. Each expedition was an uncanny reflection of the other, during neither of which did he actually set foot on his island, and of which he never wrote his own account, copying it instead from a French translation of an unauthorized account of Cook's third voyage, written by John Rickman. But he was sufficiently a victim of scurvy to write a firsthand account of the disease, *Reflections sur le scorbut* (1782), an affliction that coincided in his case with various and furious extravagances, such as deserting his consort not once, but twice; carrying his mistress, Louise Seguin, on the second voyage; publicly accusing his senior officer de Cheyron of trying to seduce her; kicking her in the stomach in front his crew and then trying to throw her overboard; and, when he failed in that attempt, flinging two valuable chameleons of de Cheyron's into the ocean instead (Boulaire 1997: 104–13, 143–45; Brossard 1970: 1.478; Delepine 1998: 113–15; Lamb 2005: 1–9). There is no apparent point of reconciliation between his paroxysms of jealousy and rage and his formal account of the disease that impelled him to indulge them, although it appears that his worst excesses coincided with a bad outbreak of scurvy on his ship, *Le Roland* (Delepine 1998: 115).

Trotter discusses an example of jealous rage very like Kerguelen's: "Mr Farquhar, surgeon of the *Captain*, mentions a very strong case of jealousy that happened in that ship.... The wife [of the officer concerned] was young and remarkably handsome; and these fits were often caused by other men paying her attention" (Trotter 1804: 3.364). Although he doesn't specify the presence of scurvy, he gives the story in the context of the symptoms of nervous debility that generally accompany it, such as imaginary fears, sudden dejection, "sighings, moanings, sudden laugher and crying" (3.362). He goes on, "The state of mind which attends this condition of body is difficult to be described ... the temper is soured with trifles, becomes fretful, irritable, and passionate.... This gloomy turn of mind often assumes a temporary derangement of the intellect; and not only forgets all its old attachments, but shews the utmost signs of dislike to those who had been most dear" (3.364). In this labile condition, the scorbutic imagination is especially prone to nostalgia, a condition first formally named as such by Banks when the *Endeavour* was becoming scorbutic on the way to Batavia, now adjusted by Trotter to "scorbutic nostalgia" (Banks 1962: 2.145; Trotter 1792: 45). Banks noticed how, in this state of feeling, an emotion would be rapidly supplanted by its opposite when the ship turned for home, "The sick became well and the melancholy lookd gay" (Banks 1962: 2.145), a sequence Trotter noticed in reverse as his patients awoke from dreams of home to bitter reality: "In dreams they are tantalized by the favourite idea; and on waking the mortifying disappointment is expressed with the utmost regret, with groans, and weeping, altogether childish" (Trotter 1792: 45, 44). Several times, he notices their morbid susceptibility to slight impressions, resulting in tears, withdrawal, or interminable and pedantically exact descriptions of feelings arising "from any ruffle of temper occasioned by changeable weather, or other slight causes" (44). This susceptibility springs from a tenderness of nerves often observed in scorbutic cases. Sounds and smells could cause them bliss but also agony, or even kill them. Their imaginative lives were therefore very active even if they were not pleasant—much more so than their bodies. Scurvy provides, William Falconer believes, "a remarkable instance of the influence of the passions of the mind" (Falconer 1791: 124).

The neuroscientific explanation for such a strange affective counterpart to physical decay is as yet incomplete, but so far it offers this intriguing hint: As well as synthesizing collagen, ascorbate produces the chemicals crucial to the healthy action of the nervous system. It is especially important in sustaining the supply of dopamine and serotonin to the brain, responsible for regulating neuronal excitability (Harrison and May 2009). When these sensory modulators fail to work, the ear will hear too much, the palate taste, and the nose smell, more than can possibly be pleasant, and the eye become inordinately sensitive to light and color. These uninhibited sensations are commonly called "sensory phantoms," the offspring of unregulated excitations in the cranial nerves, but actually they are intensifiers, carrying more information from the senses to the brain than it can handle, so that fragrance becomes disgusting, light blinding, and music deafening. The adjacency of scurvy to the legendary scandals of Bligh, Cook, and Kerguelen is simply to be observed at this point in my history. They serve to show that excitement arms the patient against the more languid tendencies of the disease and encourages an oscillation between feelings of the most opposite kinds: voluptuousness and terror, disgust and ecstasy, desolation and satisfaction, exuberance and languor. The naturalist Anders Sparrman landed at Dusky Bay with scurvy at the same early stage as that which Wales found so awkward to express. In his journal, he remembers how he shot some ducks: "The blood from these warm birds which were dying in my hands, running over my fingers, excited me to a degree I had never previously experienced.... This filled me with amazement, but the next moment I felt frightened" (Sparrman 1944: 49). It was prior to this, on the first loop into the Antarctic seas, that Cook saw on his rigging the most beautiful ornaments of ice he had ever seen, and no sooner was he charmed by the sight than he was terrified at the thought of the danger in which it put his vessel (Cook 1777: 1.37). Shortly before this, the elder Forster had been studying phosphorescence, noting the paradox of what he believed was the putrescent remains of minute polyps that were nevertheless capable of bursting into "that luminous appearance which we so much admire," like the scurvy of the sea or the sea snakes of *The Rime of the Ancient Mariner* (Forster 1996: 59). In this he echoed Boyle, who was fascinated by phosphoric light not least because it lent such glamour to rotten fish and stale piss (Daston 2001: 312–13). Again and again in the vicinity of scurvy, beauty and horrified disgust are wrapped up in the same bundle.

Let me summarize then the various impediments scurvy puts in the way of any satisfactory firsthand report of its effects. First, there is the confusion caused by its appearance as it impersonates other diseases or is impersonated by them, apparently singular only in its multifariousness. Second, there are the difficulties of diagnosis and prognosis owing to inconsistent patterns of its attack, sometimes striking at the joints, sometimes at the gums, sometimes at the bowel or stomach, sometimes at the mind, and then seeming to disappear (as often it arrives) for no better reason than a change of mood. Third is the uncertainty concerning the efficacy of available remedies, which sometimes seem to work and at other times to fail. Fourth are the punctilios of the navy that teach officers to be discreet about any outbreak of scurvy on their ships, easier to manage because of the crowd of ailments it mimics, one of which can easily be identified as the present malady: rheumatism, for example, or prickly heat. Fifth is the factor of the distance and isolation that so often define the situation of scurvy, whose worst rampages take place when the ship is removed from any neutral observer capable of authenticating what has happened. Sixth is the penetration by scurvy of all aspects of that remote situation, comprehending not only the body but also the ship and the elements that surround it, finally infecting the sensorium that is reacting so energetically and imprecisely to its environment. Seventh, there is the spectrum of vivid feelings, ranging from intense pleasure to uncontrollable unease, brought into play by an illness capable of provoking feelings, words, and actions so extraordinary they inevitably compromise the authority of the person who is giving an account of them, or using them to give the account.

Take, for example, Kerguelen in his short treatise on the subject, so much at odds with the violence of his antics near Kerguelen Island. Like Trotter, he lays his emphasis on fresh vegetables and ripe fruit, mentioning Lind with approval and observing how the very sight of the cure can actually effect it: "On remarque que la vue de ces fruits releve l'esprit abbattu des scorbutiques expirans" [One observes that the sight of fruit revives the sunken spirits of the expiring scorbutic patient] (Kerguelen-Tremarec 1782: 214). Observe also how accurately he seems to understand the extreme nature of the illusions that swept him into such astonishing improprieties: "Et souvent leur imagination échauffée par la douce illusion d'un songe les transport dans des jardins charmants & dans des vergers delicieux" [And often their imaginations, warmed by the sweet illusion of a dream, transport them to charming gardens and orchards] (213). At no point is Kerguelen capable of joining the voice of a medical analyst of this malady to the memoir of a victim of it. The last we hear of him, he is sailing for the third time toward his island, so repulsive to him that twice before he refused to land on it, and twice so urgent was his need to be away that he left his consort in the lurch. Now he is commander of a ship filled with goods intended as the foundation of a new society, a utopia like that his mentor Philibert Commerson claimed to have seen on Tahiti. Whenever there is an attempt to unite steady observation with uneven experiences, the professed candor of the reporter sits awkwardly with the turbulence of what he is reporting; and then a corresponding friction develops in his relation to the reader, as Wales demonstrates. It was shortly after his witnessing a scorbutic, blood-red ocean that Byron met the Patagonian giants, men seven feet tall with frames proportionately large, who dwarfed him and his crew. It was a claim that had been made before by mariners, but given the scientific agenda of these voyages of the 1760s, it excited nothing but hilarity at home, including a brilliant mock-utopia from Horace Walpole. There is, therefore, always a possibility that a narrative of scurvy might itself exhibit signs of the disease, either being dulled by the bleakness and the pain, or brightened by its hallucinations. Two celebrated scorbutic voyages of discovery, Anson's to Juan Fernandez and Bougainville's to Tahiti, attracted witty and skeptical "supplements," one by Gabriel Francois Coyer, the other by Denis Diderot. In them, the extravagance of scorbutic reactions contributes to utopias of pure pleasure, one based on bibelots and romances, the other on erotic abandon (Coyer 1752; Diderot 1992).

On the other hand, when Dampier pauses to give the exact hue, consistency, and volume of a cloud, or Pascoe Thomas spends time specifying the tints on the wing of a little red bird, it is likely that what is being recalled is not just a hallucination, but a sensation so intense that the description of it exhibits the startling precision of a specimen drawing. When a combination of tedium and incipient scurvy causes Cook to complain repeatedly of the desolation of what he sees, or Forster to find the noise of drunken mariners perfectly intolerable, there are traces of the black and obsessive mood in which, as Trotter reports, the scorbutic victim exhibits "a selfish desire of engrossing the sympathy and attention of others, to the narration of their own sufferings" (Trotter 1804: 3.62). But there is also a revelation of the incidents and emotions of lives on ships that seldom makes its way into journals. Very acutely, Trotter points out that this insight is "a new subject in nautical medicine," one in which "the hero and the infant unite." The staunchest tar "may therefore be allowed to exemplify the same extremes of the great and the little, in his composition, in his pleasures, his passions and diseases" (3.61). This provokes an intriguing uncertainty in the testimony of the afflicted. By the time of his landing at Dusky Bay, Cook had become quite uncertain of his capacity to identify a fact. He confesses of the place, "We now saw it under so many disadvantageous circumstances, that the less I say about it, the fewer mistakes I shall make" (1.67). Trying to organize the violent see-saw of his feelings as the *Resolution* made its way once more toward the ice, and he grew ill with scurvy, the elder Forster gave up thinking about natural history and started listing his symptoms instead of his specimens: the pains in his neck and jaw, toothache, a swollen cheek (Forster 1982: 3.447).

Within the circuit of diseased observers who become childishly exact in the account of their sickness, or blindly obsessive about imagined scenes of elsewhere (on his second arrival at his horrid island, Kerguelen became obsessed with the charms of Atongil Bay in Madagascar [Delepine 1998: 113]),

Trotter places surgeons themselves, particularly their mates, who seem to have been especially prone to nostalgia (Trotter 1804: 373). For example, when Banks notices his exemption from the onset of nostalgia among the crew of the *Endeavour*, only to confess two days later that he, too, is longing for home, surgeons were immediately implicated in the conditions causing the disease they were inspecting and treating. So ardent was his pursuit of the facts of scurvy that William Stark, Pringle's protégé, deliberately contracted it (Lloyd and Coulter 1961: 3.316). Being an enthusiast for malt, he eschewed citrus juice on principle and lost his life; but even if he hadn't, who is to say that he would have been any more successful than Wales and Kerguelen in bridging the gulf between the vividness of his feelings and the requirements of a credible experimental account. I have often wondered whether Trotter's fascination with the temperamental side of the disease reflects his attempt to achieve something like an impartial statement of what had once been his own disorienting passions. Who knows how closely surgeons such as Perry on the *Endeavour*, Hutchinson on the *Dolphin*, and Ettrick on the *Centurion* participated in the symptoms of which they were giving an account, and whether their preferences for a systematic theoretical analysis had not something to do with a desire of stabilizing what was for each individual too turbulent and terrifying an experience.

Trotter on the other hand, and maybe Lind too from his days afloat, opted for empiricism in the hope that lonely facts might eventually coalesce into a shapely narrative. For this to happen, there was going to have to be some experimental value to the morbidly enlarged sensations caused by the scorbutic disturbances of the synapses. Was it possible, Trotter must have wondered, for scurvy to act as its own magnifying glass, revealing exceptional beauties and fathomless terrors hitherto unavailable to minds settled firmly within the norms of daily consciousness? Were its extravagances the flourishes of an occult legislation and harmony? There were good reasons to doubt that this might ever be the case, but empiricists had little choice but to support the possibility in what amounted to a conjectural defence of factuality, since sensations were their bedrock of knowledge. *Nihil in intellectu quod non prius in sensu*. From whatever angle they approached scurvy, sensations were the guides, fallible or sure. If the ear hears too much, and an eye blinded by all the light it lets in, then some benefit must accrue from the pressure of supererogatory information, something more than just noise and dazzle. In this respect, the history of scurvy sets a scene analogous to a larger stage occupied by Cartesians and Neoplatonists on the one hand, and by Epicurean materialists and the Royal Society on the other. The former regarded empirical knowledge as the residue of an assault from external contingencies, a humiliating subjection to the accidental and meaningless impressions made by out-ward things. The latter knew the weaknesses of the senses and had a choice: Either they could improve their performance by means of prostheses such as microscopes and megaphones, or they could sit down with senses suitable to our place in the chain of being, whether blunt or keen, content to take facts as facts rather than signs of imminent order. It is with that battle, and the stakes riding on it for those trying to clear up scurvy, that the next chapter deals.

# CHAPTER TWO

# Effluvia

## PART I

Bacon is only Epicurus over again.
—William Blake, *Annotations to Reynolds's Discourses* (1808)

The line of Hippocratic thought renewed by Thomas Sydenham and Herman Boerhaave, which emphasized the part played by the environment—airs, waters, and places—in the preservation of health, dominated both theoretical and empirical enquiries into scurvy. The importance of air, whether considered as the circumambient element or as the amount of oxygen in the blood, led by way of MacBride's notion of the fixed air in malt to the pneumatic theory of Beddoes, espoused with certain qualifications by Trotter. Innovations in marine ventilation, such as Samuel Sutton's air-extraction machine, endorsed by Richard Mead, were understood to clear the atmosphere of infection and the threat of scorbutic contagion, at least among those who believed it was a contagious disease. Systems for purifying water were believed to be equally important, although less practicable, but the drinking of fresh water played a prominent part in scorbutic fantasies and in the treatment of victims once they were back on land. Place, of course, was a critical factor, whether it was the absence of the homeland amidst the blank monotony of sea and sky, or its artificial substitution by means of baths of local soils aboard ship, or the luxuriant vegetation and waterfalls of a landfall that promised either a quick death from overexcitement or an instant cure. The Hippocratic environmental approach generally relied on two principles: contagion and putrefaction. Effluvia emanating from filthy quarters or carried in air overloaded with moisture and salt entered the body and poisoned it, disturbing its equilibrium and accelerating its susceptibility to decay. In *The Rime of the Ancient Mariner*, the dice game played for the parched and starving sailor is won by Life in Death, not Death itself. Scorbutic victims provided dramatic examples of putrefaction in the living body when their skin turned black and they began to emit a foul and (as it was thought) infectious stench from their sores and their mouths. Often the loss of flesh from the jaw and other parts of the body made a shocking sight. Dr. Buzzard reported to the Admiralty Enquiry into the Nares Expedition, "Sometimes the disorganisation of the flesh is sufficient completely to expose the bones" (RCA 1877: 198). Of all the routes taken by germs lurking in the environment, the mouth provided the most commodious way both in and out of the body. Bad air, rotten food, fetid water, and even lime juice (as some experts thought) cooperated with all the other effluvia that multiplied in the rotten spaces of a ship to turn living flesh into carrion.

We have seen that the proponents of a scorbutic etiology comprising defective food, infection, and contagion were leagued against those who believed that its cause was nutritional, a deficiency in food, air, or water rather than a taint. According to the latter school, sailors were destroyed not by what entered the body but by what didn't. Buzzard put the case with admirable clarity to the Nares Enquiry: "I have arrived at the conclusion that scurvy is a peculiar state of mal-nutrition, supervening gradually upon the continued use of a dietary deficient in fresh vegetable material, and tending to death after a longer or shorter interval." He was convinced that fresh vegetables and raw meat possessed "certain vital or living properties which are more or less destroyed in cooking" (RCA 1877: 191, 182). Although everyone agreed that once it had arrived, scurvy was exacerbated by the combined pressure of a fetid atmosphere, damp clothes, poor rations, and cramped quarters, its immediate or proximate cause, according to the doctrine of deficiency, lay not in mephitic effluvia but in the want of the vital ingredient that fresh vegetables and unsalted meat never failed to supply. That is to say, there was no inimical foreign substance invading the body whose expulsion was necessary to recovery, rather there was a mysterious quality in food that the organism must absorb if it was not to sicken. The model of physical health, according to the hypothesis of poisonous effluvia, was based on a balance of the humors disrupted by the interception of bad matter; whereas the model appropriate to deficiency was much more dynamic, involving a perpetual flow of particles in whose tides the body and nerves were immersed, acting and reacting with them. Trouble brewed only when the flow was stemmed.

"A Humane Body," Boyle reminded the reader of his essay *The Great Efficacy of Effluviums* (1673), "ought not to be look'd upon meerly as an aggregate of Bones, Flesh, and other consistents, but as a most curious and living Engin, some of whose parts, though so nicely fram'd as to be very easily affected by external Agents, are yet capable of having great Operations upon the other parts of the Body, they help to compose" (Boyle [1672] 1999: 7.267). Thus far, Boyle and Hobbes were of one mind, for the latter described the arrival of a sense impression and its sequel in the same way, "So when the action of the same object is continued from the Eyes, Eares, and other organs to the Heart, the reall effect there is nothing but Motion, or Endeavour; which consisteth in Appetite, or Aversion to, or from the object moving." In the event of appetite or pleasure predominating, the motion acts as "a Corroboration of Vitall motion, and a help thereunto" (Hobbes 2004: 40). Accordingly, the living engine of the body is altered for the better by incoming effluvia, the receptivity of its organs enhanced, and the overflow of their energy discharged back upon the world. A "knock or thrust from without" (as Cudworth calls it contemptuously [Cudworth 1731: 99]) is the beginning of a creative reaction to the environment, an outward flow that makes sensory impressions more vivid and exact in what Thomas Willis calls the irradiation, triumph, and ovation of the animal soul.

This doctrine of effluvia provided a model for those who understood sensation—and life itself—as a flux and reflux of matter in which losses are being perpetually made up, provided that new matter of the right sort keeps coming in. Boyle was an eminent example of how the atomism of Epicurus and Lucretius was making converts among philosophers, experimental scientists, and physicians on both sides of the Channel during the seventeenth century. Epicurus was mediated to the world chiefly via Lucretius's *De rerum natura*, a poem exerting a powerful influence on the work of Rene Descartes, Pierre Gassendi, Marin Mersenne, Thomas Hobbes, John Evelyn, Lucy Hutchinson, Margaret Cavendish, Thomas Willis, and Walter Charleton, among others, who became variously attached to the idea that all phenomena are the result of atoms moving through a void, deviating infinitesimally from the straight line into spirals and vortices of various curvatures that briefly cause bodies to take shape before being dissolved into their original particles (Wilson 2008). Newton's formula for fluxions provides a very apt illustration: "The continuous motion of a point generates a line, the motion of a line generates surface, and the motion of a surface generates a solid" (Guicciardini 2009: 171). From this perspective, everything subsists in a state of modification because all things, whether they cause or suffer change, are in a state of motion: all properties are really accidents (not primary but secondary qualities as Locke would say) and all events are contingencies whose connections with each other are never necessary (as David Hume was to emphasize). In the sixth book of *De rerum natura*, the mobile relations of matter to matter defy prediction or management, like the shapes of clouds on a windy day or the flashes of lightning in a storm. There were numerous elaborations of this basic thesis of flowing atoms, and predominating in the following pages are the ones first articulated by Willis and Charleton, who applied the model of Epicurean materialism to their observations on scurvy.

Lucretius's notion of the drift and flows of effluvia did not exclude infection. In his philosophical poem, he describes vividly the plague in Athens, listing symptoms which would not look out of place in an account of scurvy ("a mind disordered in all this sorrow and fear, a gloomy brow, a mad and fierce look, ears also troubled and full of droning, quick pants or deep breaths rising at long intervals" [Lucretius 2006: 583; 6.1182–86]). This pathological sidelight on the cyclic aggregation and diminution of bodies was understood by Willis and Charleton to arise not from corruption but from a blockage that hinders the free flow of matter inward and outward. Lucretius taught that in daily acts of ingestion and defecation, or in more tumultuous activities such as oratory and sex, something is alternately gained and lost by the bodies involved. The voice abrading the larynx as it leaves the throat may feel like an injury, whereas the emission of sexual fluids gives keen pleasure. Contrariwise, the entry of effluvia may be so violent as to cause discomfort, such as the brightness of the sun penetrating the eyes (Lucretius 2006: 303, 4.325–27), or it may be made so gently and variously that it produces nothing but delight, like the sunlit awnings in a theater that stain the audience with their colors (283, 4.75–80). It follows that there is an active as well as a passive mood to all sensation, and in each organ of sense there is a faculty of receiving and emitting energized matter. While the eye may be struck with the species or effigies of things—the rinds and films that dart incessantly from their superficies and leave their physical traces on the retina, the skin, or the ear—the opposite motion equally well may be induced. Objects striking the eyes of jaundiced persons, for instance, change their hue "because many seeds of this greenish-yellow colour stream out from their bodies to meet the images of things" (303, 4.335–36). Sexual desire is aroused by the blows of images "bringing news of a lovely face and beautiful bloom," yet it is consummated when the genitals send their seed "towards that whither the dire craving tends" (357, 4.1045–46). Ovid gives a Lucretian turn to his description of Ajax's suicide in the thirteenth book of *Metamorphoses*: "Into his brest (not wounded erst) he thrust his deathfull swoord. / His hand to pull it out ageine unable was. The blood / Did spout it out" (Ovid 2000: 331 [13.473–75]). This hydraulic flow inward and outward is observable in all aspects of a material existence.

Robert Boyle and Sir Kenelm Digby were fascinated by the challenge of confirming experimentally the truth of these propositions about the actions and reactions of effluvia. When Boyle noticed, for example, that under a microscope the color green is revealed to be a mixture of minute grains of blue and yellow, he conjectured that color is not a pure effluvium, but a compound of latent tints called out by the angle of light—not a property of the object, therefore, but an accident of its placement. He ran a number of experiments to show that green, the color that fascinates him most, is like a chord in music, blue united with yellow in an unstable compound (Boyle 1670: 44–85). He went on to show, by using colored glass and white cards crossed with colored ribbon, how these blends can be refashioned by the eye if the retina is taught to hold a tint and then print it on another, and what extraordinary depths and diapasons of green these experimental superimpositions are capable of yielding.

For his part, Digby was fascinated by nutrition and the mutual changes it induced. "Observe the sustenance we take; which that, it may be once part of our body, is first changed into a substance like our body, and ceaseth being, what it was" (Digby 1658: 8). Yet in his astonishing allegory of the bean plant as a willful monad operating anarchically in a state of nature, following "its own swing" and inclination for dominion by hoisting itself up over rival growth, the idea of food simply adding to mass is adjusted in favor of a crucial catalyst whose presence will cause a plant to flourish out of all proportion to its seed and whose lack will cause it to forsake its natural course, to dwindle and perish (idem 1669: 211; see Schmidgen 2013: 66–77). The presence of this vital principle causes limitless exuberance; its loss results in terminal collapse: it is, in effect, the principle of flow, what Buzzard called the living principle of food and motion. The sap of the plant is not just water, Digby concluded, "There must be something else enclosed within it, to which the water serves but for a Vehicle." And this he defines as "a balsamic Saline juice," or a "nitrous salt" (222–23), that is never so much part of the plant that its presence cannot be detected, and yet which acts as the motive power of the plant's organic independence.

This idea was developed by the most intrepid Epicurean of the age, Walter Charleton, in two treatises, *The Natural History of Nutrition* (1659) and later in his *De scorbuto* (1672). In the former, he begins with the same basic materialist proposition as Digby, that "to nourish is to substitute such and so much of matter, as was decay'd in the parts, namely flesh, nerves, veins, arteries, &c" (Charleton 1659: 2). Incorporated food is indistinguishable from what it replaces. Like Digby, Charleton soon stands back from this bald assertion, "As if Nutrition were nothing else but a selection and attraction of fit aliment; and that there were not required in every part a concoction, assimilation, apposition and transmutation" (9). He divides the substitution of matter into the renovation of the blood ("fewel of the vital flame") and what he calls "the instauration of parts," the repair of the fibrous, membranous, and nervous materials of the body (11, 59). The substance that accomplishes this latter task he calls the *succus nutritius*, or nutritive sap, a colorless balsam resembling the white of an egg that he has observed leaking from cut nerves. He associates this balsam with the cement that heals broken bones and with spermatic fluid. It is distilled out of food by the glands of the loins, notably the pancreas, stored in the brain and spinal fluid, and distributed through the body by means of the nerves (169). At no point, he affirms, does the system of the nerves overlap with the circulation of the blood. Thus in cases of paralysis, the muscle decays not through want of blood, but through a deficiency of the *succus nutritius* supplied by the nerves. Of all the organs enriched by this amazing albumen, a fluid that is itself the source of flow, the brain is the most important and the most retentive of what it possesses. Since the chief office of the brain is, according to Charleton, to govern a complex transfer between matter outside and mind inside, it is the brain that is responsible for the reflexive act that studies incoming effluvia and dispatches its own supply to embrace and transform objects. He says, "We see with our eyes, but do not by them perceive that we see, but by the mediation of another internal sense, or sensitive organ, the Brain, by which we judge of all objects offered to the External senses" (Charleton 1659: 122). Without the *succus nutritius* summoning this prime organ to perceive what is seen, Boyle's experiment on the color green would be impossible, for the eye alone would have no power to judge or change it.

Charleton improved the idea of the nutritional sap out of Harvey's work on circulation; but when he imported it into his discussion of scurvy he established his own division between the two systems of circulation, of the blood and the nervous juice. It was the latter to which he attended, along with his fellow specialist on nerves and the brain, Thomas Willis. Like Lucretius, they both allowed a place for contagion in cases of effluvial transfer from polluted bodies to healthy ones, curdling and souring the contents of the nerves until, in cases of paralysis, Willis says, the nerve is "discoloured like filth or dirt" (Willis 1681: 102). "Scurvy passes from infected people to healthy ones, transmigrating from one to the other by subtle effluvia, which alter and pollute the receiver's blood and succus nutritius by fermentation," wrote Charleton in a momentary desertion of his theory of disease as blockage. These particles, he says, cause a feculence that terminates in paralysis, stupor, and convulsion (Charleton 1672: 53). On the other hand, Willis and Charleton allowed for something like Digby's "swing" in the sensorium, a charge of reactive energy powered by an agent that cleanses and refreshes the nervous juice. The source of this energy was the pancreas, acting like a still or limbeck to purify the *succus nutritius* circulating in the nerves, for "if you examine the substance, temper, sweetness, lightness, whiteness, and softness of this said gland, you will easily realise that in no way is it inconsistent with the particular excrement to be removed from the said nerves" (104). Degeneration of the *succus nutritius* is explained by Charleton not only as an infection, then, but also as the result of a blockage preventing this cleansing operation of the nervous fluid by the pancreas. And blockage brings about a cessation in the flow of spirits—an impediment to the transfer of energy such as that promoted by Digby's nitrous or saline sap of a plant—inevitably preventing the renovation of "the dewy liquid of the nerves": "Therefore whenever it has ... lost its necessary qualities of good and praiseworthy nourishment, it will inevitably happen that all the spermatic parts (and indeed from these almost the entire body is constructed) not only slowly wither away ... but are also burdened with an unpleasant, incongruous and irritating *succus*, as a daily growing load' (42, 149, *marginalium* 238).

Thus Charleton anticipates Willis's account of how the depleted nervous juice reacts with the brain, nerves, spirits, and organs to provoke scurvy. But in several key respects, Charleton's idea of the *succus nutritius* is prophetic of later developments. Like Digby's nitrous salt or Willis's "latex" (Willis 1684: 173), it names something riding in the vehicle of food that is deployed by the body's organs as a catalyst for nutrition. Without it, the organism is "incapable of the daily repair of the substance of its parts" (Charleton 1672: 247). But it also forms a reserve of energy that may be beamed outward as "swing." In this respect, Charleton's *succus nutritius* looks forward to Jean Grimaud's *sucs nourriciers* which, in his *Memoire sur la nutrition* (1789), supply energy to a living machine that is constantly consuming itself. Grimaud's sap is distinct from the engines of the nerves and the blood in being inorganic, an external and in itself nonnutritive substance that penetrates the body but is not of the body (Grimaud 1789: 33; see Williams 2012: 212). In different ways, Grimaud and Charleton anticipate the biochemical solution to the enigma of scurvy by supposing an extra element in food that has nothing to do with mass but which is critical to the operation of the brain and nerves in the process of acquiring, sustaining, and renewing both mass and energy. Charleton is unique in supposing further that the system of nerves is sealed off from the circulation of the blood, a phenomenon now well understood by neuroscientists presently studying the function of the choroid plexus and the very high concentrations of ascorbate in the spinal fluid, the brain, and the eye (May 2013: 95–105).

Charleton's atomism allows him a central position in the debate that lasted at least two centuries concerning the dynamics of the reception of sensations and the production of ideas, an issue of considerable moment to theories of mechanism and to the practice of empirical observation, and one bearing distinctly on the history of scurvy's relation to the protocols of experimental knowledge. For devout Epicureans such as Boyle, it was axiomatic that the stream of effluvia made a physical impact on the organs of sense and that perception involved not only the print but also the material consequences of that original impression. Nothing exhibits this process more dramatically than sight, for according to Lucretius, the effigies, films, or rinds fly off from the surface of objects to lodge successively on the retina: "Where we turn our sight, there all things strike upon it with shape and colour" (Lucretius 2006: 297; 4.243). Nowhere in Lucretius's work is there any suggestion that perception is an analogy or a sign of an object: it is consubstantial with it, unmediated, a blow that is received and returned. There was, therefore, a direct correlation between the operations of the senses and the contents of the imagination; and reason itself, in sorting these traces of material impacts, was inspecting the traces left by coalescent and evanescent effluvia. Boyle hoped that one day it would be possible with the aid of microscopes to study their actual trajectories (Boyle [1672] 1999: 7.281).

There was no more ardent advocate of empiricism than Lucretius because of what he understood as the self-evidence of sensation: Nothing can be of greater credit than the senses, he declared: "Will the ear be able to convict the eye, or the touch the ear? Will the taste of the mouth again refute the touch, will nose confound it, or eye disprove it?" (Lucretius 2006: 315)—or as Lucy Hutchinson has it, "Can th'eare, the sight denie? / Shall th'eare, or tast, the feeling sense oppose? / Or shall the eie, dispute against the nose?" (Hutchinson 2012: 1.251; 4.508–10). George Berkeley was no materialist, but he was quite sure that the senses corroborated each other, especially the eye and the finger. Hume's *Treatise of Human Nature* (1738) begins by laying down the self-evidence of sensation as a fundamental principle of empiricism: "Those perceptions, which enter with most force and violence, we may name impressions; and under this name I comprehend all our sensations, passions and emotions, as they make their first appearance in the soul. By ideas I mean the faint images of these in thinking and reasoning" (Hume 1978: 1). And by images, he meant *images*: "That idea of red, which we form in the dark, and that impression, which strikes our eyes in the sun shine, differ only in degree, not in nature" (3). Erasmus Darwin was adamant in defense of this model of perception, telling Beddoes, "If you allow an idea of perception to be *a part* of the extremity of nerve, of touch, or sight stimulated into action, that part must have figure, and that figure must resemble the figure of the body acting on it" (Stock 1811: Appendix 6, xliii). Associationists such as

David Hartley, Joseph Priestley, and the early Coleridge were not quite of this opinion, believing that the transactions between the nerves and the brain took place as vibrations rather than as the direct impact of matter on the sensorium, but they insisted that even the most abstract conceptions could be broken down into the simple sensations of which the first ideas were copies, now lodged in the cerebellum as miniature vibrations (vibratiuncles) of the original impressions. If green was like a chord in music for Boyle, then so was a complex idea signified by words such as honor or unity for Hartley; both could be analyzed into their constituent sensations (Hartley 1810:1.9–35; 330–35).

The divisions that ramified between Cartesians and leading empiricists during the seventeenth and eighteenth centuries did not chiefly emerge, as one might have expected, from debates about the degree of correspondence between an impression and an idea. Descartes's twelfth rule of perception laid it down that the figure received from a sense impression is transferred to the brain as an analogy of what it represents, "For there is nothing amongst bodily things exactly like this force" (quoted in Lezra 1997: 110). Although he and Locke maintained very different views about the origin of ideas, they agreed that they are not still lifes of the things that excited them. "There is nothing like our Ideas in the Bodies themselves," said Locke (Locke 1979: 137; [II, vii 14–15]). As an Epicurean, Hobbes was quite sure that sensation was purely a process of matter and motion operating between the outside and the inside, "As wee see in the water, though the wind cease, the waves give not over rowling for a long time after; so also it happeneth in that motion, which is made in the internall parts of a man, then, when he Sees, Dreams, &c. For after the object is removed, or the eye shut, wee still retain an image of the thing seen, though more obscure than when we see it" (Hobbes 2004: 150). Notwithstanding his use of the word "image," suggestive of a mimetic relation between the two, and his largely materialist account of sense perception, Hobbes agrees with Locke this far, "The object is one thing, the image or fancy is another" (14). Margaret Cavendish denied that one sense was capable of corroborating another: directly confronting Lucretius, she declared in an accidental pun, "The nose knows not what the eyes see." She also rejected the notion of sensation as imprint—"It is not the real body of the object which the glass presents ... only figures or patterns" (Cavendish 2001: 46, 51). Gottfried Leibniz and later Buffon agreed, "There is no resemblance between sensations and the Objects which produce them.... Internal sensation is totally different from its cause" (Cavendish 2001: 51; Wilson 2008: 157, 218; Buffon and Daubenton 1791: 2.356).

Thomas Stanley, a publicist of Epicurean materialism, said that the eidola and effluvia of objects glide into the senses and that "we receive something from without, which causeth the sensation itself," but he wasn't about to call any of the senses mere channels or porches through which images pass intact to the mind (Stanley 1687: 886–87). More to the point, neither Digby nor Charleton wholly dissented from the Cartesian position. They believed that the senses were merely the medium by which a certain excitation reached the brain, where it was formed into an idea by its "innate and proper Faculty" (Digby 1669: 349; Charleton 1657: 93). Willis was closer to Lucretius on this issue when he talked of impressions distinctly represented in the brain, "marked with the type or shaddow of the Objected thing ... as upon a white wall" (Willis 1683: 59), but he was still a long way from notions formed by a direct material impact. The most powerful advocate for the purely conventional function of the figure or sign of sensation was Thomas Reid, who founded his common sense philosophy upon the principle that they had nothing physically in common. He was adamant in his refusal of the "unphilosophical fiction of images in the brain" because he was certain that "no sensation can resemble any external object" (Reid 1997: 121, 176). His illustration concerning what he calls the "geometry of visibles" is precisely the opposite of Lucretius's and, incidentally, Newton's: "The geometrician, while he looks at his diagram, and demonstrates a proposition, hath a figure present to his eye, which is only a sign and representative of a tangible figure.... These two figures have different properties, so that what he demonstrates of the one, is not true of the other" (105–6). The geometrician, that is to say, is working with a symbol, like an $x$ or a $y$ in algebra. However, it is possible that Reid insisted on this division between ideas and sensations not only to introduce us to a world of signs but also to remind us of a time in the development of human cognition when we enjoyed a more vivid and immediate relation to the impressions of things themselves; and if we are ever to recover it, then "we must accustom ourselves to attend to them, and to reflect upon them, that we may be able to disjoin them from ... the qualities signified or suggested by them" (61–62). By losing the language of geometrical abstraction, a world of sensuous self-evidence might return.

It was clear however that a powerful party of empiricists wished, like Cartesians, to avoid the humiliation of suffering the blows of matter, and the symbolic transfer of impression into idea softened or even avoided a predicament Lucretius wasn't at all concerned with moderating: "We see that the blows take upon our body," he said, "exactly as if some object were striking us, and giving us the feeling of its own body outside" (Lucretius 2006: 297,

4.239–43). But Baruch Spinoza was horrified by this assault from the environment that turned a perception into a bruise: "Weakness consists in this alone, that man allows himself to be led by things which are outside him, and is determined by them to do those things which the common constitution of external things demands" (Spinoza 1993: 164). Ralph Cudworth rejected entirely any epistemology that depended on such intrusive violence and preferred to believe that thought belongs to "the Awakening and Exciting of the Inward Active Powers of the Mind" (Cudworth 1731: 99). Digby tried to show that Descartes placed the business of impressions in a civil light, where things gave "a blow to our exterior Organs," much as a visitor might use a door knocker, or a flapper use his bladder among the philosophers of Swift's Laputa: "Our soul and mind hath notice, by this means, of every thing that knocks at our gates" (Digby 1669: 346). Clearly Locke's distinction between primary and secondary qualities and Hobbes's between the original and its image were designed to emphasize the etiquette of entry and lodgment.

Few in this debate wished to entertain, or at least be seen to entertain, the idea of an action without an agent or of an event without an architect. Looking back to the seventeenth century, William Hazlitt was sure that this was not a guarantee that empirical materialism could make. Despite the distinction he drew between object and fancy, Hobbes, for instance, was, in Hazlitt's opinion, a man who "resisted all impressions but those which were derived from the downright blows of matter"; consequently, "the external image pressed so close upon his mind that it destroyed the power of consciousness, and left no room for attention to any thing but it self" (Hazlitt 1904: 11.28–29). Eventually, Coleridge renounced associationism on the grounds of its blind and involuntary subjection to random impressions, in which "images act upon our minds, as far as they act at all, by their own force as images" (Coleridge 1930: 1.129).

One way or another, the totality of the impression of the eidolon or image of the thing itself, obliterating any margin for judicious reflections of the rational soul, haunts the work of experimental science and voyages of discovery. Like the elephant in the room of knowledge, it makes the sheer perception of a fact look no different from the delusion or fiction Descartes hypothesized in the *Meditations* (1641) when he decided to treat all sensations as if they were enchantments. Later in the eighteenth century, as we shall see, it constituted the experience of induced and adventitious ecstasies as well as the symptomology of nervous diseases; and with increasing frequency, it acted as the muse as well as the subject of poetry. Whether feared as the loss of self-control or welcomed as the liberation from all constraints (like the "swing" of Digby's willful bean plant), the imperative demand for exclusive attention made by the thing itself heralds a nervous commotion in which the creative as well as the passive tendencies of the human mind seem equally to be canvassed. It provokes the swift interchange of inward and outward impulses typical of the sublime, and also the state of possession characteristic of the reveries Erasmus Darwin called diseases of volition. For nowhere in either the dilation or contraction of spirit that follows the "knock or thrust from without" (Cudworth 1731: 99) seems there to be any room for mediation, reflection, or choice.

The organ at the center of this vortex is usually the eye, which exhibits the same indiscriminate and morbid alertness Charleton alludes to when he says that under some emergencies and distractions the soul crowds itself into a single sense. Everything becomes sight because no distinction remains between the image that comes in and those it seems to awaken: the object and the fancy of it, as Hobbes would say, become identical, and the distinction on which Reid's "geometry of visibles" is raised, falls down. Then the soul "undergoes various metamorphoses, and is invested in strange apparitions, and confused with delusory images" (Charleton 1670: 28–29). All the organs of sense were vulnerable to this kind of overload. One of the most astonishing episodes in the history of madness was provided at the end of the eighteenth century by James Tilly Matthews, who believed himself the victim of a team of seven operatives working an "Air-Loom" or effluvia machine that, by directing a stream of vile matter at him ("stinking human breath—stench of the sesspool—gaz from the anus of the horse"), were able to control his voice, heighten or extinguish his senses, and manufacture the "event-workings" of his life (Jay 2012: 150). This state of involuntary excitement was induced artificially by laughing gas at Beddoes's Pneumatic Institute; and, of course, it had a pathology all its own in the morbid distortion of sense impressions in the later stages of scurvy.

## PART II

> Science appears but what in truth she is,
> Not as our glory and our absolute boast,
> But as a succedaneum, and a prop
> To our infirmity.
> —William Wordsworth, *The Prelude* (1805)

Science plays a paradoxical role in the history of scurvy, for on the one side, it contributes to the increase of the disease by advancing the art and machinery of navigation to the degree that voyages become too long for the human organism to sustain and, on the other, it promises instruments so ingeniously conceived and cleverly constructed that the measure of time and space is made—as far as possible—certain, so that voyagers, go as far as they might toward the Terra Incognita, would always know where they were. Health might not be secure, but knowledge was; for as Thomas Sprat put it, the mariner has "eyes in all parts … to receive information from every quarter of the earth" (Sprat 1667: 20). But what if the eye were scorbutic, overloaded as Charleton describes, "invested in strange apparitions, and confused with delusory images"? What could correct its misinformation?

The first answer was food of that special sort containing the unanalyzed essence Digby called a salt, Charleton a sap, and Willis a latex, each in their own way getting as close to a solution of the problem of scurvy as was possible at this stage in the history of brain chemistry. However, the more accurately the problem was described, the more the nutritional debate about scurvy edged into terrain bounded by theological issues, for defenders of the deficiency thesis had discovered a latent weakness in the human constitution that led inevitably to corruption and death if it were not strengthened by a supplement not naturally a property of the body or the animal soul. The school of infection supposed, on the contrary, a human organism naturally healthy, vulnerable only to the influence of exotic toxins. These opposing views lent piquancy to the allegorical dimensions of discovery in the South Seas, for in the centuries dividing Magellan's achievements from Cook's, it was widely believed that it was the last possible location of a terrestrial paradise, and frequently a landfall was cast in these terms. De Quiros meant to establish a New Jerusalem in Vanuatu; Cook thought he had found one in Tahiti, where men and women went naked without shame and found their food growing conveniently on trees, no need of the sweat of their brows. The question was whether paradise was to be imported or found, and the answer bore directly on the topic of inherent versus acquired corruption. De Quiros meant to institute a perfect world, while Cook was quite sure that he was the agent of a historical force destined to ruin the one he had discovered. Nowhere was this force more destructively spent, or the allegory of sin more literally applied, than in the settlement of Australia as a British penal colony. It was based on the belief that the vitiated criminal mind, finding its remedy in the medicine of reformatory prison discipline, would at length take its rightful place in a land "comparable to the Garden of Eden." Yet savage corporal punishment combined with overwork on a diet almost entirely deficient in the vital ingredient of vegetables and fruit led to scenes of such dreadful misery they were, an impartial witness observed, "more to be dreaded than death itself" (Backhouse 1843: xlviii–xlix, lxi).

When the archangel Michael conducts our forlorn parents from Eden in Milton's *Paradise Lost*, they are made aware of three things. The first is that their expulsion is necessary because human flesh, once as pure as spirit, is now like a festering lily, breeding a "distemper" that must be eliminated from the garden. The second is that the world is all before them, "no despicable gift," yet sufficiently unlike Eden to be suitable for grossness such as theirs. Their job is to explore, appropriate, and people it; so the third thing is that they need equipment suitable for such tasks, beginning with better eyes, for their sight has been dimmed by sin. Michael supplies them with herbal spectacles consisting of euphrasy and rue, by means of which the extent and future of the world begins to be opened to them. It is soon clear that time, which has just begun, has little pleasant in store, and yet the probation of human life on earth is to be consummated with the salvation of the just. So the whole history of human dominion of the earth is controlled by two coordinates—the extensiveness of space and the probative intensity of time— seasoned with a paradox: namely, if desire of knowledge caused the Fall, what part is it to play in colonizing the earth? Will human grossness merely stain it, or might it be possible for the senses to be purged and corrected, just as Adam's eyes have been by Michael, in order to yield knowledge as pure as human flesh once was?

Milton provides the axis of sensationist empiricists represented by Hooke and Boyle with a myth of discovery that is friendly to its priorities: the improvement of the sense organs for the sake of experimental knowledge. At the same time, it is suitable for the implications of Charleton's and Willis's researches, inasmuch as accurate observation serves a soteriological as well as an epistemological purpose, by neutralizing a congenital deficiency in the organs and the minds of the discoverers. Robert Hooke, whose horological inventions helped so much in the search for an accurate measurement of longitude, wrote: "And as at first, mankind fell by tasting of the forbidden Tree of Knowledge, so we, their Posterity, may be in part restor'd by the same way … by tasting too those fruits of Natural Knowledge, that were never yet forbidden" (Hooke [1665] 2003: vii). For Hooke, redemption and knowledge are cut from the same cloth; but so (it seems) is sin, for in the fruit of knowledge it begins and ends. No matter how many devices Hooke and his colleagues invented to further this project, there was a human infirmity no machine could entirely prop up, nor any medicine cure reliably at sea. Scurvy made obscenely obvious what God had already noticed about the corrupt human body: that it bred putrescence and death out of itself, not infected but self-consuming, with symptoms growing more grotesque as the disease progressed: "The mouth and gums, that grew proud flesh in foyson / Till gangrene seemed the blood to poison, / Gangrene that carried foul and ful-some taint, / Spreading infection through the neighbouring air" (Camoens 1881: 5.81 [Burton's trans., 2.620]).

In Hooke's dispensation scurvy, like sin, was not an unequivocal impedi-ment to experimental discovery, so much as another branch of it: it was, for instance, *revealed* as an infirmity peculiar to humans that everything waiting to be *found* in the world at large might mend. Or, rather more paradoxically, it was the sinful element of knowledge that knowledge itself was destined to expel. In Hooke's Baconian replay of knowledge as salvation, the conse-quences of the desire for forbidden knowledge are to be eradicated by those flowing from the unforbidden sort; but where is the standard by which they are to be distinguished? How is sinful fruit to be distinguished from the re-demptive sort? Bacon's program for physical immortality, for example, was potentially heretical since it aimed to do without the mediation of Christ. To this extent, the universality of the disease "discovered" in the process of dis-covering the world might be expected, like the salvific sin of knowledge itself to participate to some degree in the cognitive labor it appeared only to hinder. In this chapter and the following ones, I shall argue that scurvy, discovery, and the taking possession of strange lands formed a secular triad operating in the same relation to its parts as sin, knowledge, and redemption in the theo-logical sphere. How exactly could this be?

Well, the purpose of the various prostheses built by Hooke was to improve the imperfect operation of the senses—hygroscopes for the nose, telegraphs for the ear, microscopes for the eye, megaphones for the voice. These were intended, just like Michael's application of euphrasy and rue to Adam's eyes, to expedite the discovery of "new Worlds and Terra-Incognita's" (Hooke [1665] 2003: xvi). Bacon placed his scientific utopia in the same Terra Incog-nita, filling Saloman's House with experimental machines designed as helps for the organs of sense. Although not included by Steven Shapin in the di-verse places of experiment (shops, cellars, coffeehouses, palaces, and college rooms [Shapin 2010]), ships were larger examples of the same thing, being both a platform for experimental discoveries as well as machines designed to bring remote lands more fully under the dominion of the eye and the other four senses. But if this enterprise were to succeed, the eye, no less than the nose, the palate, the skin, and the ear, had to be working at full pitch. To en-hance their susceptibility to impressions, Hooke advised his colleagues to approach experiments armed not only with new machines but also with a temperamental inclination for novelty. Even the simplest and most routine materials ought to be approached in a spirit of wonder, "as if they were the greatest Rarity," while the investigator was advised to "imagine himself a Person of some other Country or Calling, that he had never heard of, or seen the like before" (Hooke 1969: 61–62; cited in Daston and Park 1998: 315). By situating the mind in the pretence of a Terra Incognita, the real one is fetched closer, and we "bring Philosophy from words to action" (Hooke [1665] 2003: xxvi).

Hooke's belief in the efficacy of sensory prostheses was not unanimously admired, and the reason lay in the division that has already been examined between philosophers favoring a direct link between impressions and images, and those who supposed the brain to perform a complex relay between a sensation and its representation as an idea. Hooke was manifestly of the former party. As far as he was concerned, any device capable of increasing the distinctness and force of an impression was bound to get closer to what Sprat called "the realities of things" (Sprat 1667: 26). Locke and Cavendish belonged to the latter school, taking a dim view of machines that enlarged the capacity of the senses beyond the level of gross normality needed for consensual witnessing (Schaffer 1998: 83–89; Golinski 2005: 133–36). The danger was conceived to be twofold. In the first place, the experience of an artificially enlarged power of sight or hearing was disturbing and isolating. If our ears were more finely tuned, we should be hearing things all the time, the fall of a leaf pounding in our ears; or if our eyes were more acute, seeing lice the size of swine rooting among a beggar's rags (Swift 1995: 115). Without the safeguard of a limited receptivity to things, our lives would be made intolerable, there would be no rest, no power of discrimination. Such a witness would live "in a quite different World from other People. Nothing would appear the same to him, and others" (Locke 1979: 303; [II, xxiii, 12]). Locke saw no advantage at all in the kind of foreignness Hooke recommended to the experimental scientist as the temperamental alpha and omega of discovery. As for Cavendish, she said his microscope could never penetrate the surface of things in order to find out the secrets of their constitution; it merely disarranged the distances, textures, and angles that made them familiar or comely, revealing instead the immodesties, moles, and hairs that cause the maids of honor in Brobdingnag to appear so repulsive to Gulliver. She and Locke are hostile to the unusual, whereas Hooke takes the unfamiliarity of a fact to be the standard of its value. Had they been asked for a solution to the problem of scurvy at sea, they would have said that it lay within the compass of what we are allowed to know, not what we aspire to find out.

In his *History of the Royal Society* (1667), Thomas Sprat was confident that soon "there will scarce be a ship come up the Thames, that does not make some return of experiments" (Sprat 1667: 86). Even buccaneers such as William Dampier, whose heads were filled more with thoughts of treasure than of truth, made serious attempts to describe and delineate the products of the new lands they discovered and to produce specimens of their inhabitants. Dampier brought back Jeoly, a tattooed native prince from Mindanao, reported to be the "just Wonder of the Age" (Douglas 2005: 34)—not a *bare* wonder merely reported, but a *just* one that could be observed and studied. It was in this scientific spirit that Dampier and many of his brethren made memoranda of weather, currents, and soundings, even sketching the looming of capes and promontories, imitating the experimental attention to particulars recommended by the Royal Society. They likewise rendered their discoveries in the close, naked, and natural choice of words most apt for particulars of facts—the same sort of prose later adapted by Defoe and Swift for their fictional voyages. Dampier warns his reader how exact he means to be ("I have been exactly and strictly careful to give only True Relations and Descriptions of Things … a Plain and Just Account of the true Nature and State of the Things described" [Dampier 1939: lxviii), and so does Gulliver ("I could perhaps like others have astonished thee with strange improbable Tales; but I rather chose to relate plain Matter of Fact in the simplest Manner and Style" [Swift 1995: 261]). But it is strange how easily the methods and discourse associated with the sober and punctual interrogation of nature slide into the action and language of fiction, a clandestine marriage with a remarkable future.

The navigations of the Indian and the Pacific Oceans authorized by the French and British governments in the mid-eighteenth century, and subsequently by the Spanish and the Russians, took the experimental role of the mariner to a new dimension. Unlike Anson, who had been sent out as a species of naval buccaneer, licensed to pillage the Spanish while probing possible trade routes between the islands of the South Seas and China, picking up what scraps of information where he could, Cook, Bougainville, Bligh, and J. F. de Galaup de La Perouse were officially engaged on scientific projects, set out in advance in their instructions. On his first voyage, Cook, for example, was to measure the transit of Venus from a position on the earth's surface previously thought impossible; on his second, he was testing newly invented chronometers to see if they offered a solution to the problem of calculating longitude. Throughout his three voyages, he was confirming the nonexistence of the Great Southern Continent and (apparently) of the Northwest Passage, favorite ideas of geographical theorists such as Alexander Dalrymple and Arthur Dobbs (Lamb 2001: 90; Williams 2009: 98–115). He and his colleagues carried with them as supercargoes experts responsible for accurate collection and measurement of data, together with artists whose job was faithfully to represent those novelties that were not transportable. A great deal of specialist equipment was necessary for this work: Larcum Kendal's and John Arnold's chronometers, Gowin Knight's azimuth compass, and John Bird's astronomical quadrant.

According to the elder Forster, the multifarious experimental duties of the cruise became a sort of joke among the crew of the *Resolution* during the second voyage:

> We dined this day upon beef, which our crew called *Experimental-beef*: to understand this, I only will observe, that every thing which our Sailors found not to be quite in the common way of a man of war, they called *Experimental*. The beer made of the Essence of beer, or Malt they called *Experimental beer*, the very Water distilled from Sea Water by Mr Irwin's method was *Experimental Water*, Mr Wales, the Astronomer, Mr Hodges, the painter, Myself & my Son were comprehended under the name of *Experimental Gentlemen*. (Forster 1982: 3.309–10)

It is evident from this account that the emphasis on experimental enquiry was internally directed as well as externally. Besides acting as the instrument of research, the ship was often the scene and object of it. While new seas were traversed and landfalls were being made on strange islands, the proper fuel for the labor needed to make these discoveries was quite as urgent an object of enquiry as any other, and one in which everyone on board was involved. Antiscorbutic food and drink were critical elements of the whole enterprise, for experimental voyages to the ends of the earth relied more and more on a dependable standard of health and perception. It might be said that the premier experiment of all the others conducted on the *Endeavour* and the *Resolution* was the search for a food capable of preventing scurvy and, in the event that the malady did occur, a medicine fit to cure it. Although estimates of the scorbutic rate of mortality vary, there was no doubt that it ranked as the premier occupational disease of the great maritime era (Carpenter 1986: 253).

Despite its frequent iteration, the claim that Cook conquered scurvy is not true. None of the experiments conducted either on the *Endeavour* or the *Resolution* provided a remedy; in fact, Lloyd and Coulter believe Cook's endorsement of concentrated malt ("without doubt one of the best antiscorbutic sea-medicines yet found out" [Cook 1776: 403]) put the search for a preventive back by fifty years. The clinical trials conducted by William Perry aboard the *Endeavour* were plainly misleading, and the institutional support for MacBride's claims for malt, led by Sir John Pringle, was biased by a combination of thrift and theory. On the *Resolution*, the early stages of scurvy were endured by the experimental gentleman as well as Cook himself, although it would seem, from the guarded manner in which William Wales announces his own symptoms, that it was not politic to say so. But sin, let us say, was on the lookout, determined not to be ignored. It is perhaps no coincidence that problems with Arnold's chronometer—on one occasion it refused to be wound up, and on another the lock had been forced and jammed—occurred on an expedition where the experimental gentlemen in charge of them succumbed to scurvy, leading to a public squabble once Wales and the two Forsters were back in London, each accusing the other of a want of scientific candor. The younger Forster reached for metaphors of scorbutic nastiness in his rebuttal of Wales's charges, "hatched like a basilisk in the damp and noisome cabbins of the *Resolution*" (Cook 1961: 175–76, 798, clii). A coincidence of scurvy, a thwarted scientific enterprise, and a disintegrating ship were frequent enough: Davis's ship had been eaten up by serpents as his men died; Carteret's crew were too sick to repair an ill-found vessel; similar crises forced Flinders and Nicolas Baudin to interrupt their circumnavigations of Australia. However, the failure of a chronometer on Cook's second voyage is significant since it was a machine designed to bring to absolute perfection the measure of the lapse of time—a calculation that Locke typically believed was a subjective experience, being as long or short as the train of a person's ideas. Like Boyle's experimental air pump leaking air into a vacuum intended to be total, or an experimental ship going rotten in the stem, or the experimental malt that failed to cure scurvy, or the potentially immortal living engine of the scientist's experimental body sickening on long voyages, no timekeeper, not even John Harrison's, was able to overcome a limitation inseparable from its design and operation—namely the effects of friction. Even the most delicately constructed escapement was incapable of telling a time that never needed resetting. The same impediment—or sin—infested every other experimental attempt at eliminating the defects of engines.

Rather than experimentalists confronting infirmity as an obstruction merely to be removed, there was a possibility, with scurvy at any rate, of treating it as something like an ally, for scurvy had its own way of supplying the defects of dull senses and imperfect attention. It was well known to induce stupor in its victims, but shrewd observers such as Thomas Trotter had observed that in some cases it rendered the cranial nerves inordinately susceptible to noises, smells, tastes, textures, and sights. Sudden sounds, such as the report of a musket or a cannon, could kill scorbutic sailors. Even pleasant stimuli, such as a drink of fresh water or a long-awaited taste of fruit, might provoke a seizure and put an end to their lives. In his *Omoo*, Melville recalls how scorbutic seamen would be troubled by the scent of flowers (Melville 1847: 64). When Bernardin de St. Pierre landed on Mauritius badly afflicted with scurvy, he was thrown into an ecstasy of disgust and alarm at the smells of the trees and shrubs (Bernardin de St. Pierre 1800: 66). Often, as we have seen, the sensation passed the frontier from pain to pleasure, or vice versa.

The scorbutic eye was particularly absorbed by sights, so much so that vision seemed to envelop the viewer and turn the orb of the individual organ inside out, as in the passive voice adopted by Pedro Fernandez de Quiros in his description of a festival he organized after he and his scorbutic crew reached Vanuatu: "There were seen amongst the green branches so many plumes of feathers and sashes, so many pikes, halberds, javelins, bright sword-blades, spears, lances, and on the breasts so many crosses, and so much gold, and so many colours and silken dresses, and many eyes could not contain what sprung from the heart, and they shed tears of joy" (de Quiros 1904: 1.261). Spectacular novelties such as coral grew more wonderful for Flinders as scurvy heightened the impression, turning dangerous animate rock into fascinating simulacra of fresh food: "We had wheat sheaves, mushrooms, stags horns, cabbage leaves, and a variety of other forms, glowing under water with vivid tints of every shade betwixt green, purple, brown, and white" (Flinders 1814: 2.88). But when he recollected this was an armed illusion that could tear out the bottom of his crazy ship, he sped away from it.

Of what use might these lonely and largely incommunicable intensities be to science? Before attempting an answer, I want to categorize them, for they are nothing if not various. It is clear that they are felt at first as an affliction, either because of bodily suffering, revulsion, or a disorienting sense of isolation. The morbid excess of the sensations serves to make the situation incomparably worse, such as no tongue can declare. "It is not perhaps very easy for the most fertile imagination to conceive by what our danger and distress could be increased" (Carteret in Hawkesworth 1773: 1.405). Annihilated by misery, there is nothing left to witness except occasionally the feeling that life has turned into an unpleasant dream or fiction. However, the same situation—like Boyle's emphatical green—may at a different time or under a different perspective excite sensations of pleasure quite as powerful as the former painful ones. When this alternation happens rapidly, a dramatic combination of delight and disgust, or of alarm and attraction, keeps opposite emotions in play, circling around opposite foci, something like an affective ellipse. Under certain circumstances, the sight of a desert island enchanted Pascoe Thomas, and under others, it plunged him into misery ("a Situation ten times worse to me than at any other in the whole Course of the Voyage" [Thomas 1745: 38, 149]). On one hand, the senses are found in symphonic company with each other, while on the other, they produce nothing but dissonance. When the cravings of appetite, stimulated by a dream and regarded by Trotter as the premier symptom of scurvy, are contradicted by waking to the distressing emptiness of a ship still at sea, then the imagination intensifies a loss; but when what is imagined or dreamt really takes place, and fantastic food and liquid materialize, then delight knows no bounds.

If sensation is the basis of knowledge, continuous with the ideas it generates, as Hume supposed, and not merely a prompt for the signaling system of the brain, as Reid maintained, then some principle of sensory reflexivity is needed in cases of extraordinary sensitivity if the totality of an impression is not to swamp cognition. Synesthesia and ellipsis—occasions when intensity causes sensations to collide or blend—provide this kind of balance: a fluid and brief confirmation of the resemblance of different motions and feelings. Willis called it dilation or irradiation; Charleton (perhaps after Hobbes) named it corroboration. As it receives "the strokes of all sensible things," Willis says, the sensitive soul will "leap back and recede into it self;" but having absorbed the impression, there is a reflux of energy that "by an emanation[,] as it were a certain vibration of the Spirits, exerts or puts forth its virtue and force of acting" (Willis 1681: 95). In blushing for shame, for example, he notes a very rapid alternation of these two movements, consistent with the impulses of modesty and self-defense: "Concerning this Passion, 'tis observable that when the Corporeall Soul being abashed, is enforced to repress its Compass, she notwithstanding being desirous, as it were to hide this Affection, drives forth outwardly the Blood, and stirs up a redness in the Cheeks" (idem 1683: 54). Relying on Lucretius, Charleton referred his idea of corroboration to erotic love, when fierce desire prompts the soul to cherish the image of a dreamt or fancied object, and the heart to send auxiliary animal spirits into the brain to "corroborate the Idea of this Desire, as that whole brigades of them may be from thence dispatched into the Organs of the Senses, and into all Muscles, whose motions may move especially to conduce to obtain what is so vehemently desired" (Charleton 1670: 109).

Willis illustrates the voluptuous possibilities of corroboration as follows: "We imagine the Drinking of excellent Wine, with a certain Pleasure, then we indulge it; the Imagination of its Pleasure is again sharpened by the taste, and then by a reflected Appetite drinking is repeated. So as it were in a Circle, the Throat or Appetite provokes the Sensation, and the Sensation causes the Appetite to be sharpened, and iterated" (Willis 1683: 49). Here, it is plain that the most powerful and pleasurable link established between the imagination and the reality of things occurs when there is a reciprocal flow between the image of a thing powerfully desired and the experience of seeing, tasting, touching, and consuming the very thing. Trotter noted the remarkable shift from the misery of want to the ecstasy of delight: "The patient in the inveterate stage of the disease seems to gather strength even from the sight of fruit: the spirits are exhilarated by the taste itself, and the juice is swallowed, with emotions of the most voluptuous luxury" (Trotter 1792: 141–42). John Mitchel, an Irish political prisoner en route for Tasmania aboard a scorbutic transport, wished never to forget the "brutal rapture" with which he devoured six oranges when the ship landed in Pernambuco. There is a gentler example of corroboration when the parched Ancient Mariner wakes from a dream of drinking to find his thirst quenched by the rain falling on his bare skin: "Sure I had drunken in my dreams, / And still my body drank" (Coleridge 2004: 177, ll. 304–5). If the sin of scurvy (eating fruit so abundant it allowed the gene synthesizing ascorbate to mutate) is not to kill its victims, it is not, in the first place, the abstract fruit of knowledge that will save them, but rather the immediate experience of fruit, seen and tasted.

Boyle stands out in the field of science as much as Trotter in medicine as a resolute empiricist, reluctant to proceed beyond the observation and evidence of a fact. His air pump was in some respects the antitype of James Tilly Matthews's "air-loom," for instead of concentrating effluvia, it was a machine designed to empty a vessel of every trace of matter; but his experiments were pursued in search of "a little vital Quintessence," the analogy in air to what Digby called "a hidden food of life" sought by himself, Charleton, and Willis in the process of nutrition (Wilson 2008: 77–78). Like Newton, Boyle disdained to sport with hypotheses, and like Bacon, his mode of enquiry was inductive, proceeding from no a priori positions in its production of facts. So it is interesting to see how Boyle makes his way from something akin to the total event of scorbutic intensity to his version of sensory corroboration.

In a remarkable account of the suspense experienced by the human witness when confronting facts of uncertain provenance, Boyle turns to fiction in order to explain his feelings:

But in the Book of Nature, as in a well-contriv'd Romance, the parts have such a connection and relation to one another, and the things we would discover are so darkly or incompletely knowable by those that precede them, that the mind is never satisfied till it comes to the end of the Book; till when all that is discover'd in the progress, is unable to keep the mind from being molested with Impatience to find that yet conceal'd, which will not be known till one does att [*sic*] least make further progress. And yet the full discovery of Natures Mysteries, is so unlikely to fall to any mans share in this Life, that the case of the Pursuers of them is at best like theirs that light upon some excellent Romance, of which they shall never read the latter parts. For, indeed ...

there is such a Relation betwixt Natural Bodies, and they may in so
many ways (and divers of them unobserv'd) work upon, or suffer from,
one another, that he who makes a new Experiment, or discovers a new
Phenomenon, must not presently think, that he has discover'd a new
Truth, or detected an old Error. For ... he will oftentimes find reason to
doubt, whether the Experiment or Observation have been so skillfully
and warily made in all circumstances, as to afford him such an Account
of the matter of fact, as a severe Naturalist would desire. (Boyle 1671:
118–19)

Boyle's sense of the experimental situation resolves itself into an uncommon
parallel between scientific observation and the reading of a genre of fiction
by which he had all his life been fascinated: romance. Like the permutation
of the passages of war and love in the fiction of La Calprenede and Madeleine
de Scudery, the mobility of the observer's position in an experimental situation has nothing to do with probability, or with social consensus. Romance
(unlike the novel and unlike the instrumental fictions of representation favored by Boyle's antagonist, Hobbes) neglects the predictable series of a
recognizable world in favor of events without precedent, parallel, or sequel,
inevitably occurring in an unnavigable space: stunning and absorbing, and
embarrassing to every principle of accountability. Henry James called romance a fantasy of unrelatedness. It deals with "the kind of experience ...
disengaged, disembroiled, disencumbered, exempt from the conditions that
we usually know to attach to it" (James 1947: 33). As a great expert in romance points out in Charlotte Lennox's *The Female Quixote* (1752), its
heroes are highly unpredictable because they are not bound to any system of
action, "triumphing not only over all natural avow'd Allegiance, but superior
even to Friendship, Duty, and Honour itself" (Lennox 1973: 321). Having
attempted a romance of his own (*The Martyrdom of Theodora and Didymus*
[1687]), Boyle is willing in some degree to entertain under the auspices of
a suspended judgment the possibility that romance is hospitable to "a new
kind of fact that signified nothing at all" (Daston 1991: 100). While Boyle's
use of fiction bears some resemblance to the sheer improbability of Descartes's hypotheses (a world invented by a djinn, machines walking around
dressed in clothes), it is not instrumentally deployed on behalf of some higher
principle of cognition. For Boyle, romance is a source of instances of the perplexity of someone compelled by a phenomenon awaiting corroboration,
and in this respect, it stands closer to the kinds of fictions referred to impa-
tiently by Trotter and Beddoes in their work on nervous diseases, as we shall
see in the fifth chapter.

If this scene of perplexity represents the zero degree of experimental satisfaction, equivalent to the isolation and confusion of the scorbutic singleton, then Boyle evolves his response to it, as we have partly seen, by means of his study of the effects of effluvia as they make their way into the body and through its organs, before being projected outward in the form of perceptual energy. In his analysis of this flow, he outlines the same reciprocal action of recession and dilation favored by Charleton and Willis. But when it comes to sensations of extraordinary strength, Boyle's explanation is remarkable for locating the cause of preternatural sensations in the body's reaction to disease. There is the man who having recovered from the plague could smell an infected person before any tokens of the distemper were visible; another who was able, after a severe inflammation of his eyes, to see colors in the dark; and another, a physician who, falling sick of a fever, discovered afterward he could overhear whispered speech at a great distance (Boyle [1672] 1999: 7.268, 282). The organs of the living engine of the body evince a susceptibility deriving directly from an infirmity possibly grievous in itself but, like an inoculation, productive of happy results. Were Boyle's view of the compensations of disease to be transferred to the stark polarities of deficiency versus defect in the debate over scurvy, he would be seen inclining somewhat equivocally to the former, maintaining a strict proportionality between the impact of the knock or thrust from without and the reactive energy of the sense organ. Regardless of the absence of a nutritive sap or the presence of a bacterium, however, he has detected a refinement of the senses owing not to the addition of a vital essence, but to a radical adaptation of the organism triggered by illness. In terms of recession and dilation, this agrees with what Charleton says about corroboration, except that he never ascribes it to scurvy or any other indisposition. For his part, Willis understands that the corruption of the nervous juice provokes spasmodic "inordinations of the animal Spirits" and, specifically in cases of scurvy, refracted and disturbed irradiations, "like the sun in misty air," along with "cracklings, or convulsive explosions" of the animal spirits (Willis 1681: 127; 1684: 184, 61). But we have to wait until Trotter's *Observations on Scurvy* (1792) for an account of the disease that deals at length with its excitements as well as its languors and tries to explain them as symptoms of scurvy: examples of patients "gathering strength even from the sight of ... fruit: the spirits exhilarated by the taste itself, and the juice ... swallowed, with emotions of the most voluptuous luxury" (Trotter 1792: 141–42). Boyle, as it were, affords sin an active part in the economy of experiment, for even the failure of an enquiry leads to suspense that is not unenjoyable, while sickness, far from proving a distraction from the experience of things may, at least under certain conditions, focus it strangely. So

Boyle's skepticism is not like Locke's or Cavendish's, for he welcomes any enhancement of the sensory encounter with the world as much as his friend Hooke, but as to the source and tendency of its salvific virtue, he has a different opinion.

So I want to offer some examples of experiences, optical, auditory, and olfactory, that illustrate what Boyle discovered from his studies of effluvia. Here is de Quiros in Vanuatu, after terrible experiences with scurvy while voyaging toward the place he names the New Jerusalem, celebrating the festival of Corpus Christi in a weird travesty of chivalric ritual:

> There were seen amongst the green branches so many plumes of feathers and sashes, so many pikes, halberds, javelins, bright sword-blades, spears, lances, and on the breasts so many crosses, and so much gold, and so many colours and silken dresses, that many eyes could not contain what sprung from the heart, and they shed tears of joy. (de Quiros 1904: 1.261)

The burden of the optical overload has to be distributed among what is seen, so that seeing and feeling are recorded in the passive voice as operating on a multitude of eyes independent of persons, as if the pressure of impressions had turned the scene into a kaleidoscope of active colors and textures, circulating so freely there are no viewpoints left to claim them, for the human eyes are all dazzled and blinded by tears. Every actual observer dwindles away, like the scorbutic Forster on the *Resolution*, leaving light and color sporting in a prospect of disembodied looking. A scorbutic shift to the passive in a very different key is found in Kerguelen's narrative of the discovery of his island. He stood off it in such a ferment of pathological disgust that he could not bear to encounter it directly in his person, so just as he chose a deputy for the discovery, he assigns one for the report. He quotes a garbled account of his achievement from a translation of John Rickman's journal of Cook's third voyage, then turns to the reader as a reader himself, declaring, "On voit par cet extrait du Journal Anglois, que j'ai decouvert cette isle, que je nommerai Isle de Kerguelen" [One sees from the extract of this English journal that I discovered this island that I shall name Kerguelen Island] (Kerguelen-Tremarec 1782: 31). Never can a discovery have been more oblique or unwilling, the indicative being resumed only after the first person of the discoverer has been authenticated as someone else's third.

Heading for New Holland, Dampier gazes at the sky with the close attention Lucretius recommends in his sixth book until it begins to operate on him as if he were its instrument, eliciting from him emotions of different keys and pitches according to each chromatic variation:

> The Night before, the Sun set in a black Cloud, which appeared just like Land; and the Clouds above it were gilded of a dark red Colour. And on the Tuesday, as the Sun drew near the Horizon, the Clouds were gilded very prettily to the Eye, tho' at the same time my Mind dreaded the Consequences of it. When the Sun was now not above 2 deg. high, it entered into a dark Smoaky-coloured Cloud that lay parallel with the Horizon, from whence presently seem'd to issue many dusky blackish Beams. The Sky was at this time covered with small hard Clouds (as we call such as lye scattering about, not likely to Rain) very thick one by another; and such of them as lay next to the Bank of Clouds at the Horizon, were of a pure Gold Colour to 3 or 4 deg. high above the Bank: From these to about 10 deg. high they were redder, and very bright; above them they were of a darker Colour still. (Dampier 1939: 76)

The detail is out of all proportion to the significance of what is observed. Dampier has been carried across the boundary that divides factuality from pleasure, and this pleasure has no maritime, social, or subjective value. The sky seems to be painting itself, and his moods respond to each change of hue. In his *Histoire Naturelle* (1750), Buffon supposes Adam in the same situation as Dampier when he sees the light, hears birdsong, and eats grapes for the first time, exclaiming, "All this fills me with astonishment and transport. Till now I had only enjoyed pleasures, but taste gives me an idea of voluptuousness" ["Quelle saveur! Quelle nouveauté de sensation! Jusque-là je n'avois en que des plaisirs; le goût me donna le sentiment de la volupté"] (cited in Douthwaite 2002: 74).

When Bernardin de St. Pierre landed at Mauritius, suffering badly from scurvy himself, his principle sensations for some time were those of disgust. The offensive smell, look, and taste of the vegetation and its fruits overpowered his senses. He reports coming across a plant resembling a red July flower whose blooms smell unpleasant, and a hedging plant whose fruit is always sour. *Bois de canelle* and stinkwood smell horribly of human excrement. The tree called *mapou* has a disagreeable taste with a bark that inflames his throat when he carelessly puts a piece in his mouth. Even when a flower smells sweet, it is only a trick of proximity: "The veloutier … exhales an odour, that at a distance is agreeable, less so when you draw near, and quite close is perfectly loathesome" (Bernardin de St. Pierre 1800: 66). There is a pigeon whose flesh tastes sweet but will soon throw you into convulsions if you swallow it; and there is a butterfly with dust on its wings that will blind you (65–79). It is hard to say where the novelties of natural history end and the repugnance of an overactive sensibility begins, but it is certainly a landscape where all action is lodged in the flora and fauna, and all passion in the visitor.

In *Paradise Lost*, Milton compares Satan's journey to Paradise with da Gama's off Mozambique. That was where the Portuguese got scurvy, but for his part, the infernal voyager is cheered by the scents of flowers from the shore: "Now gentle gales / Fanning thir odoriferous wings dispense / Native perfumes, and whisper whence they stole / Those balmie spoiles" (Milton 1958: 77 [IV, ll. 156–59]). Here is Melville describing how painfully this self-active air affected a scurvied whaleman in the South Seas:

> The Trades scarce filled our swooning sails; the air was languid with the aroma of a thousand strange, flowering shrubs. Upon inhaling it, one of the sick who had recently shown symptoms of scurvy, cried out in pain, and was carried below. This is no unusual effect in such cases. (Melville 1847: 64)

And here is James Colnett off Patagonia, his men down with scurvy, experiencing a modern version of the sirens of the *Odyssey*:

> About eight o'clock in the evening an animal rose alongside the ship, and uttered such shrieks and tones of lamentation so like those produced by the female human voice, when expressing the deepest distress, as to occasion no small degree of alarm among those who first heard it.... I conjectured it to be a female seal that had lost its cub, or a cub that had lost its dam; but I never heard any noise whatever that approached so near those sounds which proceed from the organs of utterance in the human species.... We had one man who was literally panic struck by the appearance and cries of the seal.... If we had remained twenty-four hours at sea, he would not have recovered. (Colnett 1795: 169–76)

This is a dramatic example of what Coleridge meant when he talked of sounds shorn of the sense of outness—"what a horrid disease every moment would become" (Coleridge 2002: 1.1307 f.34).

These instances of morbidly acute sensation bear directly on what Trotter says of the mortal danger of irritated nerves, namely that "the living body possesses the faculty, if I may call it so, of receiving impressions, and retaining them, even to the hazard of its destruction" (Trotter 1807: 199). If sounds can haunt a person to madness, and the odor of flowers agonize the man who smells them, then the relation between object and subject, conceived by Locke as governed by powers operating at regular pressures via an insensible medium (Locke 1979: 142 [II, ix, 25]), is really much more variable, immediate, and perilous than he believed, liable to plant very different ideas of things in the brain than those customarily associated with seeing, smelling, tasting, and hearing things, where the object is one thing and the idea of it quite another. According to the situation in which the impressions are received, the level of sensation can rise to a point where what is felt is indistinguishable from the impact of the object itself, and as Spinoza says, the victim is entirely governed by the common constitution of external things. In fact, Trotter's insight introduces, like Boyle's, an inkling of what A. N. Whitehead calls the totality of the bodily event when a force is unleashed capable of carrying us to "depths beyond anything which we can grasp with a clear apprehension" (Whitehead 1967: 149, 92). When such an event occurs, then agency is discovered in the whir of reciprocal material action that Boyle compares to hydraulic exchange and which, in its first cycle, might possibly be overwhelming. But when pleasure comes, it is transcendent: as in de Quiros's New Jerusalem, where seeing takes place around a multitude of dazzled eyes; or in Dampier's sky, which paints itself on his eye.

As this debate developed in the eighteenth century between those ready to acknowledge some value in enhanced sensations and those attached to the norms established by the limits of human faculties, it is surprising how often the rhetoric of those on the normative side of it recruited as wild conjectures actual scenes of agony or delight that had been experienced and witnessed at sea. For example, Francis Hutcheson floats what he thinks is a set of improbable hypotheses concerning the morbid excess of sensation, justifying the ways of Providence to his reader by ridiculing the extravagance of what Trotter had actually seen and heard as a surgeon:

> Would we allow room to our Invention, to conceive what sort of Mechanism, what constitutions of Senses and Affections a malicious powerful Being might have formed ... [h]ow easy had it been to have contrived some necessary Engines of Misery without any use; some Member of no other service but to be matter of Torment; Senses incapable of bearing the surrounding Objects without Pain; Eyes pained with the Light; a Palate offended with the Fruits of the Earth; a Skin as tender as the Coats of the Eye.... Human Society might have been

made as uneasy as the Company of Enemies.... Malice, Rancour, Distrust, might have been our natural temper. (Hutcheson 2002: 119)

Hutcheson is listing as impossibilities the sensory phantoms incident to scurvy; yet Trotter had witnessed the effects of dilated eyes, ears deafened by tinnitus, bad tastes in the mouth, and all the other enormities of scorbutic prostration. He had observed the sequel, too: "the temper ... soured by trifles, fretful, irritable, and passionate," and sometimes elevated by delight (Trotter 1807: 3.362). In the same vein as Hutcheson, in a passage in his *Essay on Man* (1733) satirizing the ideal of sensory prosthesis, Alexander Pope introduces as an improbable conjecture the very scene Melville witnessed on a whaleship:

Say what the use, were finer optics given,
T'inspect a mite, not comprehend the heaven?
Or touch, if tremblingly alive all o'er,
To smart and agonize at every pore?
Or quick effluvia darting through the brain,
Die of a rose in aromatic pain? (Pope 1787: 2.50 [1, ll. 195–200])

## PART III

The Situation, like all others, has notions of her own to put into the brain.
—Laurence Sterne, *Tristram Shandy* (1759–67)

I mean to spend the rest of this chapter studying the primary states of scorbutic excitement, leaving the reflective or corroborative ones to the next. This will involve an extended engagement with an environment not susceptible to measurement and an individual inside it whose pathological alertness is inversely proportionate to any faculty of will or expression. The autonomy necessary for action and speech was menaced at the ends of the earth by impressions which, already too novel to be comfortable, were redoubled by an innate infirmity. As his ship drifted past a paradise island abundantly supplied with the coconuts, bananas, and limes his crew were crying out for, but which their scorbutic prostration made it impossible to reach, John Byron explains, "This is a situation ... in which reason cannot preserve mankind from the power which fancy is perpetually exerting to aggravate the calamities of life" (Byron in Hawkesworth 1773: 1.93). Under circumstances similarly tantalizing, Pascoe Thomas complained, "It made our Situation appear ten times worse to me than at any other in the whole Course of the Voyage" (Thomas 1745: 149).

In the South Seas, the fundamental problem of the discoverer's *situation*, a word of great importance in the history of scurvy, arose from the uncertain relation of time to space. Until Harrison's perfection of the chronometer, an instrument that had a complicated history of its own in Cook's second and third voyages, longitude could be calculated by abstruse methods based on logarithmic tables, and only then if the ship was stable enough to take a reliable observation of the moon. While latitude was a straightforward reckoning, without longitude it was impossible to say where on that line the ship lay. Mariners, therefore, were never sure of where they were unless they reached a mapped landfall, but often as not, its coordinates were incorrectly given. Anson lost many men on the *Centurion* by taking a wrong turn on Juan Fernandez's line of latitude. Until the traverse of the ocean could be exactly correlated with duration, one's location was, as the Portuguese sailors called it, the *punto de fantasia*, the imaginary position. Here an ancient conundrum of philosophy was renewed, not only by the extent of the latest explorations but also by the strong Epicurean vein in experimental science.

Plato and Aristotle had called space and time phantasms, purely abstract concepts that arrived at a specification only by relativities: happening to be in such a place at such a time. This was a proposition Lucretius emphasized because it confirmed his claim that history was an assemblage of what he called "eventa" or accidents, actions happening to occur then and there, without any ontological solidity. "Time also exists not of itself," he wrote, "but from things themselves is derived the sense of what has been done in the past.... Nor may we admit that anyone has a sense of time by itself separated from the movement of things" (Lucretius 2006: 39, 1.459–63). As for the infinity of space, it is penetrable only as far as we can imagine the bodies that move within it (177, 2.148–157). "Verum positura discrepitant res," he announces: "Position marks the difference.... When the combinations of matter, its motions, order, position, shapes are changed, the thing also must be changed" (175, 2.1019–22).

In the section called "Concerning Position or Situation" of his *Philosophical Arrangements* (1775), James Harris explains that position or situation arises "from the relation which the distinction of *parts within* bear to the distinctions of *place without*; and it varies, of course, as this relation is found to vary." The corollary he draws is this: "The fewer of these internal distinctions any being possesses, the less always the number of its possible positions. As it possesses more, its positions increase with them" (Harris 1841: 345). It all seems very straightforward, except for the degree of variation. Might it be possible to suppose that the number of internal distinctions among the parts of a single being could be so few as to occasion a radical disproportion between them and those of the "place without," and what would happen then? Could we suppose further that the being with fewer positions is a human one, bereft of options and initiatives, and possessed by feelings suitable to that state of privation, while the situation is the beneficiary of a multitude of the available positions? This is what I have supposed with respect to scurvy in the previous chapter, noting that there is (for example) no proportion at all between what Kerguelen actually did under the influence of the disease and what (and how) he wrote about it in his treatise and journal. That is to say there is no relation between the nature of the experience in the first person and his knowledge of it in the third, the "parts within" and the "place without." Laurence Sterne amused himself with this idea in *Tristram Shandy* in the scene where Tristram, his ideas passing "gummous" through his pen from the "parts within," decides to liven them up by going to the glass to shave himself. An action exerted on his own body by his own hand in a "place without" is expected to yield him ideas or positions; but even if it doesn't, then "the Situation, like all others, has notions of her own to put into the brain" (Sterne 1992: 506 [IX, xiii]). Caught between a stimulus arising from an image of his head and a set of possible positions offered by the arrangement of the mirror's frame, the soap, and the razor within a room, he pits the ideas he wagers will emerge from his own brain against those arising from the situation—or, since he has already personified it, Situation.

I will now attempt two maritime parallels with Harris's and Sterne's approach to the extreme relativity of situation. In the journal of the voyage of the *Bounty*'s launch from the scene of the mutiny to the Dutch settlement at Coupang in Timor, the word "situation" looms large. It marks the difference between the "flattering situation" of a naval commander who has brought his task to a successful conclusion and that of a man reduced to impotence, thrust into an "unhappy Situation" in regard to which "very little reflection will show our miserable situation" (Bligh 1937: 1.131, 2.122). His situation goes from bad to worse—perilous, wild, deplorable—until the implied relation to a compassionate reader, the last positional option, breaks down: "It is a difficult thing to believe in our situation" (2.159), he says, much in the same tone adopted by William Wales when he complains of the scurvy, as if he had gone beyond the real world and was isolated at that moment in a nightmare, haunted by spectral figures who try not to look at each other, the sight is so unpleasant ("Our appearances were horrible, and I could look no way but I caught the Eye of someone" [2.165]). In Coupang, Bligh rehearses the events of this awful navigation, starving in an open boat on wintry seas, and he finds that the images of each successive phase still transfix him, even now when they have reached land, for an imagined observer would be hard put to choose what was more startling, "the Eyes of Famine sparkling at immediate releif [*sic*], or their Preserver horror Struck at the Spectres of Men" (2.227). Lodged in the mind and body of one of those "spectres" and totally at the mercy of the situation, Bligh confesses, "I appear to be contemplating a Dream rather than a reality with the Facts staring me in the Face" (2.229).

However, in the course of the voyage, especially after they had pierced the Great Barrier Reef, he made successful efforts to retrieve the situation by finding small amounts of fresh food and by feats of cartography and hydrography that would have been remarkable in a fully manned vessel of the proper rate. Thus, he ends up redeeming what the mutiny had cost him: his reputation as a skilled seaman and an equitable commander. At these moments, Bligh "calculates his situation": "the situation was four leagues from the Main, and we were on the N Westernmost of four small Keys" (2.224, 2.195). When he writes, "My situation being so low I could see nothing of the Reef towards the Sea," he doesn't mean he is blinded by despair but that his position relative to the reef gives him no aperture through which to view it. So "situation" goes through a kind of loop whose beginning and end is an adjustment of the relative positions of inward and outward factors, and at whose extreme extent there is nothing but illusion, tumult, isolation, and pain.

If we set Bligh's adventure in the launch alongside Apsley Cherry-Garrard's trek to Cape Crozier with Bowers and Wilson, likewise fraught with starvation, incipient scurvy, and the imminent probability of death, then it becomes possible to guess what is involved when the position of the eyewitness is overwhelmed by stresses so severe and numerous that all opportunities of calculation are lost, and the situation starts turning into a personification (Situation) with a monopoly on all available ideas. He begins his memoir with a testimony to the annihilating power of the Antarctic: "Nothing is more striking about the exploration of the Southern Polar region than its absences" (Cherry-Garrard 1994: xlix). The expedition to collect freshly laid Emperor penguin eggs, thought to contain the secret of the missing link, was undertaken during the polar winter, with practically no light. There were three principal privations driving each of the three individuals involved (Cherry-Garrard, Wilson, and Bowers) into the mighty arms of the Situation. First was a diet of pemmican, biscuit, butter, and oil, with no supplements of seal meat or penguin, which at best gave them half the calories they needed to fuel the enterprise and none of the necessary vitamins. Second is the unintelligibility of the landscape that is visible. At the end of the lava precipices of which Cape Crozier is formed, there was the "Pressure," their name for the giant wrinkles made of coastal pack ice, forced up by the action of the ocean against the cliffs of the coast, over which it was scarcely possible to draw a sleigh. "The tumult of pressure which climbed against them showed no order here. Four hundred miles of moving ice behind it had just tossed and twisted those giant ridges until Job himself would have lacked words to reproach their Maker" (272–73). Third is the terror provoked by the presence of a peril one cannot see or (what is worse) formed of nothing but absence. From the lip of a crevasse, "we could see the nothingness below" (130); and in a blizzard that overtakes them while trying to pitch their tent "we could see literally nothing" (175). This nothingness haunted Cherry-Garrard all the way to Kensington, where he went to deposit the eggs they had taken so many risks to obtain: "I don't know nothing about no eggs," the custodian at the Natural History Museum told him (305). The cumulative effect of starvation, chaotic masses of ice and rock, and the total loss of any sense of relative position is a state of mind and feeling he cannot express other than as a weird passivity: "I cannot write how helpless I believed we were to help ourselves, and how we were brought out of a very terrible series of experiences.... I always had the feeling that the whole series of events had been brought about by an extraordinary run of accidents, and that after a certain stage it was quite beyond our power to guide the course of them" (296). If Sterne's Situation thinks, Cherry-Garrard's acts; and as for Tristram's hypothesis of his own image installed in a place beyond himself, Bligh finds it experimentally proven in his own case when he is told by the boatswain that of all the horrible spectral faces in the boat, his own was undoubtedly the worst (Bligh 1937: 219).

Locke thought all extreme sensations useless to the scheme of knowledge because they disrupt our calculations of space and time. He said, "From such points fixed in sensible Beings we reckon, and from them we measure out Portions of those infinite Quantities ... which we call Time and Place. For Duration and Space being in themselves uniform and boundless, the Order and Position of things, without such known settled Points, would be lost in them; and all things would lie jumbled in an incurable Confusion" (Locke 1979: 199 [II, xv, 5]). The idea of inhabiting such confusion was impossible for him to imagine without assuming a total alienation from identity and common sense. A century later, Coleridge attempted a synthesis of this doctrine of position or situation in his *Treatise on Method* (1817), arguing that the mind's calculation of the relation of things (Locke's "points fixed in sensible Beings") sets everything in order vis-à-vis the subject. Method retrieves the senses from subjugation to sheer impressions of the "machinery of the external world" (Coleridge 1995: 1.634). If passions are the "state of undergoing" whatever the situation decrees, a merely involuntary and passive reception of the stream of impressions, then method regains for the mind the reflective consciousness upon which understanding and knowledge rely. "Take away from sounds etc. the sense of outness," he said, "what a horrid disease every moment would become." He saw experience and books as filters of the intensities that disturb and disorient our understanding of "outward reality": they are "necessary no doubt, if only to give a *light* and *shade* in the mind ... for all being vivid = (the whole becomes) a dream" (Coleridge 2002: 2.1422; 2.2526). Yet of all the Romantic poets and thinkers, Coleridge is most adept at evoking the state of undergoing, when the will is suspended, and the senses settle on an object and are absorbed by it: "The Eyes quietly & stedfastly dwelling on an object not as if looking at it or as seeing anything in it, or as in any way exerting an act of Sight upon it, but as if the whole attention were listening to what the heart was feeling & saying about it" (2.3027.11.93).

There were many situations in which this kind of reverie or fixation occurred in the eighteenth century, where the joint sense of duration as time and of space as position was suspended, and the imagination was compelled by whatever was seen, felt, heard, or remembered to react as if the here and now of an impression excluded all other points of reference. Such was the experience of second sight in the Highlands of Scotland, of those addicted to the reading of fiction, of voyagers hallucinating in places no European had ever seen, of self-experimentalists inhaling laughing gas at the Pneumatic Institute, of solitary walkers in France, and of opium eaters in England. As the sense of oceanic displacement combined with a scorbutic trance, John Mitchel, in the nineteenth century, gives a singular account of it: "My utter loneliness in this populous ship amidst the strange grandeur of the ocean, and for so many days ... all my life is the seeing of the eye only" (Mitchel 1864: 93). The Ancient Mariner remembers a starker arrangement of forces surrounding visual paralysis when "the sky and the sea, and the sea and the sky / Lay like a load on my weary eye" (Coleridge 2004: 175, ll. 250–51).

I want to explore two situations where this sort of fixation occurs during the pursuit of experimental knowledge. These are the ship and the laboratory, places where knowledge of facts is garnered and processed, yet where the coordinates of time and space were at risk of being lost; for when facts are not seen but seeing—staring observers in the face—they begin to strike the mind as preternatural and incomparable, as if they were happening in another realm; and then the relation between the parts within and the place without becomes very indistinct, at least from the human point of view. The condition is that of Etienne Bonnot de Condillac's statue when it first begins to feel: "It cannot differentiate between a cause within and a cause without.... The statue knows of no other state than that in which it finds itself.... It cannot make the difference between imagining a sensation and having one" (Condillac 1930: 8, 14, 18).

We need to be clear, therefore, how new developments in naval medicine sort with the examples of effluvia and situation we have been handling thus far. "Parts within" include the nerves, brain, animal spirits, reason, memory, and imagination, and the parts without are the material facts that modify the inward parts (including the skin): air, gas, temperature, aliment, shapes, colors, noises, odors, surfaces, hands, faces, and missiles. In the meetings between the two zones, effluvia are circulated, sense impressions received, passions refracted, mass eroded or restored, distances in time and space calculated, the memory filled, the imagination activated, physical reactions prompted, and some form of energy or matter sent out from the sensitive soul to envelop the earth, "this light, this glory, this fair luminous mist" (Coleridge 2004: 307, l. 62) Under normal conditions, the distinctions belonging to these two positions are mutually reinforcing, but if the parts within should abjure their loyalty to the place without, or the place without become so inhospitable, inaccessible, or inscrutable as to block the interchange of sensations and reactions, then the motions of the inner life of the sensorium yield either to an outward force that is for the moment irresistible and unintelligible or to an impulse supplied entirely from within. At this juncture, facts, having no position, strike so hard upon the senses they commandeer every scrap of attention; or else position is usurped completely by the imagination, which proceeds to fill the phantasms of space and time not with accidents or secondary qualities but with its own fictions. So it is important to distinguish between situation and its personification, Situation, for the one may have ideas to impart to the brain, but the other provokes the extremes of paralysis and failure to recognize what is what.

For at least a century before the scientific expeditions of the mid-eighteenth century, the exaggerations, errors, and special pleading of sailors under the dominion of Situation was often categorized as romance. Woodes Rogers, who earns a place in the history of fiction as the rescuer of Alexander Selkirk, the castaway on Juan Fernandez who supplied Defoe with the idea for *Robinson Crusoe* (1719), complained of the romantic accounts of buccaneers such as Dampier, invented for no better purpose than to conceal their own ignorance and poltroonery. Yet Jacob Roggeveen, the Dutch navigator saw no difference between Rogers's fallible account of antiscorbutic plants on Juan Fernandez and Dampier's, for both were utterly unreliable, "devised and represented after the fashion of romances" (Roggeveen 1970: 81). William

Betagh accused George Shelvocke (whose journal was one of the inspirations for the shooting of the albatross in *The Rime of the Ancient Mariner*) of writing "a wild story full of abominable romance and vain glory" (Betagh 1728: 109). Horace Walpole mocked George Anson's voyage as the feat of "Admiral Amadis," and of the commodore's decisive treatment of the agents of the Middle Kingdom in Canton, he joked, "Admiral Almanzor made one man of war box the ears of the whole empire of China (Walpole 1937–83: 9.55, 35.284). When Byron returned with his tale of Patagonian giants, Walpole reaffirmed what Peter Heylyn had proposed of the Pacific a century earlier, namely that "writers of Romance have a new field opened to them" (idem [1766] 1964: 203).

Here is the description of Dampier's skipper wounding himself with a fishhook, made by a man so sunk in a situational trance he is unable to perceive the relative importance of what he is looking at, or to gauge the drift of feelings the scene arouses: "One Time our Captain after he had haled in a good fish, being eager at his Sport, and throwing out his Line too hastily, the Hook hitched in the Palm of his Hand, and the Weight of the Lead that was thrown with a Jerk, and hung about six Foot from the Hook, forced the Beard quite through, that it appeared at the Back of his Hand" (Dampier 1729: 2.20). It is an event without any bearing on another, a fact without meaning; but as an impression, it could not be more vivid or exact. It is as if Dampier were describing an experiment whose sole focus is the datum it manifests, unrelated to any other point in space or time. Here is another example from Baudin's account of his voyage round Australia, notable (like Dampier's) for the heavy incidence of scurvy: "On our way back to the boats, I killed a snake that was about 2 feet long. There did not seem to me to be anything particular about it; its markings were a cindery grey with white points and its belly was completely white. None of its movements made me suspect that it could be dangerous. When I killed it, it was half way down a hole and was attempting to get right in. I tried to pull it out after it was dead, but it came apart in the middle" (Baudin 2004: 314). What on earth is the reason for commemorating such an unremarkable event: a short snake, dull markings, not dangerous, and trying to make itself scarce?

In a private moment on the Great Barrier Reef, Cook briefly considered the value of the discovery of facts as they concerned himself alone, regardless of their public utility, confiding, "Were it not for the pleasure which naturly [*sic*] results to a Man from being the first discoverer, even if it was nothing more than Sands and Shoals, this service would be insupportable especially in far distant parts, like this, short of Provisions and almost every other necessary" (Cook 1955: 380). Pleasure in this sort of private observation is correlated with his exasperation with facts that mean nothing. Of the penguins in the deep south, he says, "We had now been so often deceived by these birds, that we could no longer look upon them, nor indeed upon any other oceanic birds ... as sure signs of the vicinity of land" (idem 1777: 1.53). When Forster succumbed to scurvy in the same latitudes and began to despair over his slender collections of specimens, as well as how little he had found out of public value, he felt quite distinctly the shrinkage of his scientific self as his passions became increasingly ferocious and uncertain ("I wither, I dwindle away" [Forster 1982: 3.447]). In these examples, the social definition of the eyewitness as a competent agent is weakened either by a spontaneous indulgence in pointless observation or by the recognition of the subject's fading powers. Either way, the force of impressions subverts the translation of facts into exchangeable property. At these moments, Dampier, Baudin, Cook, and Forster are forced, as Whitehead says, to know away from and beyond their personalities, and to suffer the compulsion of a fact whose relation to the situation is undisclosed and therefore unattached to any other (Whitehead 1967: 88–89).

Time and again, and often in more dramatic circumstances, the witness will own only that nothing can be owned of such an experience because it is singularly shocking or incomparably voluptuous. "Such Confusion cannot be imagin'd by any who were not Eye-witnesses of it" (Campbell 1747: 14); "The fragrance of these green valleys, brought off by flaws of wind at intervals, was truly delicious, and a person that has at no time enjoyed, can scarcely be able to conceive with what delight we received it, after having been for a length of time at sea; it actually seems to take hold upon the feelings in such a manner as to reanimate the whole system" (Fanning 1989: 86). In either case, the force lies with the phenomenon, not the passive recipient, and readers are excluded from everything but the notice of their exclusion. Here is Richard Walter at Juan Fernandez: "It is scarcely credible with what eagerness and transport we viewed the shore.... Those only who have endured a long series of thirst, and who can readily recall the desire and agitation which the ideas alone of springs and brooks have at that time raised in them, can judge of the emotion with which we eyed a large cascade of the most transparent water" (Walter 1838: 111). Notice that the common experience of thirst doesn't provide an entrée here to the emotions of desire and agitation accompanying the actual observation of water, only to an idea that such emotions are likely to be intense in such circumstances. Walter is shifting the register from what might be commonly imagined to be true (as in a novel) to what presently and absolutely resists any normative judgment, as in the discontinuous episodes of romance. Carteret does the same: "It is not perhaps very easy for the most fertile imagination to conceive by what our danger and distress could be increased" (Carteret in Hawkesworth 1773: 1.405). Like Walter, he uses the first-person plural to define a sensory experience that belongs scarcely even to the single individual who reports it, sprawling as he is in the grip of a passion portable neither within the community of the ship nor over the page. As de Quiros reports in parallel circumstances, "There was not much good fellowship, owing to the great sickness and little conformity of feeling," emphasizing the loneliness of scorbutic intensities even though many people are simultaneously in the grip of them (de Quiros 1904: 1.105).

Although this form of expression is used so often in journals as to seem trite, its familiarity ought not to disguise how it patrols the boundaries of the communicable and the consensual only to desert them. The predicament of such reporters is one of unlimited loneliness, bordering on the isolation of the person hypothesized by Adam Smith, who is burdened with "the misfortune to imagine that nobody believed a single word he said, [and therefore] would feel himself the outcast of society" (cited in Shapin 1994: 12). This is the predicament of Coleridge's Ancient Mariner who, in losing the community of his ship and of living things, speaks of unparalleled loneliness to an auditor who has no means of apprehending what he is talking about ("This soul hath been / Alone on a wide wide sea: / So lonely 'twas, that God himself / Scarce seemed there to be" (Coleridge 2004: 186, ll. 597–600). Primo Levi dreamt he was home after his metamorphosis in Auschwitz and that he had begun to tell his family of the vile things he had seen and suffered, only to find that they are paying no attention to him: "A desolating grief is now born in me, like certain barely remembered pains of one's early infancy. It is pain in its pure state, not tempered by a sense of reality and by the intrusion of extraneous circumstances, a pain like that which makes children cry" (Levi 1996: 60). Frederick Douglass heard the sound of the loneliness of that pain in the songs sung by the slaves on the Lloyd plantation, which he compared to the singing in unison of individuals, each cast away on their own desert island, each uniquely miserable (Douglass 2003: 76).

These instances of morbidly acute sensation have none of the elliptical energy of the flux and reflux of the animal spirits mentioned by Charleton, Willis, and Boyle. They conform instead to the fixation of what Willis calls "an Idol in the Brain" (Willis 1683: 50), typical in his opinion of the lovelorn, but discussed in the next chapter under the heading of "nostalgia." Such obsessive harboring of a reiterated image, with no outlet or relief because of the impossibility of saying what it is like, bears directly on what Trotter says of the mortal danger of irritated nerves, namely that "the living body possesses the faculty, if I may call it so, of receiving impressions, and retaining them, even to the hazard of its destruction" (Trotter 1807: 199). If sounds can haunt a person to madness, and the odor of flowers kill the person who smells them, then the relation between object and subject, conceived by Locke as governed by powers operating at regular pressures via an insensible medium (Locke 1979: 142 [II, ix, 25]), is really much more variable, immediate, and perilous than he believed, liable to plant very different ideas of things in the brain from the system of cognition that holds the object to be one thing and the idea of it quite another. In fact, Trotter's insight introduces a sketch of what Erasmus Darwin called "diseases of volition," common in reveries that divorce the patient from any commerce with reality, such as the lady who asked the company to notice how her head had fallen off and rolled to the corner of the room, where it was presently being attacked by a small dog. These are examples of what Whitehead calls the totality of the event; Tristram Shandy a "Situation"; Cherry-Garrard the experience of nothing; and Bligh a confusion of facts and dreams.

So a Situation need not necessarily be terrible, but it will be total; and as scientists became more alert to the importance of sensibility in the matter of witnessing natural phenomena, especially the process of self-experimentation in which the observer and the object are the same, the idea of raiding the total event for a bodily sensation not one's own became attractive. The glorious prospect of being fully folded into a natural phenomenon sent Coleridge on punishing walks over the Lakeland peaks, risking his health and his limbs so he could know the "total feeling worshipping the power & 'eternal Link' of energy" (Coleridge 1956: 1.638; quoted in Holmes 1989: 291). Natural historians were notable exponents of this torturous yet delicious entry into the life of things. Having said farewell to his dying horse, and finding himself naked and starving some five hundred miles from the nearest European settlement, Mungo Park experienced a rare insight into the exquisite organization of a plant: "At this moment, painful as my reflections were, the extraordinary beauty of a small moss, in fructification, irresistibly caught my eye.... I could not contemplate the delicate conformation of its roots, leaves, and capsula, without admiration" (Park 2000: 227). Whether his early experience of scorbutic disgust at the plant life of Mauritius had prepared Bernardin de St. Pierre's sensibility for this intense aesthetic reaction, there is no doubt of his ability to distribute himself pleasurably among the microscopic actions and reactions of the natural world, and to let his imagination do the job of a microscope. Using a strawberry plant as his starting point and Bernard de Fontenelle as a guide, he supposed vast colonies of tiny insects on each leaf and then considered how the giant flower might strike the eye of each miniature witness: "Each part of the flowers must offer them spectacles of which we have no idea. The flowers' yellow anthers, suspended from white filaments, present them with a pair of golden joists balanced on columns more beautiful than ivory; the corolla, vaults of ruby and topaz of an incomparable height; the nectars, rivers of sugar; the other flowering parts, cups, urns, pavilions, domes" (Bernardin de St. Pierre 1804: 1.110; cited in Stalnaker 2010: 81).

Dampier was a favorite of Bernardin de St. Pierre because he would always make an attempt at the indescribable: his account of the look and taste of a plantain, for example, which in shape, he compared to a sausage; in texture and color, to butter in winter; and in taste, to a mixture of apple and Bon Chretien pear (Dampier 1999: 282; cited in Stalnaker 2010: 90). Bernardin de St. Pierre was a favorite of Alexander von Humboldt for the same reason, namely that he could enter into the nature of things on their terms rather than his own, enlarging the sense of what Humboldt called "plant geography" at the expense of the symmetry of his own mental architecture and the amenity of his surroundings. Generally, Humboldt sought this sympathy with animate nature by means of pain, hooking his deltoid muscle via a galvanic circuit to a frog's leg, putting mosquito venom into his open wounds, suffocating himself in mines to estimate the toxicity of gases, and shredding his feet and reducing his clothes to rags in order to appreciate the physiology of the Andean waxwood palm (Dettelbach 2005: 45–47). By means of self-inflicted agony equivalent to scorbutic extremities in distant seas, Park and Humboldt were able, like Dampier and Bernardin de St. Pierre, to orchestrate the intensities of confusing isolation, sensory overload, and a dwindled self into a description as exact, marvelous, and unrelated as an episode in romance.

The history of scurvy actually intertwined with these radical departures in experimental science when Thomas Beddoes published his book, *Observations on the Nature and Cure of Calculus, Sea Scurvy, Consumption, Catarrh and Fever* in 1793. Animated by Joseph Priestley's work on air, specifically his isolation of oxygen in 1774, Beddoes believed that experts in chemical medicine were about to revolutionize the treatment of diseases. Not just oxygen, but all sorts of gases were locked up in compounds, awaiting the pneumatic chemist capable of releasing them and exploiting for therapeutic purposes "principles of more extensive influence than even gravity itself" (Priestley 1790: 1.x; quoted in Jay 2012: 9). Oxgyen now took the place of the vital principle Boyle had searched for in the air pump, the invisible source of energy equivalent to Charleton's sap, Willis's latex, and Digby's nitrous salt. The task was to analyze the various constituents of air in different places and then to adjust the mixture by adding what was missing or extracting what was harmful. By means of self-experiments with hyperoxygenation, Beddoes had come to the conclusion, for example, that pulmonary consumption was attended with too great a presence of oxygen, and that a thinner and drier air was a suitable treatment. Contrariwise, he believed that scurvy was owing to a deficit of oxygen, the joint effect of oceanic air overloaded with moisture and salt from outside and miasmas from inside the ship, all contributing to the kind of suffocation he identified as the prime symptom of scorbutic decline. This was the single theoretical account of scurvy that Thomas Trotter was prepared to accept as a basis for medical intervention. However, he and Beddoes differed about the source of oxygen, whether it was released into the organism by the digestion of fresh fruit and vegetables (as Trotter believed) or whether it was conserved by proper ventilation (as Beddoes thought). This debate rehearsed in some respects the disagreements between the disciples of MacBride, who aimed by means of doses of malt to supply the loss of "fixed air" (carbon dioxide) in the bodies of seamen, and those who simply wished to ameliorate the temperature, odor, and humidity of their daily intake of the surrounding air for the sake of comfort and general health—in effect an offshoot of the argument about deficiency versus toxicity. Cook represented for Beddoes a magnificent example of ventilation, while Trotter praised him for the importance he attached to regular refreshment. But the point labored by Beddoes and understood by Trotter was that air was a congeries of different gases, all of which could be produced in the laboratory and thus distinguished as "factitious airs." Pure oxygen was a factitious air obtained by dripping acid on manganese, hydrogen by heating zinc in sulphuric acid, carbon dioxide by adding water to red-hot chalk (Jay 2010: 100).

Impressed by Helenus Scott's success in curing syphilis with nitric acid, Beddoes was on the lookout for a factitious air combining the virtue of the acid with that of oxygen when an American chemist called Samuel Mitchill came up with two possible gases that had first been identified by Priestley. These were nitric oxide, whose russet fumes were toxic but which we now know is necessary in the dilation of blood vessels, and nitrous oxide, a factitious air quite colorless and apparently innocent, but to which Mitchell assigned very sinister properties. If oxygen was vital air, the principle necessary for combustion and life, nitrous oxide was the opposite, the active principle of contagion, fever, and decay. If the ancient doctrine of putrefaction needed a factitious air of malign potency, this was it. Mitchill named it "septon" and apostrophized it in heroic couplets as the seed of all infections and contagions to which the human body was vulnerable. Since he understood scurvy as a contagious disease, Mitchill believed that in septon he had located the cause of scurvy as well as of yellow fever, typhus, and dysentery. Mitchill's discoveries were publicized in Britain by his associate Winthrop Saltonstall, whose book, *An Inaugural Dissertation on the Chemical and Medical History of Septon, Azote, or Nitrogene* (1796), was reviewed by Beddoes in the *Monthly Review* of the same year (XX: 490–93). Mitchill's work on septon was excerpted for an appendix to Beddoes second edition of *Considerations on Factitious Airs* (1796). For his part, Trotter included in the second edition of his *Medicina Nautica* (1804) an extensive account of "the Mitchellian Doctrine of septic Fluids," together with extracts from Saltonstall.

Intrigued by a compound of nitrogen and oxygen with such allegedly dangerous properties, Humphrey Davy subjected it to a series of experiments and found none of them. However, in the course of his examination he developed a method of producing larger volumes of the gas, and when he breathed it, his reactions were those of someone whose sensations had been preternaturally enlarged: his eyes dazzled, his hearing became very keen, and his muscles were so charged with energy he began to rush around the laboratory shouting loudly with glee. This moment of excitement inaugurated a series of public self-experiments with nitrous oxide at Beddoes's Pneumatic Institute, undertaken by eminent intellectuals such as Robert Southey, Richard Edgeworth, and Thomas Wedgwood. Davy records his own reactions to the gas in *Researches, Chemical and Philosophical, Chiefly Concerning Nitrous Oxide* (1800), where he returns frequently to the phenomenon of sensory overload: "I imagined that I had increased sensibility of touch: my fingers were pained by anything rough.... I was certainly more irritable, and felt more acutely from trifling circumstances.... My visible impressions were dazzling and apparently magnified.... When I have breathed it amidst noise, the sense of hearing has been painfully affected even by moderate intensity of sound" (Davy 1800: 464, 487, 491). The reactions of his fellow self-experimentalists were more or less the same: Southey's sense of taste and hearing became uncommonly quick. M. M. Coates reported that he seemed "to feel most exquisitely at every nerve." Beddoes said his sense of smell was sharper, and that generally "I felt as if composed of finely vibrating strings" (508, 532, 544–46). The coupling of septon with scurvy had been proved a mistake with respect to contagion, but it had been dramatically confirmed with respect to nervous excitement.

More remarkable than these testimonies to enlarged sensibility is the sparse contribution made by the one person from whom we might have expected a degree of Humboldtian precision: Coleridge. The single detail he volunteers is lodged in the bare phrase: "in great extacy" (517), together with the perfunctory recollection of feeling "more unmingled pleasure than ever before" (518). Davy explains on his own account why the description of such astonishing feelings was destined to be so scant. The first reason is the loss of any sense of relation in terms of space or duration; for although he claimed to be "perfectly aware of my situation," he confessed of the same experiment, "I lost all connection with external things.... I existed in a world of newly connected and newly modified ideas" (487). The second reason was that memory either failed or became unhinged: the sensations had vanished by the time he came to write them down, or his ideas came in such vivid and unusual streams they made no sense (479). The third reason was that his susceptibility to impressions and the disorder of his ideas made it impossible to articulate them in sensible words at the time they occurred because they were so singular. Davy wrote: "When pleasure and pains are new or connected with new ideas, they can never be intelligibly detailed unless associated during their existence with terms standing for analogous feelings" (495). Although Davy had decisively shown by means of extensive and rigorous chemical analyses of the gas that Mitchill's assessment of its mephitic power was mistaken, he had found it capable of inducing changes in the nerves and emotions so extreme he believed they might conclude in what he called "laesion of organisation" (467)—not quite septon's feast of living flesh that Mitchill had poetically invoked, but close enough to what Charleton, Willis, and Trotter had suspected of nervous irritation, namely, that if it went far enough it could damage or destroy the organism, beginning with the brain (Trotter 1807: 199).

In one of his most exhilarating experiences of the gas, Davy's sight and hearing were so tender that he felt himself in the grip of "perceptions totally novel" and cried out, "Nothing exists but thoughts! The world is composed of impressions, ideas, pleasures and pains" (Davy 1800: 488–89). Evidently, he could no longer distinguish between events taking place in his brain and the circumstances of the room. The predominantly pleasurable sensations of his confusion do not disguise his being adrift in a pure situation, whirling in an environment that is seeing, hearing, and thinking for him, like de Quiros's fantastic procession or Dampier's wonderful clouds. Bligh and Cherry-Garrard found themselves in the same place, but in agony, so there was nothing charming about the novelty of their new worlds. Davy wanted to believe that he had internalized his gaseous Terra Incognita and so mastered it. In effect, he was claiming the experience as a reverie such as Erasmus Darwin defined in his *Zoonomia* (1801), where the relation to the ordinary world of facts made communicable by analogy is temporarily at an end. But for Davy, an important difference was to be observed: his imagination didn't seal him off from that world but instead entirely supplanted it with one of its own creation.

As Mike Jay points out, here at the limit of sensory excitement, Davy believed he had discovered that there was no difference between absorbing the impression of an object and imagining it. Like Condillac's statue, he could not tell the difference between passivity and activity, "between a cause within, and a cause without" (Condillac 1930: 8). At this pitch of sensation, both were the same, and the illusion it generated for Davy is explained by Mike Jay as follows: "Reality itself was constructed in the mind, from the information delivered by the senses: [Davy's] culminating experiment had proved, as nothing ever had before, that an altered sensory and mental frame had the power to generate an entirely different universe" (Jay 2010: 199). If we compare this with Locke on the alien and incommunicable world of hypersensitivity, or with Bligh in Coupang, alarmed by the gulf between his ghastly countenance and the real world—"I appear to be contemplating a Dream rather than a reality with the Facts staring me the Face" (229)—we see how much the loneliness and pain of such a trance is elided in Davy's account. Indeed, it can scarcely be called an account since, lacking all analogies, his ecstasy remains entirely his own. There is nothing that it is like being Humphrey Davy under the influence of laughing gas—unless, like partygoers in Britain at the moment, you take some yourself. The kind of corroboration studied by Charleton, Willis, and Boyle, where the impact of things and their subsequent irradiation by the mind form the two foci of a working ellipse, is missing from these scenes in the Pneumatic Institute.

Jan Golinski supposes that Davy's experiments with nitrous oxide included a radical project of self-fashioning, involving two personae, the passive and poetic receptor, and the punctual and active scientific operator. Rather than floating beyond his own personality, Davy was doubling and reinforcing it. Golinski relates this bifurcation to the two characters in Davy's last work, a reverie called *Consolations in Travel* (1830), where the action is divided between the narrator Philalethes, who describes several life-threatening experiences similar to those Davy had suffered in the course of self-experiment, and "the Unknown," who intervenes with good advice and on one occasion saves him from drowning (Golinski 2011: 26). The division is like the one so often found in the poetry of Wordsworth, where the exceptional experience of the self is framed by an older version of the same self who is capable of understanding its importance as a memory, but quite incapable of saying what it was like at the time. Davy had warned that detailed ideas and intelligible language are not available to someone so excited as to be deprived of analogies, and the price paid for retrospection is to know how far removed it is from that immediate and present intimacy with the sensation of the thing itself. So the moment is only retrieved by Wordsworth with a statement of the impossibility of conveying what it was like: "I cannot paint what then I was" (*Tintern Abbey*, ll. 75–76), followed by a kind of descriptive embarrassment, a circling of the scene as if the necessary analogies might spring out from behind a pool, a rock, or a thorn.

As we have seen, the same rhetoric of expressive incompetence is deployed in defense of the immediacy of the pains and pleasures overwhelming the isolated individual in a distant place: I cannot tell because I was mad, or because I can't remember, or because what I said at the time was nonsense, or because I don't have words now that are adequate to convey it, or because you were not there with me, or because you have never suffered or enjoyed a situation like that one and don't care if you never do. The tussle takes place between a self that cannot mediate between the passion and the audience, and another self that makes something of putting that difficulty into words: for example, Davy's extravagant gestures under the influence of nitrous oxide compared with the sober account of all the tests he ran on the gas; or Kerguelen's irrational and violent behavior under the influence of scurvy compared with the treatise he wrote on the disease. The failure to recall an event in terms adequate to the pressure of the original excitement may help to fashion an articulate and reflective self, but at the cost of abandoning the other presence to invisibility and silence. On the other hand, it may be that the self abandoned to pleasure or pain inside an unnavigable situation preserves something by a silence or a noise that the other self spoils in the attempt to make it articulate. Thus Coleridge's poem "Constancy to an Ideal Object" (1828) turns a reverie inside out, so that what was only passionately and immediately known to the first-person singular becomes the delusion observed by a third. In the next chapter, I will explore the range of reflective possibilities available to scorbutic memoirists as they try to reach a public dwelling on the other side of the Situation.

# CHAPTER THREE

# Nostalgia

If what you love is absent, yet its images are there ... the sore
quickens and becomes inveterate by feeding, daily the madness
takes on and the tribulation grows heavier.

—Lucretius, *De rerum natura*

The discussion in the last chapter of pathologically present sensations,
emotions, and images showed how closely the ecstatic intensities of
scurvy resemble others (reverie, the reading of certain kinds of fiction, the
inhalation of laughing gas, laboratory encounters with singular facts, etc.),
where what is imagined is perceived as a phenomenon with a real existence,
or conversely, where what is experienced as a fact strikes the mind as so singular it appears like a fiction or a dream. It remains to examine the qualities
and gradations of these ecstasies in order to judge what difference, if any,
distinguishes the enlargement of Humphrey Davy's sense impressions under
the influence of nitrous oxide from the extraordinary susceptibilities of scorbutic patients, and whether the extreme fits of reverie discussed by Erasmus
Darwin have anything in common with the delusion common among mariners that the blue ocean has turned into a green landscape. To do this, I want
to begin by emphasizing two things.

The first is the nervous component of scurvy, a disease that many of its
students point out effloresces as much in emotional and psychological forms
as bodily ones and, in its latter stages, luridly confounds the two—"People's
minds ... became as loose and unsteady as their teeth from scurvy," as Georg
Steller puts it in his account of Vitus Bering's voyage (Steller 1988: 115). In
order to probe as exactly as I can the links between a disturbed or uncommonly excited state of mind and a depraved condition of the body and its
appetites, I shall place a specific set of scorbutic delusions alongside two conditions often associated with them but which are epidemiologically distinct.
The first is the pure form of nostalgia, the pathological longing for home,
considered independently of the maritime longings for a landfall with which

it is often associated; the other is calenture, a thoroughly maritime disorder exhibiting hallucinations closely resembling scorbutic fantasies and nostalgic obsessions, yet without any apparent cause in nutritional deficits or homesickness, for it can occur within a week of putting to sea.

The second emphasis will advance what I have already claimed for the importance of the word "situation" and the meanings it attracts of a total and unframeable environment in which what normally would have been perceived as a set of distinct positions, periods, and options available to a rational subject are transformed into an event without relative temporal or spatial dimensions. Such confusion leaves the perceptions and feelings as powerful as they are disorganized, and the mind is made prey to impressions whose origin might as easily be the imagination as the external world. When no reflective distance separates the individual from what seems like an atmosphere of images and excitements encircling the senses, then colors flash and sounds reverberate with an awareness of what Coleridge calls their "outness" diminished if not entirely lost. Then things are seen and heard by virtue of a power with no necessarily phenomenal origin, and images can act by their own force as images. The effects such a vortex makes on the brain are extremely vivid, and either they reveal the passive and overwhelmed plight of the ego, quite sunk in what is befalling it—what Willis calls the falling down of the whole soul—or, the very opposite, they bespeak a union with the world at once anticipated and enjoyed by the imagination, a feat Willis described as the soul's ovation and triumph. How far these two tendencies are congenial to the experience and expression of the three varieties of nostalgia that I am about to discuss—whether they simply represent various degrees of alienation from the sense of outness that stabilizes a situation, or whether on the contrary they are quite distinct, each with a dynamic peculiar to itself—this chapter shall settle as its final business.

Thomas Trotter was the first naval physician explicitly to consider the link between scurvy and homesickness as more than fortuitous when he described the symptoms of "scorbutic nostalgia" in 1792, deploying what he called "the second species" of William Cullen's nosological definition of the disease (Trotter 1792: 145, 45). In the lethargy and depression following the onset of scurvy, he noticed, along with the usual early symptoms of ghastly countenance, aching limbs, swollen gums, and hemorrhages of the hair follicles, various degrees of yearning for food, liquid, land, and home, often culminating in fits of hopeless tears. "I consider these longings as the first symptom and constant attendants of the disease in all its stages." He went on to specify them: "The cravings of appetite, not only amuse their waking hours, with thoughts of green fields, and streams of pure water; but in dreams they are tantalized by the favourite ideas; and on waking, the mortifying disappointment is expressed with the utmost regret, with groans, and weeping, altogether childish." No physician, he warned, ought to overlook such emphatic "cravings of nature" (44, 35).

Like other medical men with an interest in the psychological and sensory accompaniments of scurvy, such as Thomas Willis and Walter Charleton in the previous century, and James Lind and William Falconer in the eighteenth, Trotter wanted to know more about the reciprocal actions of the scorbutic body and the disturbed mind. Eventually, he was to devote a whole book to the topic of nervous disease. With his colleagues Falconer and Thomas Beddoes, he was exploiting the advances made in the study of nervous irritability by Albrecht von Haller, William Cullen, Robert Whytt, and John Brown (Harrison 2010: 62). Like them, he paid close attention to the passions of his patients, fascinated by the contribution they might make to the speed of organic disintegration, as well as to the destruction of the personality. Of rugged sailors bathed in childish tears, he observed, "The hero and the infant here unite," an astonishing factor in "a new subject in nautical medicine, unfolding such peculiarities in the constitution of officers and seamen, that it would be unpardonable to pass it by" (Trotter 1804: 3.362).

His interest in the affective aspect of the disease was aroused when working as a surgeon on the Liverpool slave ship *Brookes* in 1783, bound for the Gold Coast and Antigua. It was not a job he relished, but he was assiduous in attending to the human cargo. At night many of the Africans, having woken to find themselves elsewhere than the scene of their dreams, made "an howling, melancholy noise, expressive of extreme anguish" (*Abstract of the Evidence* 1792: 44). When he asked his interpreter to find out why, "She discovered it to be owing to their having dreamt they were in their own country again, and finding themselves when awake in the hold of a slave ship" (ibid.). Trotter was convinced that this passionate nostalgia was linked to the severe outbreak of scurvy that troubled the ship during its lading as well as its voyage. Like the dejection of pressed men, homesickness among slaves seemed to hasten the onset of scurvy and then to aggravate it, by which time the absence of familiar food and the craving for familiar places were blended as factors both in the scorbutic condition of the patient and its delusional accompaniments (Trotter 1792: 63). Trotter was perpetually aware of the novelty of this kind of analysis and the strangeness or foreignness of its object, "attended with many singular phenomena never before isolated," for which (he concedes readily) an adequate terminology does not exist (50–63; 1804: 3.364). "Scorbutic nostalgia" amounted therefore to a nomenclatural breakthrough—as far as I am aware, he was the sole inventor and employer of it—

but if it was to be anything more, then the alliance of the two diseases needed more specification.

Trotter was prepared to offer some. What he had done was to modify two entries in William Cullen's *Nosology* (1793), a systematic table of illnesses organized on the principles of a Linnaean taxonomy. These were genus 86 *Scorbutus*, and genus 105 *Nostalgia*. Neither of Cullen's outlines of these diseases was particularly original. Trotter thought his assigning cold weather as a proximate cause of scurvy was wrong, and there was nothing new about listing nostalgia as an urgent desire to revisit the homeland. However, Cullen had placed nostalgia in the class of Locales, his name for distempers of the body, and then in the order of Dysorexiae ("False or defective appetites"). That is to say, he was not putting nostalgia in the class of nervous diseases, nor was he assigning it to the order of delusional longings, but treating it as a physical complaint accompanied by difficulties in the consumption of food, such as *anorexia* (eating too little), and *pica* (eating what is not food; Cullen 1800: 162–63). Moreover, he offered two kinds of nostalgia, *simplex* and *complicata* (164), allowing homesickness some degree of kinship with other diseases in the same class or order. This caused some astonishment among his colleagues, but his biographer John Thomson pointed out that Cullen was determined to mark the boundary between delusions and appetites arising from organic morbidity, and those owing to a derangement of the mind (Thomson 1839: 2.65). For his part, Trotter could not be blind to the resemblance between scorbutic symptoms (craving for food sometimes alternating with a voluptuous enjoyment of it) and the order in which Cullen had placed nostalgia, where preternatural appetites wore the appearance of the pathological sensitivities typical of scurvy, such as *polydipsia* (unquenchable thirst) and *bulimia* (unappeasable hunger). So he chose to unite nostalgia with scurvy as a complex instance of two diseases: nostalgia, which Cullen regarded in some important respects as a nutritional disorder, and scurvy, which Trotter was convinced was nutritional, too. He did this knowing that scurvy could mimic a whole hospital of infirmities while at the same time (in his view) remaining "of so singular a nature that no disease seems analogous to it ... by any concourse of symptoms or method of cure" (Trotter 1792: 106). Indeed, he proclaimed that "in forming a diagnosis of Scurvy there is but little danger of confounding it with any other disease" (42). In proposing the mixed case of scorbutic nostalgia, he had two motives for compromising what seemed like an unequivocal position on the singularity of scurvy. The first was to take the opportunity of elaborating the testimony he had given a Parliamentary Select Committee on the Slave Trade the year before the *Observations* was published, where the relationship between the nervous and somatic elements of nostalgia was introduced, but left unclear; the second was to recruit a powerful ally in the debate about deficient versus defective nutrition (a futile gesture, as it turned out).

Like Cullen, Trotter was airing what must have struck his colleagues as a paradox by presenting symptoms proper solely to scurvy as typical of another disease, but reference to his etiology helps clear up some of the contradiction. It was based on four degrees of distance between cause and symptom. Furthest away and of least concern was a pretended cause, such as the humoral theory of scorbutic putrescence entertained by many of his colleagues, a conjecture unconfirmed by observation. Remote causes were next, divided between occasional and predisposing: occasional causes being found in the state of the ship (damp, unventilated, cold, dirty) and predisposing causes lodging in the temperament of the individual who might be melancholy, laboring under a sense of injustice, or just longing for home. Next was the exciting cause, and Trotter was convinced, as were Cullen, Cook, Blane, Anson, Hulme, Nelson, and other influential medical and naval personnel, that this was a shortage of fresh food, especially vegetables and fruit. Finally there was the proximate or immediate cause, of which Trotter thought everyone including himself was ignorant, asserting "the proximate cause of Scurvy is still to be sought for" (124). So how exactly did nostalgia contribute to the scorbutic emergency on the *Brookes*?

As Trotter expands his account, it becomes evident that scurvy already had a footing in the vessel before it set sail, with at least seven people dead of it on the African coast before the course for Antigua was set. In his account of the outbreak, Trotter initially wants to blame the Duncas, people from marshy coastal land who took no exercise, grew very corpulent on a diet of corn mash, and brooded over their misfortunes; however, he said in his testimony that it was those of an "exquisite sensibility," particularly among the women of the Fantee nation, who were most prone to nostalgia and, by implication, to scurvy, too (56, 62). To help us understand how he was developing his diagnosis of scorbutic nostalgia, Trotter marks a distinction between two mental aberrations that were assumed to have an effect on the body, *hypochondriasis* and *idiosyncrasy*. A hypochondriacal patient will suffer physical symptoms arising from a disturbed state of mind. Trotter mentions dyspepsia as a case where mental anguish acts as an exciting cause of a physical symptom (72). But with idiosyncrasy, this is not so, Trotter believes, for it acts only as a predisposing cause: "I shall not ... attempt to explain any symptom that may be said to arise from idiosyncrasy" (43). As for the physical symptoms of the brooding Duncas, the exciting cause is to be found in their diet and their idleness; they are behaving merely "as if hypochondriacal" (46).

While insisting that no idiosyncratic predisposition can produce scurvy without an exciting cause, he concedes that "where some are afflicted before others, we ought, certainly, to assign it to idiosyncrasy, or peculiarity of temperament." But it is never to be forgotten that "when the body has been subjected to the diet of which Scurvy is the immediate Effect, it will make its appearance, though in a longer time, without any predisposition whatever" (67). So if diet was the exciting cause with the Duncas, and temperament only a predisposing cause with the Fantee women, the exciting cause in both cases—that is to say, of scurvy and nostalgia—was food. Although Trotter doesn't quite spell it out, the question is really whether nostalgia is acting as a predisposing cause of scurvy or as one of its symptoms, since scurvy was well established before the *Brookes* set sail for the Leeward Islands. He is quite prepared to concede that misery may weaken resistance, but since the disease was already on the march, the identification of its effect on the nerves and emotions is a more important task for him than a medical speculation about the likelihood of nerves and emotions contributing to the sickness.

So what attracts his attention are those cravings, lamentations, and tears whose neglect would be unpardonable in a physician, not because they are symptoms of nostalgia, but because they are symptoms of scurvy that have never before been analyzed as such. "I consider these longings as the first symptom and the constant attendants of the disease in all its stages" (44). He never leaves it in doubt that these longings and cravings refer to the food and drink whose absence acts as the exciting cause of scurvy, nor does he concede that idiosyncratic yearnings for home have anything more than a tertiary relation to the primary symptom. No sooner does he instance the cravings for green stuff than he adds a physical symptom, emphasizing that the want of the right victuals produces bodily effects of which extravagant emotions are merely the outriders: "Around this time the colour of the face is changed" (45). Between the cravings of Trotter's first symptom of scurvy and the account of nostalgia given in the entry in the *Encyclopédie méthodique* (1782–1832) cited by Starobinski, there is a perfect reverse symmetry, for in the latter, the sadness of the patient, the haggard look, glassy eye, lifeless countenance, and torpor are accompanied by physical correlates such as an uneven pulse and fever, secondary indications of the indisputable presence of a fatal nostalgia (Starobinski 1966: 97). By producing a nervous disorder as the exciting cause of a physical breakdown, the author of the entry, Philippe Pinel, lays his emphasis squarely contrary to Trotter's, where nutrition is the cause of a bodily affliction and alterations in temperament are merely symptomatic of that organic disorder.

One of the first experimental identifications of nostalgia as a disorder of the mind possibly in league with scurvy was made by Joseph Banks on the *Endeavour* during a voyage allegedly free from scurvy but during which he was obliged at least twice to dose himself with inspissated citrus juice specially prepared for him by Nathaniel Hulme. On the second occasion, with blisters in his mouth, he noted that all the crew but himself, Solander, and Cook were "pretty far gone" with homesickness, a condition "the Physicians have gone so far as to esteem a disease with the name of Nostalgia" (Banks 1962: 2.145). Three days later, he succumbed himself when he remembered Dampier sailing the same seas, and "this thought made home recur to my mind stronger than it had done throughout the whole voyage" (ibid.) Charles Darwin was to find himself in this condition off the coast of South America sixty years later, overwhelmed by memories of the friends of his youth (Beer 1996: 22–23). James Watt believes that the motifs on a dinner service commissioned by Anson in China at the conclusion of his scorbutic Odyssey through the Pacific are hieroglyphs of homesickness, consisting of scenes from Juan Fernandez and Tinian interspersed with domestic images of dogs, sheep, the Eddystone Light, and Plymouth Sound, "betraying the nostalgia which commonly accompanied vitamin deficiency" (Watt 1998: 577). That these images were set in the glaze of dinner plates, destined to be covered with the food once so passionately desired and of which they had proved so happily to be the harbinger, is a singular triumph of design—nostalgia in reverse, as it were—an adjustment Trotter would have appreciated. In Banks's case, nostalgia appeared to last only as long as his other symptoms of scurvy, dissipated (it seems) by Hulme's medicine.

In 1788, Falconer had published a prizewinning essay, *On the Influence of the Passions on the Disorders of the Body*. In it he gave scurvy pride of place as a remarkable instance of psychosomatic transfers, telling the story out of James Lind about the siege of Breda, where the Prince of Orange cured his scorbutic men with a placebo pill, an incident Trotter found incredible, but which Lind had justified as an authentic lesson in "the wonderful and powerful influence of the passions of the mind on the states and disorders of the body" (cited in Falconer 1788: 85; Trotter 1792: 123). Hulme's account of this event (in which the siege of 1627 seems mixed up with the one of 1637) is quite contrary to Lind's and Falconer's, for he interprets it as a tragic allegory of nutritional deficit: "Had the states of Holland provided the city of Breda ... with plenty of [fresh food], it is more than probable, that so many hundreds of their bravest men would not have died piecemeal by the scurvy: but would have held out till the prince of ORANGE could have come up to their relief"

(Hulme 1768: 72). Falconer for his part went on to discuss nostalgia as potentially fatal, using material from Johannes Hofer, the Swiss physician responsible for identifying homesickness as a disease and giving it its official name. Then he makes an observation of his own that tied nostalgia to physical evidence of scurvy ("livid or purple spots upon the body") and to the nutritional problems that may have caused it, but to which it also gave rise. First of all stating that all alterations of mood affect the body to some degree, he turns to scurvy and says "perhaps this is the only endemic disorder of which we have any knowledge ... in which mental affections are *specifically* hurtful" (Falconer 1788: 92). By this, he means that nostalgia is hypochondriacal, not idiosyncratic, and that Trotter's first symptom of scurvy—nervous collapse consequent upon a depleted constitution—is quite the reverse, actually provoking the petechiae and lesions that Falconer interprets as symptomatic of mental distress.

Here Falconer and Trotter part company, for Falconer was installing idiosyncrasies as exciting (possibly even proximate) causes of the disease. In drawing the line at Lind's story of the Breda placebo, Trotter was insisting that the cure of an established scorbutic condition could not merely be imagined into existence, or banished by a change of mood. Something substantial would have to be added to hope if an organic amelioration were to occur; likewise, the absence of something equally solid would need to act as the exciting cause if a predisposing cause were to herald and possibly accelerate a fatal outcome. In order to become scorbutic, the nostalgia of the Africans on the *Brookes* had to be owing to a change in their diet, most likely the corn mash they were made to eat on the coast (at first), followed by the refusal of food so common among the homesick once the ship was under way. What hurries the patient into scurvy is not misery itself, but the alteration in diet caused by it, or coincident with it.

The first difference between scorbutic nostalgia and pure nostalgia emerges here, since there were plenty of alleged cures of nostalgia obtained simply by the promise of returning home, provided the physical symptoms of self-starvation had not become too severe (Dames 2001: 33–34). But in spite of the fact that scorbutic decline was far from steady, often interspersed with false signs of recovery, there is no reliable evidence that scorbutic nostalgia was ever cured merely by encouragement, promises, or good news. Once established, its excesses appeared "as if hypochondriacal" (Trotter 1792: 46), but they were in fact geared to images of the colors, foods, and liquids directly associated with the remedies the body needed.

The opposing views of Falconer and Trotter were representative of a difference between French and English physicians. Pierre Barrère performed a series of autopsies on the bodies of French soldiers who had died of nostalgia. In almost all of them, he found proof of the fact that extreme nostalgia was a fatal disease. "Cette pensée continuelle de revoir son Pays cause d'engorgemens au cerveau, des tremblemens, des roideurs aux Membres ... d'où suit la mort" [The perpetual thought of returning home causes blockages in the brain, trembling, stiffness in the limbs ... from which death follows] (Barrère 1753: 26). Again and again, he discovers the physical proof of nostalgia in the blockage of arteries and ventricles, clotted with blood as black as ink (14). D. J. Larrey found the same swollen blood vessels in the brains of men dead from nostalgia, but like Barrère's evidence, it was consistent with the vascular damage typical of scurvy and therefore typical of a nutritional deficit rather than a constitutional melancholy (Dames 2001, 30; Brown 1788: 2.347). A military contemporary of Larrey and Barrère noticed that patients talking obsessively of home lost their appetite and found it difficult to digest the little food they were able to eat, suggesting that nostalgia was indeed a predisposing cause of scurvy; but as for its role as an exciting or proximate cause, it was most probable that the damaged blood vessels and dark blood ascribed by Barrère to obsessive longings for home confirm a diagnosis of scorbutic decline. Gilbert Blane noticed in sailors dead of scurvy "large effusions of coagulated blood into the cellular membrane" (Blane 1799: 486). Giving evidence to the enquiry into the disastrous Nares expedition, Dr. Buzzard identified as specifically scorbutic those symptoms the French assigned to nostalgia, pointing out that autopsies of the brain revealed areas "gorged with very dark fluid blood, or coagula, and ... sanguineous effusion into its substance (RCA 1877: 198; Malieu de Meyserey 1754: 105). However, that is not what either Barrère or Falconer claims. They say there is a direct link between the nervous symptoms of diseases such as nostalgia and signs of physical deterioration, particularly in the blood vessels, as if (Barrère seems to suggest) the heart and the brain were literally suffused with grief.

Unless physicians espoused J. J. Scheuchzer's theory of barometric pressure as the problem behind homesickness among the Swiss (Hofer 1934: 383; Starobinski 1966: 88), nostalgia had really only one exciting cause—being away from home—which was not clinically measurable. The only sure remedy of return was likewise not medically verifiable. So in the first instance of what I have termed "pure nostalgia," there is evidence of a temperamental refusal of the patient's present circumstances but not, as Kevis Goodman suggests, a specifically physiological component, that is an exciting or proximate cause in the shape of "a somatic revolt against forced travel," unless scurvy is to be installed as something more than the "less likely accomplice" of nostalgia (Goodman 2008: 196, 204). In scorbutic dreams, on the other hand, the cravings of appetite may produce illusory scenes of domestic pleasures, but these frame vivid and exact images of the food the stomach needs, as if the imagination were urgently sending pictures to the mind on the body's behalf of what it needs: green vegetables, fresh fruit, clear water. Home is no more than the alibi of the nutritional remedy. Of the scorbutic visions of sailors, James Lind wrote, "What nature, from an inward feeling, makes them thus strongly desire, constant experience confirms to be the most certain prevention and best cure of their disease" (Lind 1753: 128). During the Admiralty enquiry mentioned previously, the question was put to Dr. Buzzard whether scorbutic dreams of fruit and vegetables were "a mental expression urging the dreamer to the choice of a food best suited to his condition," and Buzzard emphatically agreed that it was (RCA 1877: 198).

Under no circumstances do the visions of pure nostalgia conform to this model, for they are neither specific nor efficacious, as we shall see. The difference between the two conditions is always to be gauged from the reality of loss in the case of scurvy and its inexact or illusory appearance in nostalgia, each to some degree measurable when an encounter with the sought object is achieved. Trotter observed closely both the grief of those who yearned for fresh food and liquid, and the delight with which at last they beheld and consumed them. The change in mood is exactly calibrated to the imminence of real physical satisfaction, such as the scorbutic John Mitchel, thinking of oranges, carefully noting the colors of their skins, then rapturously eating them at Pernambuco (Mitchel 1864: 88). These are moments of indisputable sapid recognition when what is desperately fancied as well as needed, and what is presently being seen, tasted, and swallowed, overlap in a conjoint sensation. On the other hand, homecoming is almost inevitably spoiled by disappointment when the imagined good fails to be matched by the real thing.

Tears flow, says Willis, from "the moisturising juice of the Brain" when it still has enough power to know what it wants and to reach out for it in an "irradiation," albeit "refractedly and disturbedly" (Willis 1684: 179–84). Charleton calls it corroboration when the irradiated image is greeted by the sensation of its arrival, and then redoubled by the energy of the animal spirits ("whole brigades of them ... dispatched into the Organs of the Senses, and into all Muscles"; Charleton 1670: 109). Montaigne used the same model of corroboration in his essay "Of Experience," where he confesses he sometimes has himself woken up at night so that he might relish the pleasure of slumber "better and more sensibly" (Montaigne 1711: 3.458). Willis used the flux and reflux of the animal spirits to explain drinking a long-anticipated fine wine: "The Imagination of its Pleasure is again sharpened by the taste, and then by a reflected Appetite drinking is repeated. So as it were in a Circle, the

Throat or Appetite provokes the Sensation, and the Sensation causes the Appetite to be sharp'ned and iterated" (Willis 1683: 49). Georg Forster describes the *Resolution's* taste of fresh fish in Dusky Bay at the scorbutic conclusion of Cook's first great loop into Antarctic seas as exemplary of the action and reaction of desire and pleasure: "The real good taste of the fish, joined to our long abstinence, inclined us to look upon our first meal here, as the most delicious we had ever made in our lives" (Forster 1777: 1.124). So the still lifes of fruit and vegetables painted by the scorbutic imagination were always a tryst kept with reality, or at least with the material world, provided the victim lived long enough to enjoy it. The sequel of physical repletion, far from extinguishing appetite, reacts with it to make the sensation of taste even more delightful. Among the diseases of the order Dysoriexia, Cullen had listed *polydipsia*, or "preternatural thirst," and bulimia, "appetite for a greater quantity of food than can be digested." The scenes of repletion described above demonstrate satisfactions quite as exaggerated as the cravings that preceded them, their intensity prolonged in a variety of sensory registers. After his meal of fish in Dusky Bay, William Wales recalled, "I was entertained in bed with a serenade by the winged Inhabitants ... far superior to any ever enjoyed by a Spainish Lady" (Wales MS Journal, 26 March 1773).

Banks's thoughts of Dampier in the Arafura Sea, therefore, are likely to have had a more material connection than that of their common nationality, for in his scorbutic state he was primed to enjoy at Savu, his next landfall, tastes of exotic fruits already described in luscious detail by his predecessor, such as the durian, ripe only for a moment with a taste so rich and curious it was only just this side of repulsive. These fruits were going to put a more glorious end to scurvy than Hulme's concentrated juice, even though objectively some of them were not in the best condition: "Bad as the character is that I have given of these fruits, I eat as many as any one, and at the time thought as well and spoke as well of them as the Best freinds [*sic*] they had" (Dampier 1999: 144; Banks 1962: 2.213). To the extent nostalgia was part of his ailment, we notice that home recurred to his mind powerfully but indistinctly—did he mean London, his estate in Lincolnshire, his friends, Miss Blosset, or a permutation of them all? Apart from food, it is not certain what he was missing, or whether on his return he was particularly gratified by the reunion with it, them, or her.

In his landmark essay on the disease, Jean Starobinski generalizes the stage set of the nostalgic illusion—"sad, tender recollections, golden visions of childhood" (Starobinski 1966: 87)—while Hofer himself points out that a nostalgic mind "feels the attraction of very few objects and practically limits itself to one single idea" (cited in ibid.). Taking his lead from Hofer, Helmut

Illbruck has argued recently that the poignancy of homesickness arises from nothing else than its being sealed up in the imagination, where the impossibility of its longings ever being satisfied is the whole point, for as soon as the opportunity of satisfaction is introduced, the disease begins to disappear. Hofer had said that nostalgia "admits no remedy other than a return to the homeland" (Hofer 1934: 382), but Kant, for one, disputed the prescription, or at least the meaning of the word "remedy," when he said, "After they visit these same places, they are greatly disappointed in their expectations and thus also find their homesickness cured" (Kant 2006: 71). When Odysseus finally gets back to Ithaca, he cannot recognize it as home: "He therefore, being risen, stood and viewd / His countrey earth: which (not perceiv'd) he rew'd / And striking with his hurld-downe hands his Thyes, / He mourn's, and saide, 'Oh me! Againe where lyes / My desart way?'" (Homer 1956: 232 [13, ll. 292–96). He is the first in a long line of nostalgic voyagers who have difficulty in treating homecoming as either real or pleasant.

There is a dramatic specimen of this disappointment recorded by Robert Scott on his *Discovery* expedition to the Antarctic, when he and his party got back to their ship after an exhausting trek to the South. "How can I describe this home-coming … how our eyes wandered about amongst familiar faces and objects … how in the unwonted luxury of clean raiment we sat at a feast which realised the glories of our day-dreams; how in the intervals of chatter and gossip we scanned again the glad tidings of the home-land.… It was a welcome home indeed, yet at the time to our worn and dulled senses it appeared unreal: it seemed too good to be true that all our anxieties had so completely ended, and that rest for brain and limb was ours at last" (Scott 1907: 2.92–93). The same flatness of a long-anticipated arrival afflicted Bligh when he finally made land and ate fresh food after sailing three thousand miles in an open boat, and he feels like an unlovely spectre in an unpleasant dream (Bligh 1937: 2.229). In these two examples, one detects an unquenchable nostalgic desire not magnifying but impeding the scorbutic corroboration of the taste of food; or perhaps having already eaten, the shadow of the longing for the circumstances of food persists in the mind of the sufferer as a sort of bulimic refusal of satisfaction.

Before going any further, it is necessary to consider alongside nostalgia another disease that had a much longer history in the literature of medicine and voyaging, and this was sea-fever, or calenture, defined summarily by Samuel Johnson as "a distemper in hot climates wherein [sailors] imagine the sea to be green fields" (Johnson 1760: "Calenture"). There is a lively literary pedigree for this sort of disturbance of the brain. The episode of the sirens in the *Odyssey* is an early example, where mariners are allured overboard while listening to haunting songs sung by beautiful women, only to add their bones to the hedges made of skeletons that enclose the flowery meadows where the sirens sit and sing (Homer 1956: 210 [12, ll. 67, 236]). In the seventeenth century, the mariner who throws himself overboard into a fatal delusion of pleasure, often fringed with meadow, is a byword in the drama and poetry that lasts well into the Romantic period. Nicholas Rowe uses the image in *The Ambitious Step-Mother*, and John Dryden in *The Conquest of Granada* when Almahide tells Almanzor, " 'Tis but the raging calenture of love. / Like a distracted Passager you stand, / And see, in seas, imaginary land, / Cool groves, and flow'ry meads, and while you think / To walk, plunge in, and wonder that you sink" (Dryden 1808: 4.143 [II, iii]). In his satire on the South Sea Bubble, Swift compares the deluded bankrupt with the febrile sailor who sees "On the smooth Ocean's azure Bed / Enamell'd Fields, and verdant Trees" (Swift 1958: 1.251). Wordsworth's fullest treatment of calenture is found in *The Brothers* (1800), where Leonard "would often hang / Over the vessel's side, and gaze and gaze; / And, while the broad blue wave and sparkling foam / Flashed round him images and hues that wrought / In union with the employment of his heart, / He thus by feverish passion overcome, / Even with the organs of his bodily eye, / Below him, in the bosom of the deep, / Saw mountains; saw the forms of sheep that grazed / On verdant hills" (Wordsworth 1984: 157, ll. 51–60).

Trying to offer an intelligible account of the mutiny on the *Bounty*, Bligh naturalized the irrational impulses of calenture: he said his men were willing to risk their own destruction for "alurements of disipation [*sic*] ... more than equal to any thing that can be conceived" (Bligh 1937: 2.123). James Weddell had a man on his ship who had seen a siren with green hair sitting on a rock and heard her distinctly emitting cries "in a musical strain ... a musical noise" (Weddell 1827: 143). In his autobiographical romance, *Loose Fantasies* (1968), Sir Kenelm Digby recalls his expedition to the Mediterranean in 1628 when his ship was ravaged by a contagious fever: "But that which of all others seemed to cause most compassion, was the furious madness of most of those who were near their end, the sickness, then taking their brain; and those were in so great abundance that there were scarce men enough to keep them from running overboard, or from creeping out of the ports, the extreme heat of the disease being such that they desired all refreshings, and their depraved fantasy made them believe the sea to be a spacious and pleasant green meadow" (Digby 1968: 166). In *Moby-Dick* (1851), Ishmael issues a warning about the peril of the individual who stares too long at the ocean and loses his footing: "Lulled into such an opium-like listlessness of vacant, unconscious reverie is this absent-minded youth by the blending cadence of waves with thoughts, that at last he loses his identity; takes the mystic ocean at his feet for the visible image of that deep, blue, bottomless soul, pervading mankind and nature" (Melville 1972: 257).

The point they labor and Wordsworth transposes in *The Brothers* is the powerful urge of the victims to throw themselves into the mirage or hallucination appearing below them. Calenture can be a very active state. People under its influence know what they want and believe they know how to get it: "Those affected have a fierce look, and are very unruly, being so eager to get to their imaginary cool verdure" (Chambers 1786: "Calenture"). Digby talks of the difficulty of restraining men who were convinced that heaven on earth lay on the other side of a porthole. Even when the objective was pursued in the more sedate manner of Juan Francisco, a seaman with de Quiros, the very deliberateness of his preparations evinces the same immoveable determination to reach land evident in the actions of his more reckless colleagues. He carefully fashioned himself a tiny raft made of plank and a buoy, with two empty jars for flotation, and after paying his debts and making his will, he disappeared in search of the shore visible three leagues from the ship, and was never seen again (Kelly 1966: 1.261). Whether calenture was in some cases a purely optical reaction to the monotony of ocean views or the motion of the ship, a kind of hypnosis uninflected by a conscious longing for home; or whether it was a decision simply to leave the vessel and seek the shore, such as that made by Juan Francisco and later by Herman Melville's Tommo; or whether it was indeed a powerful hallucination associated with heat and a strong fever, such as nearly carries Tobias Smollett's Roderick Random over the side of the *Thunder*, is hard to determine. But it is clear that the victims, rather than keeping dismal faith with a memory of home, move impetuously or at least deliberately toward the sight of land, or into an exotic and seductive prospect lying either on the surface or in the depths of the sea. Drake said of his crew of circumnavigators, when they were down with the "Calentura," that the past meant nothing to them: "Yea, many of them were much decayed in their memorie" (Keevil 1957: 1.14). Many years later, Alexander Armstrong discovered the same kind of amnesia in scorbutic seamen: "I observed that the faculty of memory in the more severe cases became confused and defective" (Armstrong 1858: 42). Whatever they remember, therefore, is as much a delusion as the meadows at their feet.

Medical discussions of scurvy freely associate it with calenture. In the *Philosophical Transactions of the Royal Society* abridged by Henry Jones in 1721, "Strange Effects of the Scurvy at Paris" sits next door to "An Account of a Calenture" (vol. 5, part 1, ch. 6, xiv and xv). Ephraim Chambers excerpts the latter for his *Cyclopaedia* (1786), confirming that calenture is a delusion, "wherein the patients imagine the sea to be green fields; and if not prevented, will leap over-board ... being so eager to get to their imaginary cool verdure" (Chambers 1786: "Calenture"). In *Zoonomia*, Erasmus Darwin sees little difference between calenture and nostalgia, offering a definition that will do for both: "An unconquerable desire of returning to one's native country, frequent in long voyages, in which the patients become so insane as to throw themselves into the sea, mistaking it for green fields or meadows" (Darwin 1801: 4.82). More recently, Jean Starobinski has cited Darwin as his authority for treating calenture as the nautical variant of nostalgia (Starobinski 1966: 86).

Joyce Chaplin defines calenture along with scurvy as a symptom of "earthsickness," a somatic revolt mounted specifically against the sea rather than travel as such, expressing itself in disgust with the ocean and in an urge to return to a "whole-body experience of the whole Earth" (Chaplin 2012: 518–23; 2012a: xv). How the "whole Earth" can exclude two thirds of its surface and still be consistent with "humanity's secular self-awareness on a planetary scale" (2012: 520) she leaves unclear, but scurvy at least has a serious place in her argument about this broader conception of nostalgia as an ecumenical yearning for land in general, not just home. This agrees with several accounts of calenture where home is the last thing the impetuous maritime pastoralist is interested in. Let the sailor "but heare the call / Of any Siren, he will so despise / Both wife and children for their sorceries, / That never home turns his affection's streame" (Homer 1956: 210 [12, ll. 59–62]). Trotter noticed that for all their tearful longing for green fields, suggestive of former friends and childhood memories, his patients sometimes forgot "all old attachments, [shewing] utmost signs of dislike to those who had been most dear" (Trotter 1804: 3.364). When Wordsworth's Leonard gets back to the rural scenes he saw pictured in the bosom of the deep, he finds it "a place in which he could not bear to live" (Wordsworth 1984: 167, l. 421).

The French call nostalgia "maladie du pais" and scurvy (on at least one occasion) "mal de terre" (Chaplin 2012: 518), the first a sickness arising from a yearning for the homeland, the other from an urgent desire to be at one with any substance that is not sea, and thus very like calenture. On landing, scorbutic sailors would have their faces pressed to freshly opened earth, or even be partly buried in it, so that they might incorporate its mysterious virtue. So it didn't matter that the earth was not local, as long as it was freshly turned. Likewise for victims of calenture earth, any earth, would do, as long as it fulfilled the dream of resting on a solid element. "Mal de terre" therefore was not a longing for home but for the stuff on which homes may be reared: land covered in green growth, the elementary solace on which all bodies depend (517–23). Nevertheless, some vessels shipped soil from the home port as an antiscorbutic in the belief that only in the sod of the patria, packaged as a portable medicine, were effective effluvia to be found.

Thomas Melville, formerly a skipper of a convict transport and then a whaler, tried both remedies and found them useless without the addition of vegetables (Melville ML MS Q 36, 256). On the other hand, Richard Mead, the eminent physician, was certain there was a real efficacy in the smell of fresh earth, citing from Anson's voyage the example of a man very close to death who was revived by having his face put into newly dug soil (Mead in Sutton 1799: 119). Philip Saumarez, an officer in Anson's squadron, declared, "Nor can all the physicians, with all their materia medica find a remedy for [scurvy] equal to a turf of grass" (Saumarez in Williams 1967: 166). Sir Gilbert Blane, surely one of the most eminent authorities on the disease, agreed, recommending that the legs be buried in the soil of any available island, and the air be deeply inhaled (Blane 1799: 496–97). When Bering died of scurvy he was half-buried in earth, evidently a desperate last attempt at some relief. At all events, the sight, smell, touch, and even the taste of earth were believed in various applications to offer a cure calculated for scurvy, calenture, and nostalgia. A Russian general marching through Germany in 1733 promised that any man in his regiments suffering from nostalgia would be buried alive, a variation on the remedy of the earth-bath sufficiently terrible to keep at bay any reason for applying it (Starobinski 1966: 96).

However, it is wrong to assume that earth-bathing and its variants were universally admired and recommended, or that there was any real difference between sea-scurvy and land-scurvy. Cullen had decisively announced that scurvy was an identical disease, no matter on which element it appeared, "depending everywhere on the same causes" (Cullen 1827: 2.649). One of his epigones, describing a bad outbreak among Russian troops alluded by way of a joke to "this land sea-scurvy (if I may be indulged so whimsical a term)" (Guthrie 1788: 333). Dampier says bluntly that too rapid a shift from sea-air to land-air will kill a scorbutic mariner (Dampier 1999: 84). Nathaniel Hulme was convinced that the preference for landing scorbutic men as quickly as possible in order to get the benefit from land-air contributed to the high level of mortality on the *Centurion* when it reached Juan Fernandez, many of them dying in the boats or on the shore because they were not fit to be lifted from their hammocks. Much better rates of survival were maintained on the *Gloucester*, which had to be supplied with fresh food from boats because its crew was incapable of bringing it to anchor. "This opinion of the good effects of land-air," Hulme wrote, "may appear, at first sight, harmless in itself; yet, being adopted, it may prove of the most fatal consequence, not only to a single ship, but a whole squadron, or fleet" (Hulme 1768: 44). Twenty years later, his dreadful warning was confirmed when the First Fleet arrived in Australia and the most inveterate outbreak of scurvy anywhere and at any time afflicted the settlements in Port Jackson, Tasmania, and the penal outposts of the new colony of Australia for upwards of forty years. So it is fair to say that the earth had no efficacy other than the herbs, vegetables, animals, and fruit thriving on its surface. If they were insufficient, nothing in the soil of Australia or anywhere else could supply the deficit.

Nor is distance from home any more certain a factor in the treatment of nostalgia than earth is in that of scurvy. In her study of the severe outbreak of scurvy among the British forces occupying Quebec in 1759–60, Erica Charters notes that nostalgia was most intense among the provincial troops operating in their own territory; it was the native Canadians who "got home in their heads," not the British who were furthest away from their native land (Charters 2009: 24). The grounds of Goodman's calculation of the ratio of nostalgia to remoteness from home, and of Chaplin's correlation of the incidence of scurvy and calenture to the degree of distance from the earth, both begin to look disputable. If there is nothing in the soil of home or the substance of the earth that would account for a somatic reaction to their absence, then we are left with the imagination: Chambers's "imaginary cool verdure," the "depraved fantasy" of Digby's sailors, and what Darwin calls the "mistaking [of the sea] for green fields."

In the *Odyssey*, Homer describes all three conditions: calenture, scurvy, and nostalgia. The pastoral landscape of the sirens and the sweetness of their song spell the fatal attraction of calenture. The episode of the lotus-eaters celebrates the tryst between the cravings of appetite and the gratification of the palate, so frequently enjoyed amidst sea-celery, samphire, sorrel, and scurvy grass growing on the seashore. It is in fact Odysseus's rampant nostalgia that brings a sudden end to this scene of gorgeous self-medication. As he manhandles his famished crew back into the ship, we are told they "striv'd, and wept, and would not leave their meate / For heaven it selfe," crying like Trotter's scorbutic sailors for the loss of their dream-food. Their captain tries to win them to a more powerful object, "Nothing so sweete is as our countries earthe," he urges, mistaking their frenzied eating perhaps for an ill-judged remedy of *mal de terre*. But of course while they are cramming themselves with lotus his crew "did quite forget / (As all men else that did but taste their feast) / Both country-men and country" (Homer 1956: 154 [9, ll. 150–64]). So *maladie du pays* goes by the board too and Odysseus drags them away to disappointments more worthy of true patriots, such as failing to recognize home when finally they get there. There can be no doubt that in preventing them from continuing their feast he is acting against scorbutic nostalgia on behalf of sheer nostalgia. For as long as they can feed, the lotus-eaters may be compared with the sailors of Anson's *Centurion*, who wept with the cravings of appetite and the fear of death only to find themselves (the ones who survived being brought ashore, that is) charmingly situated on a littoral abounding in esculent vegetables where they gorged themselves back to health.

This oscillation between the absence of a desired thing and the full enjoyment of it is evident even in the scene of wine drinking described by Willis, or of fruit eating described by Mitchel. Corroboration consists in the meeting of a dream and an event that can never be fully consummated inasmuch as the turbulence of longing conspires with the extravagance of gratification to keep the encounter fluid: thus rapture is brutal and short, never lengthy and terminable, and has to be renewed from moment to moment. This forms an ellipse of loss and enjoyment that can alternate between privation and gargantuan refection typical of a long sea voyage, or it may be squeezed into the sudden coalition of delight and agony (shrieking at the smell of flowers) or into the more moderate pendulum-swing between imagining a taste and then finding it in one's mouth. Either way, the negative shadows and defines the positive. Locke is a useful commentator here:

> If it were the design of my present Undertaking, to enquire into the natural Causes and manner of Perception, I should offer this as a reason why a privative cause might, in some cases at least, produce a positive Idea, viz. That all Sensation being produced in us, only by different degrees and modes of Motion in our animal Spirits, variously agitated by external Objects, the abatement of any former motion, must as necessarily produce a new sensation, as the variation or increase of it; and so introduce a new Idea, which depends only on a different motion of the animal Spirits in that Organ. (Locke 1979: 133 [II, viii, 4])

The episode of the sirens reverses the interruption of the lotus feast, for now it is the turn of the crew to bind the intemperate desires of their captain, who would have thrown himself to his death if he could. At first sight, calenture justifies Darwin's pairing of it with pure nostalgia, for it seems to lack the nuances that organize the elliptical dance of loss and fulfillment typical of scorbutic nostalgia. If nostalgia is an idol of home lodged in the brain, a hallucination so powerful nothing real can expel it, then calenture is similarly despotic in being an absolute devotion of the will to an irresistible and fatal embrace of what is not real. But in two respects it is different, for it is lodged not in the brain, as Wordsworth (a careful reader of Darwin's discussions of reverie) is at pains to underscore, but in "the organs of the bodily eye" (Words-worth 1984: 157; l. 57). Nor has calenture anything directly to do with home, for as Circe points out of the mariner besotted with the sirens' songs, "He will so despise / Both wife and children for their sorceries, / That never home turns his affection's streame" (Homer 1956: 210 [12, 60–63]). And insofar as calenture may not always, as in Leonard's case, be a sudden and total capture of the eye and ear, either because of external restraint or the gradations by which it turns a blue sea into pasture, then it can collaborate briefly but very productively with scorbutic nostalgia.

Flinders experienced a palimpsest of a scorbutic fantasy painted on top of a calentural hallucination when, sailing off the northeastern coast of Austra-lia, he stared at the coral visible beneath the ocean: "We had wheat sheaves, mushrooms, stags horns, cabbage leaves, and a variety of other forms, glow-ing under water with vivid tints of every shade betwixt green, purple, brown, and white; equally in beauty and excelling in grandeur the most favourite parterre of the curious florist" (Flinders 1814: 2.87). All but one of these so-called flowers is an edible fungus, cereal, or vegetable. He is thinking of the exotic and the familiar at the same time, blurring the colors of a fantastic submarine garden with the shapes of ordinary victuals as if not quite sure how to orient his taste until suddenly he is aware of the danger of his situa-tion: "Whilst contemplating the richness of the scene, we could not long forget with what destruction it was pregnant" (ibid.). A measurable relation between the imaginary and the real, each adapted rather pleasurably to the demands of scurvy and calenture, is supplanted by the threat of total loss and feelings of dismay.

An alternation between homesickness and scorbutic nostalgia is to be found in the journal of Ralph Clark, a junior officer of the marines in the First Fleet to Australia of 1788. He is under the influence of competing fan-tasies that reckon, like a couple of sextants, the intervals between the place he is bound for and the one he has left behind. Observe how he swings between images of what he lacks, one anticipated (of green vegetables) and the other remembered (of sexual comfort). He is writing home to Betsey Alicia, his wife: "I would to God that we had got to Botany that I might be able to get Som Greens or other for I am much afraid that I shall get the Scurvy—oh that I was once more home to you my beloved Betsey how I would kiss and presse you to my Bosom" (Clark 1981: 82). Aside from "Botany Bay Greens" (boiled seaweed) there were very few salad supplements in the new colony, only rice, flour, peas, and salt meat—exactly the same diet that was making him worry about scurvy and later would prompt him to damn the land round Botany Bay and Port Jackson as "the poorest country in the world" (Walker and Roberts 1988: 2). Once settled there, Clark's longing for greens yield to erotic dreams of Betsey Alicia ("dreamt last night of Seeing my Beloved Alicia again in Bed and I thought that I puld her towards me"; "Dreamt that I was with my beloved Alicia and I thought that I put my hand in her breast" [Clark 1981: 54, 76]). He places her in the dream-larder where scorbutic nostalgists would keep their images of food, but the upshot is the same, namely copious tears shed for the loss of what would make him feel better if only he had it. Soon he is waking up Captain Meredith with the noise of his weeping, and he reports, "Cryd very much in my Sleep on acount I thought that two men had taking [*sic*] my belovd Alicia a Way." No greens, no Alicia, and tears begin to dominate his waking moments as well as his dreams. "I sat down and read part of the play of Jane Gray—oh what a deep Tragedy it is.... I wish that it was bed time that I might have a cry" (61, 47, 75).

It is interesting how talk of food, a tragic text, and dreams of sex are arranged in Clark's fantasy life, for even the most secure mode of wish fulfillment is destined for the same disappointments that greet his pursuit of the right kind of aliment, directing him toward the formal consolations of literature and the resource of tears as means of keeping privation within the bounds of his patience. After he dreams of Alicia's rape, imagination is found to be no better than waking life at providing him with what he wants, so instead of allowing himself to be accidentally thwarted, Clark arranges his hectic cravings as a nightly sacrament of misery before the altar of Jane Gray, "In whose blent air all his compulsions meet, / Are recognized and robed as destinies" (Larkin 1988: 59).

Captain Scott operates on a parallel axis. Having given an account of the disappointments of homecoming, he is equally illuminating about craving dreams that anticipate the disappointment of waking to an empty table by including some accident in the dream itself that spoils the feast. His companions start to dream of food not as the uncomplicated array of all they want but more like the dinners of Sancho Panza during his governorship of the island of Barataria where no sooner is a cover set in front of him than it is whipped away: "They are either sitting at a well-spread table with their arms tied, or they grasp at a dish and it slips out of their hand, or they are in the act of lifting a dainty morsel to their mouth when they fall over a precipice ... something interferes at the last moment and they wake" (Scott 1907: 2.44). So an effort is made in the conscious lives of the company to observe the proprieties permitted by the food inside each individual's stomach: "After supper, and before its pleasing effects had passed, some detachment was possible, and for half an hour or more a desultory conversation would be maintained concerning far-removed subjects; but it was ludicrous to observe the manner in which remarks gradually crept back to the old channel, and it was odds that before we slept each one of us gave, all over again, a detailed description of what he would now consider an ideal feast" (2.50). The social niceties slip into the same pattern as a dream, lasting only as long as the sensation of satisfaction and jettisoned as soon as hunger takes its place. The formal plan is too weak for the claims of appetite: it would have required the possibility of something as corroboratively real as the food on Anson's nostalgic dinner plates. Otherwise all that is left is tears, and Cherry-Garrard said he never met anyone who wept so often as Scott. When he was found dead in his tent, the tears were frozen to his face.

Trotter lays a great emphasis on tears as one of the constant attendants of scurvy in all its stages; they are one of the first symptoms no physician should neglect. Thomas Melville, a close observer of the disease, agrees. In the third and last stage of scurvy, he noticed "sudden tremors ... and very often involuntary Weepings" (Melville ML MS Q 36, 252). In an anonymous it-narrative of the early nineteenth century, *Aureus; or, the Life and Opinions of a Sovereign* (1824), there is a description of an old sailor in prison who keeps having dreams typical of scorbutic nostalgia ("I awoke, as I thought, in Elysium") and then crying uncontrollably when he awakes to where really he is (Anon. 1824: 55–63). An intriguing short narrative of starvation aboard ship, accompanied by dreams of eating followed by convulsive weeping, is to be found in *The Farther Adventures of Robinson Crusoe* (1719). The female informant recalls her "earnest Wishing or Longing for Food; something like, as I suppose, the Longing of a Woman with Child," followed by dreams of eating and the disappointment of waking: "I was exceedingly sunk in my Spirits, to find my self in the extremity of Famine.... I fell into a violent Passion of Crying" (Defoe 1927: 3.66–67). It is relatively seldom that weeping is reported of pure nostalgia, but Homer's lotus-eaters weep, and Clark and Scott too, all under the pressure of scurvy. Colnett's boatswain on the scorbutic *Prince of Wales* cried like a child when it got stuck on Banks's island in Alaska; so did Robert McLure when he lost the location of a food-dump in the Arctic (Williams 2013: 135; 2009: 303; Dames 2001: 40). The end of the last two survivors of the Knight expedition was described to Samuel Hearne by an Inuk witness: after days of looking vainly for rescue "and nothing appearing in sight, they sat down close together, and wept bitterly" (Williams 2009: 93). In these examples, desire migrates from the imagination to the organ of the bodily eye, fruitlessly reconnoitering an empty space, and it is expressed purely as a physical event, like blushing. As Willis says, tears are an effluvial radiation, reaching out for some kind of corroborative reflex that often does not arrive.

From his youth, very prone to nostalgia, Coleridge wrote a fragment about weeping that takes place inside and outside a dream, doubtless a dream of the unavailable Sara Hutchinson and similar in its way to Clark's of Betsy Alicia except that it offers no illusion of satisfaction, only of loss:

> Oft in his sleep he wept, & waking found
> His Pillow cold beneath his Cheek with Tears,
> And found his Dreams
> (So faithful to the Past, or so prophetic)
> That when he thought of what had made him weep,
> He did not recollect it as a Dream,
> And spite of open eyes & the broad Sunshine,
> The feverish Man perforce must weep again. (Coleridge 2002: 9.124)

Although Coleridge's dream clearly recalls Caliban's ("And then in dreaming / The clouds methought would open and show riches / Ready to drop upon that me, that when I waked / I cried to dream again" [*The Tempest*, 3.2.138–41]), it performs the double disappointment evident in the dreams of Scott, Clark, and Defoe's informant, where an awakening confirms what was already missing from the dream, namely a loss that can by no means be supplied, neither by imagination nor by good fortune. This is when tears are redoubled; and the question is, whether they serve to emphasize the pleonasm of loss on loss suitable for the despair of pure nostalgia, or whether they conform to the corroborative or elliptical structure of scorbutic nostalgia, where tears upon tears offer an opportunity for privative solace such as that taken by Clark when he reads Rowe's tragedy of Jane Gray in order to have a good cry. Although Coleridge wants to say that the dream is somehow continuous with the reality of the past and the future, that both show him an identical good cause for weeping, he doesn't say what the cause is, or why it should make him so upset. As it is, he divides the scene of tears between the immediacy of a dream and the comparative and reflective powers of consciousness, as if the spontaneity of his grief were quickened and complicated by reflection. He gives an outline of what he might mean in his explanation of the "double touch":

> The soul never *is*, because it either cannot or dare not be, any [ONE] THING; but lives in *approaches*—touched by the outgoing pre-existent Ghosts of many feelings——-It feels for ever as a blind man with his protended Staff dimly thro' the medium of the instrument which it pushes off, & in the act of repulsion.... As if the finger which I saw

with eyes Had, as it were, another finger invisible—Touching *me* with a ghostly touch, even while I feared the real Touch from it. What if in certain cases Touch acted by itself, co-present with vision, yet not coalescing—then I should see the finger as at a distance, and yet feel a finger touching which was nothing but it, & yet was not it ... an act of imaginary preduplication. (Coleridge 2002: 3217 f70)

It is a difficult passage, but it suggests that all imaginary endorsements of a sensation, or sensational corroborations of an image, whether these amount to an access of pleasure or a preduplication of pain, depend upon a collaboration between the nerves and mind, or one sense and another, that is productive of an emotion which is, as Erasmus Darwin would say, obedient both to inward and outward impulses. His grandson explained copious weeping as beginning with a reaction of the eye to a physical cause, triggering an association with sadness that reacts with the physical stimulus to sustain and intensify it, rather like Willis's description of wine drinking. Darwin noticed "the length of time during which some patients weep is astonishing, as well as the amount of tears they shed" as if the traces of grief were a measurable quantity (Darwin 1904: 160, 178, 157). Wordsworth describes Leonard's experience of calenture in this way: "images and hues" perceived by his organ of the bodily eye "that wrought / In union with the employment of his heart." The corollary is that even the most serene enjoyment of a pleasurable fullness is laced with these residues of not quite coincident feelings—Mitchel's "brutal rapture," for instance—that do not lodge in consciousness but rather are found flowing into it or, as Coleridge puts it, "living in approaches" and then flowing out again.

It is likely that Coleridge derived his idea of the double touch from George Berkeley who, in his *New Theory of Vision* (1709), considered the alliance between sight and touch as forming a synesthetic perception of the material world without which the arrangement of colors, light, and shade would strike us as inchoate, for "there is no solidity, no resistance or protrusion perceived by sight" unless "visible figures represent tangible figures, much after the same manner that written words do sounds" (Berkeley 1972: 75, 77). In his *Spectator* papers on the pleasures of imagination, Joseph Addison gave his full endorsement to the proposition that sight is "a diffusive kind of Touch" (Addison and Steele 1907: 6.75; No. 411). And he, like Berkeley before him and Erasmus Darwin afterward, was fascinated by Robert Hooke's claim for the tactile excitements of a prosthetic eye, namely that "the roughness and smoothness of a Body is made much more sensible by the help of a Microscope, than by the most tender and delicate Hand" (Hooke 2003: xii). Darwin said it is

always possible to "compare our ideas belonging to one sense with those belong to another" (Darwin 1801: 1.157). This discussion of the synesthetic
function of sight and touch glanced in two directions: on the one hand, toward the celebrated question proposed by William Molyneux to Locke concerning the ability of a man cured of his blindness to recognize shapes by
sight alone; and on the other, toward Descartes, who initiated the doctrine
of sensation that denied any resemblance between the nature of an object
and the sensation it causes, proposing instead that light, for instance, is not a
property of the sun but rather produced by a relay of signals taking place in
the brain (Descartes 1998: 4). According to Thomas Sprat, this seduced Descartes into an exclusive reliance on "reflexion on the naked Ideas of his own
mind" (Sprat 1667: 96), a danger which Locke, who likewise disputed the
resemblance theory of sensation, did not avoid, according to Darwin, who
said, "Mr Locke seems to have fallen into [an] ... error, by conceiving, that
the mind could form a general or abstract idea by its own operation, which
was the copy of no particular perception.... The ingenious Dr Berkley and
Mr Hume have demonstrated that such general ideas have no existence in
nature" (Darwin 1801: 1.19). Perception according to Darwin supposed a
much more direct coalition between the senses, based on what he took to be
an exact resemblance between the body, the network of the nerves, and the
structure of the medulla, so "when the idea of solidity is excited ... a part of
the extensive organ of touch is compressed by some external body, and this
part of the sensorium so compressed exactly resembles in figure the figure of
the body that compressed it" (1.151). What is seen looks like what is felt, and
vice versa.

An apt scholium on the parallel between ideas resembling nothing in
nature and the images that have no external cause, but yet are patrolled so
obsessively by the imagination in volitional diseases such as reverie and nostalgia, is supplied in those sections of the *Essay Concerning Human Understanding* (1689) quoted rather miserably by Addison when obliged to jettison his aesthetic of mutually corroborative sensation owing to the "great
Modern Discovery" that the smell or feel of a flower has nothing to do with
the flower. It arises from what Locke has to say about privation and the possibility that a positive idea may be provoked merely by the withdrawal of a
stimulus. He runs a parallel with the negative names of positive qualities,
such as "insipid" and "silence." He might have offered more dramatic examples, such as "innocence" and "immaculate." But the point having been made,
he goes on to say, "And thus one may truly be said to see Darkness," as if the
total absence of the medium of perception could trigger a perceptual event
(Locke 1979: 133 [II, viii, 5–6]). Robert Boyle had marveled at the Dutch-

man Vermaasen who could distinguish colors by touch, or the man whose sight was so fine after an inflammation that he could see colors *in the dark*, but compared with those refinements of the senses to see darkness itself owing merely to the privation of light is literally and absolutely an unexciting achievement. We shall find that Margaret Cavendish has something tart to say about this in *The Blazing World*. Berkeley's dispute with Newton about fluxions turned on the same issue, a calculus of indiscernible quantities differentiated "in an infinite Progression towards nothing, which you still approach, and never arrive at" (Berkeley 1734: 11). As for Addison, he exclaimed, "What a rough unsightly Sketch of Nature should we be entertained with, did all her Colouring disappear," yet he was forced to admit that the soul's last sight of the world, stripped of all sensory supplements, would be of just such a lightless cinder (Addison and Steele 1907: 6.64; No. 414). It seems that the oscillations between emptiness and fullness that describe the range of synesthetic perception can become, under the influence of certain principles or impulses, less ample, shrinking the ellipse into a line, and edging toward the sensory prison of pure nostalgia and extreme reverie where the patient, as Darwin puts it, "neither sees, hears, nor feels any of the surrounding objects" (Darwin 1801: 1.224).

I want to go back to the irradiations and corroborations of tears to suggest that they provide the occasion of that elliptical premonition of sensation, or its echo or after-touch, that Coleridge calls "imaginary preduplication." I want to say that they have nothing in common with pure nostalgia because the scorbutic imaginary and the calentural sensorium build no shrines to images sequestered from the world. Sailors see a green earth beneath the keel of their ship because the organ of the eye is responsible, not the imagination, and it joins what is seen to what is felt. People weep sometimes because there is something almost voluptuous in the present sense of loss. What has gone missing is always specific and bears no resemblance to the fantastic inexactitude of homesick visions, or the privative basis of Locke's pure darkness. The twin impulses inclining Clark to expect greens while remembering Betsy Alicia, or those that illuminate Flinders's joint perception of coral as food and an undersea wonderland, are experiments in the possible mutuality of different images and different senses: the domestic and the strange, the beautiful and the fatal, the felt and the seen, the full and the empty. The dinner plates commissioned by Anson are diagrams of the same thing. So the tears of scorbutic nostalgia are linked not to an inactive grief circling a purely imagined good but to a reciprocal flow between contraries that corresponds to the circulation of effluvia discussed in the previous chapter, where damage to an organ is seen to specialize and refine its operation.

In the great epics of voyaging, tears are so frequent Dryden feels obliged to apologise for Aeneas's frequent breakdowns in case he might be mistaken for "a kind of St Swithin's hero, always raining" (Dryden 1808: 14.165). The celebrated *lacrimae rerum* of the *Aeneid* flow when Dido leads him to a temple where the epochs of Troy's rise, siege, and destruction are depicted: "He stoppd, and weeping said, O friend! ev'n here / The monuments of Trojan woes appear" (idem 1811: 138 [I, ll. 680–81]). Yet Aeneas describes the sack of Troy to Dido dry-eyed, as if he has mastered the distraction he recalls: "Then, with ungoverned madness, I proclaim / Through all the silent streets, Creusa's name" (14.294). Dryden supposes that Dido was impressed by the hero's dealing thus soberly with the memory of the loss of his wife. When Odysseus on the other hand casts his mind back to these very scenes at Troy while listening to the song of Demodocus at the court of Alcinous, he cannot restrain his grief at a loss he not only witnessed but also largely organized and executed. He weeps like a child over destruction that never before cost him a tear but now, suddenly made present to his mind by the skill of the bard, provokes such a storm of tears it puts an end to the performance: "So this King / Of teare-swet anguish op't a boundlesse spring" (Homer 1956: 147 [8, 736–77]).

Aeneas, we might say, weeps politely at a predictable prompt that in a formal, monumental way recalls the tragic day of the fall of Ilium, like a gravestone. Odysseus's grief is far more complex owing to the various tributaries from which it flows. It was for example preceded by another fit of tears on the shore of Calypso's island, where he sat and thought of home, "where twas his use to view / Th'unquiet sea, sigh'd, wept, and emptie drew/ His heart of comfort" (91 [5, ll. 109–10]). Between the two scenes lies the episode of Odysseus's attempt to sail home on a raft, which for eighteen days he steers toward Ithaca until Poseidon sends a storm that wrecks the contraption and throws the hero into the sea, where for two further days he welters. In a curious anticipation of Ralph Clark's divided passion for greens and Betsey Alicia, Odysseus enters the Phaeacian court famished and homesick, his body quite emptied by the salt water that gushed from his mouth and nose as he came ashore and with a heart similarly empty of comfort. Amidst a feast set in an orchard of ripe fruits, he confesses that his emptiness wears a double aspect, "Through greatest griefe the belly must have ease / Worse than an envious belly nothing is / ... When most with cause I grieve, / It bids me still, 'Eate, man, and drinke, and live'; / And this makes all forgot. What ever ill / I ever beare, it ever bids me fill" (125; [7, ll. 306–7, 311–14]). This ingenuous confession sets the stage for Demodocus who will renew the hydraulic evacuation of salt liquid that Odysseus has just restored with food and drink, operating an ellipse whose two foci are ill and fill. As a result "ill" has now changed its referent from sickness for an imagined home lodged vaguely in the past or the future, to scenes so vivid they are for the second time taking place in front of him.

> His owne quicke powers it made
> Feele there death's horrors, and he felt life fade.
> In teares his feeling braine swet: for in things
> That move past utterance, teares ope all their springs.
> Nor are there in the Powres that all life beares
> More true interpreters of all than teares. (Homer 1956: 147
> [8, ll. 718–23])

It is useful here to remember what Boyle says about effluvia, which he was adamant did not just bounce off the surface of our organs, but entered into them, changing their relations with each other and their sensitivity to outward objects. He believed that "effluviums are of a very piercing nature" (Boyle [1673] 1999: 7.260), and it was on the basis of this axiom that conceived of the human organism as an "hydraulo-pneumatical Engin" (idem 1996: 127) constantly absorbing and expelling effluvia, a process involving not just mutual material contact between the senses and their objects, but an actual and determinate exchange of inward and outward substances. He warned his reader, "We must not for the most part look upon Effluviums as swarms of Corpuscles, that only beat against the outsides of the Bodies they invade, but as Corpuscles, which by reason of their great and frequently recruited numbers, and by the extreme Smallness of their Parts, insinuate themselves in multitudes into the minute pores of the [body]" (ibid.: 7.261). Effluvia cut their way through to the vital core and stimulate it: "They make one part of a living Engin work upon another," these parts being so nicely framed "as to be very easily affected by external Agents ... yet capable of having great Operations upon other parts of the Body, they help to compose" (7.267). "The Operation of Effluvia upon particular Bodies ... dispose and qualifie those Bodies to be wrought upon, which before they were not fit to be, by Light, Magnetisms, the Atmosphere, Gravity, or some other more Catholick Agents of Nature" (7.270). Among these great operations, Boyle records a vastly heightened reaction to sights, smells, and sounds, as we have observed: a lady who swoons at the smell of roses, a gentleman who can see colors in the dark, another who can hear whispers a long way off (7.282). In these last two cases, the extraordinary tenderness and projective power of the sensory organs were precipitated by the invasion of hostile effluvia, as it was with the man who recovered from plague and afterward could infallibly identify an infected person by a pain in his nose.

How may this be applied to the ellipse of illing and filling that causes Odysseus's tears to switch from a fairly modest stream running in the direction of home to a torrent roused by the war at Troy? Illness in the shape of extreme hunger and the emetic effects of seawater function like the plague, the smell of candle snuff, or a fever in preparing the sensorium to react with extraordinary power toward a subsequent impulse, in this case, Demodocus's powerful description of the ruin and slaughter Odysseus himself planned and wreaked. But the energy of his reaction—the vast irradiation of a remembered scene as real, with death so present that he knows as he never knew before how fearful it is, as if what he did has started to happen to him—corresponds to the elliptical movement of corroboration. The relief of filling himself with food and liquid supplies both the tearful output of the engine of his sweating brain and the fuel to make it work so mightily. The extent of the filling is the measure of his sense of ill as an event so engrossing that it makes him weep without restraint.

It is time to make some final discrimination between nostalgia proper, scorbutic nostalgia, and calenture, based on the discussion so far. In all three diseases something is missing that the imagination, the eye, or the stomach severally try to supply. In pure nostalgia the imagination alone is involved. The desire for home is the cause of the illness, but home so attenuated by the impossibility of its taking place that it cannot be represented, remembered, or realized. Unlike scorbutic craving for fruit and liquid, then, the longing of pure nostalgia keeps an exceedingly uncertain rendezvous with physical satisfaction, which may explain why tears are not inevitably an accompaniment of it. To the extent that the home of the homesick is an obsession with what isn't there, it cannot truly be mourned or wept for. Illbruck calls it by its Greek name, *pothos*, the longing for what can never be (Illbruck 2012: 13).

Lacking this double structure, pure nostalgia takes place as an extreme reverie exhibiting "inaptitude of the mind to attend to external stimuli" (Darwin 1801: 1.319), a state of mind in which all awareness of exteriority is lost and the reality of corroboration, whether positive or negative, is set at an impossible distance. To this extent, it is an unequivocal state of feeling, typical of the total loss of position suffered in extreme situations: the vertiginous horror of seeing darkness itself, for instance. Scorbutic nostalgia, on the other hand, is always doubled or mixed because it is formed out of an alliance of the imagination and the senses in which various pairs—the feeling of emptiness and the image of loss, and the idea of food and the taste of eating it—perform a quadrille, alternately joining hands and changing places. Nostalgia is dominated by a single obsessive idea that Willis calls "an Idol of the Brain" (Willis 1683: 50). Charleton says this causes the soul to crowd itself into a single sense, at which point it loses itself, subject to "strange apparitions, and confused with delusory images" (Charleton 1670: 28–29). Erasmus Darwin defines this condition as a disease of volition in which "the whole sensorial power" is so preoccupied with its own motions it precludes "all sensation consequent to external stimulus" (Darwin 1801: 1.319). With the energy of the mind centered entirely on itself, the nerves are abraded by the trapped and bound energy that can find no outlet, resulting in what Hofer calls *laesa imaginatio* (Hofer 1934: 381; Illbruck 2012: 17). August Haspel was responding to the unaffiliated and unrelated state of nostalgia when he said that its cause is a breakdown at the core of being, "that is to say, that nothing began before it and that it is, at the very beginning, all the disease" (cited in Starobinski 1966: 100). The calenture described by Herman Melville, which cancels all sense of difference between the mind and the ocean, is fatal because it includes nothing that is really there. When Humphrey Davy was able to induce an artificial reverie by means of nitrous oxide, he found himself similarly without any sense of position or duration. "I lost all connection with external things.... I existed in a world of newly connected and newly modified ideas" (Davy 1800: 487). At the same time, his memory failed or became unhinged: either the sensations had vanished by the time he came to write them down, or his ideas arrived in such vivid and unusual trains they made no alliance with the ones he already possessed (479). Like Hofer, he called this wound in the mind's organization a *laesion* (467).

It is curious therefore that a consensus of commentary on nostalgia finds it a disease conversant primarily about time rather than place. An eminent French physician insisted in the early nineteenth century that it arises solely from "memory of the past" (Roth 1993: 28). Svetlana Boym calls it a "historical emotion" (Boym 2001: 7). Jean Starobinski says it is an upheaval of the psyche owing to the workings of memory (Starobinski 1966: 89, 102). Although Helmut Illbruck describes himself as the narrator of "nostalgia's modern history," it is one founded on Hofer's and Willis's notions of a pathological ecstasy that causes damage to the cortex, "a local disease of the brain" that originates and terminates in its own wound, breeding a passion for a home that does not exist and a time that could never be or have been (Illbruck 2012: 12, 42, 56, 63). Kevis Goodman has shown how the aesthetics of nostalgia at the beginning of the nineteenth century coincided with the replacement of the craving for home with an obsession with the past. She suggests that (in Wordsworth's phrase) the *cleaving* of the mind to a single image or idea is faithful to an earlier kind of nostalgia and does in fact introduce the same coincidence of extremes (here of joining and sundering) that I have been tracing in the tearfulness of scurvy and that Ian Hacking has found typical of niche diseases such as fugue (Goodman 2010: 199–203). This is an idea I shall return to shortly.

Calenture can occur early in a voyage (within a week of getting under way [McLeod 1983: 347–50]), there is no direct nutritional reason for its onset, and lives are still lost to it at sea. It is generally agreed that it begins with an individual staring for too long at the ocean. So it is not chiefly the need of food or home that transforms the sea into a landscape, although those desires might be blended with it. We might suppose further that it is not the imagination that is engaged in a case of pure calenture, so much as "the organ of the bodily eye," in Wordsworth's phrase. Here is a splendid specimen of calenture, taken from *Moby-Dick*:

> These are the times, when in his whale-boat the rover softly feels a certain filial, confident, land-like feeling towards the sea; that he regards it as so much flowery earth; and the distant ship revealing only the tops of her masts, seems struggling forward, not through high rolling waves, but through the tall grass of a rolling prairie.... The long-drawn virgin vales; the mild blue hill-sides; as over these there steals the hush, the hum; you almost swear that play-wearied children lie sleeping in these solitudes, in some glad May-time, when the flowers of the woods are plucked. And all this mixes with your most mystic mood; so that fact and fancy, half-way meeting, interpenetrate, and form one seamless whole. (Melville 1972: 601–2)

Notice that the techniques used by Melville to deal with the problem of delivering an experience of total absorption resemble those used by autobiographers of scorbutic ecstasy: A generalized sea rover—"he"—is sympathetically linked to a generalized reader—"you"—while the "seamless whole" of the hallucination is divided between the person thoroughly engrossed by the mirage and its various analogues—prairie, a summer's day, and sleeping children with their posies by their sides—intended to bring it home to those who have never experienced anything like it. Ishmael is circling the individual in the grip of an image who cannot, on his own account, compare it with anything but itself: for him the sea is not *like* a meadow, a prairie, or a wood, it *is* all of them, clad in green and presently there at his feet. It is not a question of distance or time but of immediate access, whether he is to be on the inside or the outside of the scene of delight. Humphrey Davy pointed out the communicability of hallucinations depends upon analogy; so if "you" is to be included with "him" in the experience of calenture, analogy has to come later, inexact and congruent only with those lame expressions that mock the isolation of the sea journalist when things are very, very, very dreadful—or exquisitely delightful—and such as no tongue can declare. There are no analogies for scurvy either, as Trotter explained, nor any disease related to it "by any concourse of symptoms or method of cure" (Trotter 1792: 106); but the elliptical formation of its pains and pleasures means that there is something like a sensory reflexivity that prevents extreme feelings from being totally out of touch with the material world. Unless calenture collaborates with scorbutic nostalgia, there are no tears in it, for nothing is missing. What is wanted is there in front of you, an illusion created by an eye fatigued or damaged.

In this ecstasy of presence, there is always the danger of annihilation of which Andrew Marvell warns in *The Garden* (1681). Thinking of nothing but "a green thought in a green shade" is in effect to disappear into the "far other world and other sea" that the mind has created out of the glaucous light of a glade (Marvell 2007: 157 [ll. 45–48]). The illusion is done either in an annihilating blue or vegetable green. In this respect, calenture bids to incorporate all external stimuli as its own work, beyond analogy, where to think is to be thought and sensation is continuous with everything in view. It is like the pain that lingers in the vicinity of the moribund Mrs. Gradgrind, somewhere in the room, but she's not sure it is hers. Long after his experiments with nitrous oxide were over, Humphrey Davy would fall into reveries as boundless as Melville's masthead dreamer. One day in the Colosseum, he found himself so "perfectly lost in the new kind of sensation which I experienced, that I had no recollections and no perceptions of identity" (Davy 1851: 20). Melville knew how dangerous this state could be: "But while this sleep, this dream is on ye, move your foot or hand an inch; slip your hold at all; and your identity comes back in horror … with one half-throttled shriek you drop through that transparent air into the summer sea, no more to rise for ever" (Melville 1972: 257).

Melville's Ishmael offers the reader a sort of totemic clue in the art of mastering the totality of calentural impressions. After he gives up the hopeless task of trying to decipher the Chaldean script incised into the White Whale's brow, he talks equivocally of what it takes to "own" the whale: "For unless you own the whale, you are but a provincial and sentimentalist in Truth. But clear Truth is a thing for salamander giants only to encounter" (445). By owning the whale, he doesn't mean putting an iron in it and calling it your property, as discussed in the chapter "Fast Fish and Loose Fish," the sort of property Ahab was aiming to make when he lost his leg in the previous voyage. I think he means that owning the whale is a way of writing about it without representing or symbolizing it, but at the same time without sinking his identity into it. To accomplish this, he adapts Moby-Dick and whales in general for a variant of the calentural reverie that first absorbs the mind and then the body in a flow or drift where "every strange, half-seen, gliding, beautiful thing that eludes him; every dimly-discovered, uprising fin of some undiscernible form, seems to him the embodiment of those thoughts that only people the soul by continually flitting through it" (257). In a mood similar to this, where feeling is situational and thought is expressed in the roll and eddy of the swell, Ishmael dares to include the whale, not as an analogue or symbol but as a demonstration of how this dreaming is done. He perceives the whale's spout as the physical emanation of its incommunicable thoughts, pulsing visibly in time to the rhythm of cetaceous meditations, rising and spreading into a notion and, in moments of inspiration, illuminated by the glory of iridescence (482). A whale seen in this state of refulgent rumination is the property of no one not fit to match it thought for thought. Its poet establishes his credentials as a medium of such splendor in a congruent image of creative thinking, for while writing of the whale he reports, "I had the curiosity to place a mirror before me; and ere long saw reflected there a curious involved worming and undulation in the atmosphere over my head ... a certain semi-visible steam" (ibid.). It is his own spout, the steam of his brain-sweat rising as thoughts about whales that are thinking in spouts take visible and vaporous form. It is a fine example of transforming the singular and impenetrable intuition of calenture—"that which refuses all words ... that, which ... must be felt, be possessed, in and by its sole self" (Coleridge 2002: 3401.12 f27)—into the elliptical motion typical of scorbutic nostalgia. So vapor, liquid, and color, half tangible and half visible and all moving and changing, offer to variegate or even disturb the triumph of the organ of the bodily eye by introducing into it other angles of vision, alternative colors, and even other kinds of sensation.

I want to end this section of the chapter with another example of how this blending of calenture with scorbutic nostalgia is achieved, and first I need to discuss the vagaries of the organ of the bodily eye. In linking calenture to pure nostalgia, and offering both as specimens of reverie, Erasmus Darwin nevertheless left room for an alternative, physiological account of the disease that is much more obedient to stimuli from outside. He concluded the first volume of the 1794 edition of his *Zoonomia* with an essay by his son R. W. Darwin on ocular spectra (probably ghosted by himself), taken from the *Philosophical Transactions* of 1785. It is a thoroughgoing account of spectra (i.e., the print of shapes and colors on the retina) that distinguishes direct spectra from reverse spectra. Direct are those such as the circle of light made by a revolving firework or the glow of a bright spermaceti candle that lingers after the eye is shut, comparable with Newton's image of the sun, a sort of retinal photograph. Walter Charleton, thinking no doubt of Lucretius's picture of how theater awnings color the audience, gives the example of a direct green spectrum staining the clothes of someone resting in an arbor, like Marvell in the garden at Nun Appleton, "and this from no other Cause, but that the Images or Species of the Leaves, being as it were stript off by the incident light, and diffused into the vicine Aer, are terminated upon us and so discolour our vestiments" (Charleton 1654: 138). In the last chapter, we discussed Boyle's interest in green, which he treated chiefly as "emphatical" and subject to modification by the eye and the angle of the light, rather than an intrinsic constituent of light warrantable by touch. Coleridge disagreed, remarking of the green of plants, "Unlike all other green bodies [it] is not divided by the Prism into blue and yellow Light; but is the Prismatic Green ... the indecomponible Green" (Coleridge 2002: 3.3606.25.89). The absolute quality of green is the primary illusion of calentural reverie, so it will be interesting to see how it can be decomposed.

Colors or shapes produced by the eye that are different from those of the objects in front of it R. W. Darwin calls reverse optical spectra, the result of the organ replacing (in the case of colors) the one that is visible with the primary color on the other side of the spectrum: thus, blue is supplanted by orange, red with green, violet with yellow. In effect, the eye turns a color that has overexcited and fatigued it into its spectral opposite. Darwin repeats some of Boyle's, Newton's, and Priestley's experiments, using red silk against white paper to show how, if one gazes at it intently in a strong light for a minute or more, and then closes one's eyes, the red is changed to green. Darwin cites this as proof of the activity of the eye in modifying the stimuli of shapes and colors, reshaping and retouching its environment. This activity can result from an excess of sensibility, as when black shapes seen in very bright light turn red, or white ones turn blue. But reverse optical spectra generally arise when the eye, tired of gazing steadily at one color, goes to the yonder part of the prismatic palette for relief. It is not a deep or permanent need for something that is not there, so much as a function of the organ itself.

How precisely the sea is turned green by an eye seeking relief from blue is not altogether clear from Darwin's account. Unless it begins by seeming green (the Ancient Mariner "viewed the ocean green"), it cannot change directly to green because the reverse spectrum of blue is orange, a mixture of red and yellow that could reverse into a gentle green if there were some plausible reason for the sea looking red, yellow, or orange. There were plenty of reports of seas green and red. In a paper called "The Colour of the Greenland Sea," William Scoresby Jr. noted sharp differences between the clear blue ocean and the opaque green one in the Arctic. "The colour was nearly grass-green, with a shade of black," he reported, but sometimes it was striped with olive, and he noticed that "the ice floating in the olive-green sea was often marked about the edges with an orange-yellow stain [which] I was convinced ... must be occasioned by some yellow substance held in suspension by the water, capable of ... combining with the natural blue of the sea, so as to produce the peculiar tinge observed" (Scoresby 1820: 11). Flinders came across a red sea at Point Culver, similar to the one spotted by Byron (Byron in Hawkesworth 1773: 1.12), which he was able to confirm was owing to microorganisms: "It consisted of minute particles not more than half a line in length, and each appeared to be composed of several cohering fibres which were jointed" (Flinders 1814: 1.92). En route for Australia, John Washington Price, surgeon of the *Minerva*, spotted a very red sea filled with flakes and lumps "of a spawn-like substance," evidently krill (Price 2000: 131). Possibly the red shadow of the ship in which the Ancient Mariner beholds the sea snakes at play, "blue, glossy green, and velvet black," is beholden to J. R. Forster's work on the animalcules responsible for phosphorescence.

The most rigorously scientific account of how the sea can turn either red or green is given by Newton in the *Opticks* (1704) when he discusses the account of the different effects of absorbed and reflected light in seawater given by Edmond Halley, who on a bright sunshine day went underneath the ocean in a diving bell:

> When he was sunk many Fathoms deep into the Water the upper part of his Hand on which the Sun shone directly through the Water and through a small Glass Window in the Vessel appeared of a red Colour, like that of a Damask Rose, and the Water below and the under part of his Hand illuminated by Light reflected from the Water below look'd green. For thence it may gather'd, that the Sea-Water reflects back the violet and blue-making Rays most easily, and lets the red-making Rays pass most freely and copiously to great Depths. For thereby the Sun's direct Light at all great Depths, by reason of the predominantly red-making Rays, must appear red; and the greater the Depth is, the fuller and intenser must that red be. And at such Depths as the violet-making Rays scarce penetrate into, the blue-making, green-making, and yellow-making Rays being reflected from below more copiously than the red-making ones, must compound a green. (Newton 2003: 183)

This suggests that a vision of watery depths would be colored crimson provided the clarity of the medium and the angle of the light allowed a human eye to penetrate so far—Homer's wine-dark sea perhaps—whereas the green landscape that seduces a sailor overboard is a hallucination provoked by light sent back from the surface, where blue, yellow, and green all compound as green. However, while Coleridge was staring at Saddleback Tarn, it appeared to him "blood-crimson, and then Sea Green" (Coleridge 2002: 3.3401.13.14), suggesting that there are other causes besides reflected light, algae, and little creatures responsible for these changes in color, and that staring for a long time at the surface of the water is probably the chief one in cases of calenture.

The condition called tritanopia is the result of damage sustained by the short wavelength cones of the retina owing to the action of ultraviolet light. Anyone suffering from this disease confuses pink and orange, and blue and green. When direct and reflected light strikes an unprotected eye, for example that of a sailor staring at the ocean, blue will appear green (Mollon 2013, personal communication; Lindsey and Brown 2002). J. R. Forster, a close observer of colors and phosphoric light in the Southern Ocean, has a number of useful observations concerning the compounding of green. The only ease he found for an eye oppressed by the load of "the sky and the sea, and the sea and the sky" in the Antarctic seas were the sunsets, which he described as follows: "The setting sun commonly gilds all the sky and clouds near the horizon, with a lively gold yellow or orange; it is, therefore, by no means extraordinary to see, at sun setting, a greenish sky or cloud.... I had an opportunity to observe, in the year 1774, April 2d, in 9 deg. 30′ South latitude, at sun-setting, a beautiful green cloud, some others at a distance were of an olive-colour, and even part of the sky was tinged with a lively, delicate green" (Forster 1996: 86). The appearance of green is perhaps to be accounted for by the reverse optical spectrum activated by the attentiveness of Forster's gaze; and to the extent that yellow and orange were reflected in the ocean, he would perceive that as green too, along with the clouds and the sky.

His son Georg painted a picture of icebergs in a sea whose horizon on the right hand is illuminated by iceblink, the bright white light reflected from pack ice in the clouds above it (Fig. 6 and Plate 2). Or at least this is what announced in the title of his painting, *Ice Islands with Ice Blink*. The consensus among eyewitnesses of the event, whose authentic date appears to be 23 February 1773, as opposed to the middle of March (the Forsters' choice and Cook's), was that they were seeing a display of the Southern Lights mounting from the horizon to the zenith, accompanied by extensive phosphorescence in the sea and a remarkable iceberg, shaped like a pillar with a hollow in it (Forster 1777: 1.117; idem 1982: 2.235; Cook 1961: 94–95; van der Merwe

Fig. 6. Georg Forster, *Ice Islands with Ice Blink*. SAFE/PXD 11. Courtesy of the Mitchell Library, State Library of New South Wales. See also Plate 2.

2006: 34–45). William Wales said it was the most curious ice island he had ever seen: "Its form was that of an old square castle, one end of which had fallen into ruins; and it had a hole quite through it whose roof so exactly resembled the Gothic arch of an old Postern Gateway that I believe it would have puzzled an architect to have built a truer" (Wales MS Journal, 23 February 1773). He also reported very bright appearances of the southern lights for the latter part of February. Whatever phenomenon is responsible for the brightness in the distant sky to the right of the pictured scene, it seems to be compounded with the yellow tint of evening sunlight that contrasts dramatically with the dark cloud of indigo advancing from the opposite side of the composition. The coalition of natural light and the aurora is responsible for the specimen of a direct optical spectrum that invades the foreground of the picture, where everything is tinged with blue except that portion of the sky from which the light streams. It is likely that its brightness, echoed by the phosphorescence in the sea and by the reflected sheen of the icebergs, exhausted Georg's eye as it tried to organize a scene of white shapes arranged on a blue ground, until the retina became "insensible to white light, and thence a bluish spectrum became visible on all luminous objects" (Darwin 1794: 551). That, at any rate, is what he painted, calenture in blue rather than green.

Fig. 7. William Hodges, *Pickersgill Harbour*. © National Maritime Museum, Greenwich, London, Ministry of Defence Art Collection. See also Plate 3.

How this scene was transformed from blue to green makes a remarkable episode in the pictorial history of calenture and nostalgia, not to mention scurvy, which, by the time Georg lifted his brush, had made inroads on the health of most people aboard the *Resolution*. William Hodges painted the same two icebergs that appear in Georg's *Ice Islands with Ice Blink*, the pillar with a hole in it and a sort of ruined pyramid, probably at the same time as Georg, using oils instead of gouache. For reasons that are not known, he decided, when he arrived at Dusky Bay a month later, to use the canvas for another picture, this time of Pickersgill Harbour, the site chosen for the *Resolution*'s temporary settlement (Fig. 7 and Plate 3). The picture we see now is one of domestic peace, with a sailor returning to his berth on the ship to cook his catch of fish, the warmth of the air evident in the clothes drying on the line, and the charm of the scene radiating from a little house (actually the observatory) nestling in a glade formed by the rich green of a temperate rainforest, and all illuminated by the evening light. It would be hard to think of a

Fig. 8. X-radiograph of Pickersgill Harbour. Courtesy of the National Maritime Museum, Greenwich.

sweeter image of tranquil satisfaction, even though it could not be further from England. So nostalgia of some sort is present here; but also calenture, for even the sea has turned green, bordered with yellow.

The X-radiograph of *View in Pickersgill Harbour* (Fig. 8) shows not only the outlines of the icebergs and the southern horizon in the composition hidden beneath, it also reveals the brightness of the westering sun in the surface image to originate in the thick layer of white-lead paint used for the upper part of the pillar-shaped iceberg (Fig. 9). It shows in fact that everything strange, white, forbidding, cold, and blue in the original scene of Georg Forster's spectral calenture was not just painted over, but converted into its opposite in a formal painterly exercise in the calenture of green. It seems to be a deliberate case of iconoclasm, or more precisely "iconoclash" in Bruno Latour's terminology (Latour 2002: 14), that isn't perfectly nostalgic, for in its own elliptical way it makes a palimpsest out of a scene of privation and another of enjoyment—ill and fill. Nor does it answer the immediate de-

Fig. 9. *Pickersgill Harbour* and X-radiograph.

mands of calenture, for Hodges's brush supplants and deploys for its own purposes the effects of tritanopia and reverse optical spectra. *A View of Pickersgill Harbour* is, I should like to suggest, an example of scorbutic nostalgia where the powerful idea of absent necessities, such as warmth, food, vegetation, safety, and the color green combines with the sensations of enjoying them all by means of modified calentural effects. It is a scene that is to say of corroboration, of which Hodges evidently intended the public to know only the half.

Imagining the scene of the drowning of one of the *Centurion*'s men at Cape Horn, William Cowper in "The Castaway" impersonated the exceptionalism Trotter said was characteristic of scurvy: "We perished, each alone: / But I, beneath a rougher sea, / Was whelmed in deeper gulphs than he" (Cowper 1854: 487). I have been trying to show that what divides scorbutic nostalgia from nostalgia proper, and at the same time links it at least to the purely optical aspects of calenture, are the tears, vapors, images, and colors that breach such solitary ecstasies by contributing to the elliptical structures of corroboration, synesthesia, and preduplication. I now mean to rely on Kevis Goodman for a demonstration of how Wordsworth does this for nostalgia by taking first for his subject a person adrift in an unnavigable situation, as Cowper tries to do in "The Castaway." The experiment is not always successful because the adjustment between the isolation of the subject and the sympathy of the reader is not to be achieved by Cowper's solipsism, nor by exact representation, nor by candidly confessing, as Wordsworth does in *Tintern Abbey* (1798), that his whole project of recollection is impossible: "I cannot say what then I was." As Michel de Certeau says in his essay on utopian language, the embarrassed poet must head from "cannot say" to "can say" via "can say nothing" (de Certeau 1996: 29–47). That is to say, the difficulty of saying anything at all must develop sufficient figurative buoyancy to prevent nothing from overwhelming a minimal articulate discourse of something. Goodman finds in the pointless repetitions of a nostalgic craving—*Ich will heim, Ich will heim* [I want to go home, I want to go home]—the possibility of redeeming the involuntary nonsense of virtual tautology ("using different words when the meaning is exactly the same" [Wordsworth 1984: 594]) with the primitive ornaments of passion—iterations of the same words, their length and quantity varied solely by the surges of feeling—that were praised by Longinus and Robert Lowth, and flowed again in Christopher Smart's *Jubilate Agno* (1759–63). In the stutterings of the old sea captain of *The Thorn* and the chiming of the word "alone" in *The Rime of the Ancient Mariner*, Goodman locates the lineal descendants of Trotter's victims of "cravings of the mind" whose verbal resources were inversely proportionate to the turbulence of their feelings (Trotter 1804: 3.364). By means of the manifold figures sponsored by repetition—tautology, anadiplosis, hendiadys, and parallelism—"the effects of nostalgia are no longer just a subject of representation—they have become a defining principle of representation" (Goodman 2008: 206). That is to say that the medium installed by Melville in respect of calenture, and the elliptical modifications of scorbutic nostalgia I have been trying to analyze, are achieved for nostalgia by means of a formal transposition of "cannot say" into "can say nothing."

Wordsworth is attracted to those who have lost house and family, or to travelers heading for the unpromising prospect of homecoming, because they have refined a form of reflection analogous to that of the poet such as himself, who manifests "a disposition to be affected more than other men by absent things as if they were present; and ability of conjuring up in himself passions, which are indeed far from being the same as those produced by real events" (Wordsworth 1984: 603). Often this factitious form of nostalgia identifies as its object a person environed by memories and emotions that likewise have ceased to bear any resemblance to real events, such as Harry Gill racked by phantom chills or James Ewbank fatally deluded by a somnambulist reverie. No more than Cowper has the poet any interest in the conventional pieties of sympathy. He looks at these displaced folk as if avid for proof of their want of self-possession. "He was alone, / Had no attendant, neither dog, nor staff, / Nor knapsack—in his very dress appeared / A desolation, a simplicity / That appertained to solitude" (47, ll. 61–65). Their language likewise seems muttered or sung like an obbligato of loneliness, so that the mere sound of what once might have been articulate words takes its place as part of the scene, as much seen as heard, or as much tasted as seen. In "The Solitary Reaper," he announces quite baldly his own version of illing and filling, "I saw her singing at her work, / And o'er the sickle bending; / I listened till I had my fill" (319–20, ll. 27–29).

Sometimes the radical disorientation of the person thus represented is not susceptible to fixture, and the scene spirals round a dissolving center that generally splits into three as repetitions pile up around it: "And wherefore does she cry?— / Oh wherefore? Wherefore? Tell me why / Does she repeat that doleful cry?" The only response to these wild queries is to exhibit the location of the cries, "The spot to which she goes; / The heap that's like an infant's grave, / The pond—and thorn, so old and grey / ... / But to the thorn, and to the pond / Which is a little step beyond, / I wish that you would go" (61, ll. 86–88; 92–95; 106–8). The intention behind the ventriloquizing of this voice, perpetually demanding news that can't be had, is to emphasize the enigma of the three things: the mound, the pond, and the thorn. The technique is brought to perfection in the "spots of time" section of *The Prelude*, where each scene yields three contingent fragments expressive of the visionary dreariness engrossing the poet's mind: a pool, a beacon, and a girl; a wall, a sheep, and a hawthorn. However, he returns to them, and by means of an ecstatic geometry, or trigonometry, sets the particulars of each triad in a position and a light of its own, "The naked Pool, / The Beacon on the lonely Eminence, / The Woman, and her garments vexed and tossed / By the strong wind" (ll. 313–16); "The single sheep, and the one blasted tree, / And the bleak music of that old stone wall" (566–68; *Prelude* 11.378–79). Protecting the poet from too tumultuous exposure to these bleak and noncorroborative objects is the same art used by Ralph Clark of framing his hopeless tears for Betsey Alicia with the tragedy of Jane Gray. Compulsions get robed as destinies, "the blank and solitude of things" is irradiated by "contingencies of pomp" (*Excursion* 3.848, 4.1061 quoted in Hickey 1997: 141). The "colours and words that are unknown to man" of which the poet despaired are suddenly supplied by means of emphatic repetition, cleaving to a tongue he thought had been cleft by the inexpressible. "All these were spectacles and sounds to which / I often would repair and thence would drink, / As at a fountain" (Wordsworth 1984: 568, ll. 383–85).

These redemptions of situational blankness are made possible first of all by the distinction Erasmus Darwin drew in his *Zoonomia* between the stimuli of external objects and those of reflection, will, and association. The figure of the poet in the landscape stands in an immediate relation to the random stimuli of the scene: the sound made by the wind whistling through a wall, the pressure of the wind against the body of the woman and the poet's own, the look of a forlorn sheep caught by an eye already attuned to solitude. Synesthesia emphasizes the sort of effluvial exchange that Darwin explained in terms of sight as follows: "The immediate object of the sense of vision is light ... none of the light, which falls on the retina, is reflected from it, but adheres to or enters into combinations with the choroide coat behind it" (Darwin 1801: 1.160). This is how a poet can have his eye filled by watching a woman sing or drink the sounds of spectacles. It is also how his disposition to be affected by absent things yields to what he calls "an atmosphere of sensation" or "carrying sensation into the midst of the objects of science itself" (Wordsworth 1984: 606–7).

The rhetorical figure appropriate to these situational redemptions of the poet's physical relation to matter is not just repetition, but repetition varied as congenial effects of redoubling are required by the milieu, some of which I have mentioned. However, the one most apt in terms of nostalgia, which threatens to annihilate all real experience of objects, is the one that redeems the negative by doubling it, known as litotes and sometimes as meiosis. It generally makes an appearance where a question of the poet's "more than usual organic sensibility" is raised, either triumphantly ("I should be oppressed with no dishonourable melancholy, had I not a deep impression ... of certain powers in the great and permanent objects that act upon [the mind]" [600]); or doubtfully: "Not without reproach / Had I prolonged my watch, and now confirmed, / And my heart's specious cowardice subdued, / I left the shady nook where I had stood / And hailed the Stranger" (47, ll. 83–87). In *Tintern Abbey*, the collision between the direct stimuli of present objects and their remoter charms supplied by thought reaches a kind of crisis that only a lengthy litotes can subdue, and even then not as adroitly as may have been desired: "That time is past, / And all it aching joys are now no more, / And all its dizzy raptures. Not for this / Faint I, nor mourn nor murmur; other gifts / Have followed; for such loss, I would believe, / Abundant recompense ... Nor perchance, / If I were not thus taught, should I the more / Suffer my genial spirits to decay" (Wordsworth 1984: 133–34 [ll. 84–113]). What seems to be denied by memory and positive terms is therefore conserved, at least technically, as a nothing made glorious by reduplication.

This is, I believe, a figure faithful to fits of passion resembling those I have discussed earlier, where the oscillation between privation and figuration that intrigued Locke and Coleridge produces some sort of unstable equilibrium. Whether it is abundance or destitution that greets the mariner who makes it ashore, it is often with two negatives that he expresses the interpenetration of loss or impediment with satisfaction or delight, a sort of hesitant refining of the elliptical foci of ill and fill, sky and earth, blue ice and green landscape. When Richard Walter arrives in paradise he reports, "No valley of any extent [was] unprovided of its proper rill ... the water, too, as we afterwards found,

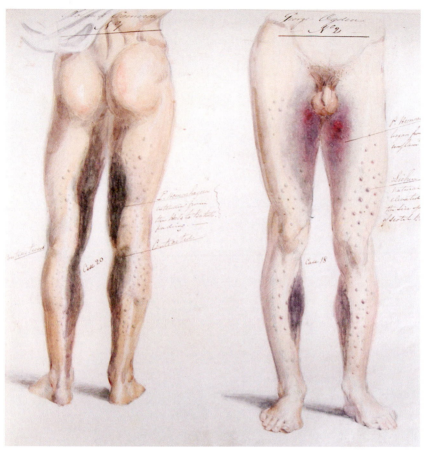

Plate 1. Henry Mahon, Scorbutic limbs. TNA, ADM 101/7/8.
Courtesy of the National Archives of the United Kingdom.

Plate 2. Georg Forster, *Ice Islands with Ice Blink*. SAFE/PXD 11. Courtesy of the Mitchell Library, State Library of New South Wales.

Plate 3. William Hodges, *Pickersgill Harbour*. © National Maritime Museum, Greenwich, London, Ministry of Defence Art Collection.

Plate 4. Nicolas-Martin Petit, *Aboriginal Dwelling*. No. 18013. Courtesy of the Muséum d'histoire naturell, Le Havre.

Plate 5. Joseph Lycett, *Aborigines Spearing Fish*. No. 2962715. Courtesy of the National Gallery of Australia.

Plate 6. Port Jackson Painter, *Aboriginal Woman Curing Her Child*. Watling Coll. 62. © The Trustees of the Natural History Museum, London.

Plate 7. Port Jackson Painter, *Grotto Point in the Entrance of Port Jackson*. Watling Coll. 7. © The Trustees of the Natural History Museum, London.

Plate 8. Port Jackson Painter, *Portrait of Balloderree*. Watling Coll. 58.
© The Trustees of the Natural History Museum, London.

was not inferior to any we had ever tasted" (Walter 1838: 111). At Dusky Bay, Cook says the same thing in the same way: "The shores and woods we found not destitute of wild fowl, so that we expected to injoy with ease what in our situation might be call'd the luxuries of life" (Cook 1961: 112). Of the Isle of Pines, Henry Cornelius van Sloetten writes, "All I shall ever say of it is, that it is [a] place ... deficient in nothing conducible to the sustentation of mans life" (Neville 2011: 30). John Mitchel writes, "Nothing [but running water] in this land looks or sounds like home. The birds have a foreign tongue: the very trees whispering to the wind, whisper in accents unknown to me ... all sights and sounds of nature are alien and outlandish" (Mitchel 1868: 257). Struck with beauty of the aurora, Scott wrote, "The suggestion of life, form, colour and movement is never less than evanescent, mysterious—no reality," as if it were empty even of nothing (Scott 1966: 208). The Chattertonian 1798 version of *The Rime of the Ancyent Marinere* is heavily marked with examples of negative hendiadys that lay their emphasis by saying twice what isn't there: a sun "ne dim ne red," a ship becalmed "withouten wind, withouten tide," a peaceful scene where "There was no breeze upon the bay, / No wave against the shore" (Coleridge 2004: 150, l. 93; 152, l. 161; 162, ll. 501–2). These litotes have in common with the standard phrase of journalistic exclusion (e.g., "No-one who has not witnessed such scenes can imagine") the double negative structure belonging to affirmations made by denying their opposites, except that it seems necessary if the experience of ecstatic recognition, experienced only for a moment and no one knows quite where, is to be communicated. It can't be poetically improved by another hand, remaining peculiar to the first-person singular, therefore litotes is not adaptable to third-person narrative: you cannot report that a person found something not unpleasant. Like scurvy, it is the singular property of a sentient individual.

Of nostalgia generally Starobinski remarks that the barrier it places between words and experience is insuperable. He sees the problem as arising first of all from "the elusive, the unobjectifiable 'object' of our search" and then from the historical distance that the over-confident modern interpreter feels sure of narrowing (Starobinski 1966: 83). "The history of emotions," he concludes, "cannot be anything other than the history of those words in which the emotion is expressed" (82). By resituating the agon of nostalgia round the idea of home rather than the past, Goodman has found an energy in nostalgic language that serves at least to add another angle to Trotter's exploration of the novelties of scorbutic nostalgia. If this depends on a close and living link between words as things and the sensations with which they are consubstantial, then it cannot be the principle of semiosis that governs Descartes's and Locke's arbitrary and colorless alliance between objects and the ideas standing for them. Like tautology and its cousins, litotes is a figure dedicated to the active implication of words with feelings and events which, while being not entirely like themselves, are at the same time not like any others.

Beddoes wrote to Darwin, "I cannot perceive any probability in the opinion that ideas resemble external objects.... I cannot at present conceive how a motion of an organ of sense can imitate extension, or colour, or any primary or secondary quality of bodies.... I am equally unable to comprehend how the stimulated part of the retina can exactly resemble the visible figure of [a] whole tree in miniature." Darwin replied, "If you allow an idea of perception to be a part of the extremity of nerve, of touch, or sight stimulated into action, that part must have figure, and that figure must resemble the figure of the body acting on it" (Stock 1811: xl, xliii). As I have pointed out, this was a debate that divided theorists of perception throughout the eighteenth century. At stake was the importance of what Locke called secondary qualities, (i.e., the immediate impression of objects upon the sensorium). If nostalgia were not to be the prison of the imagination then it was important to jettison the notion that the mind could make ideas out of nothing, no thanks to the senses—ideas of pure darkness, for instance, or quantities adding up to nothing. Scorbutic cravings and those of the homesick could be normalized only if an image or an idea could be aligned with the real and substantial excitement of an organ of sense, often the eye and the skin, and occasionally the ear, but more definitively the tongue and the throat. What seems to unite the work of Trotter with that of Cullen, and of Wordsworth with that of Darwin, is the recognition of the importance of this solid addition of outness to the exclusive interiority of the sick and the miserable. Scorbutic nostalgia was a nomenclatural novelty vindicated in medical practice, in the science of perception, and in poetry, as well as in the exigencies of seafaring.

# CHAPTER FOUR

# Australia

## Part I

How were the antipodes found out? For they were neither seen,
not heard of, nor tasted, nor smelled, nor touched.
—Margaret Cavendish, *Observations upon Experimental Philosophy*
(1666)

As we have seen, in *A Voyage to Terra Australis 1802–1804* (1814),
Flinders makes a strangely qualified promise of the narrative of the novelties of the continent where Britain had chosen to plant its latest colony: "A
history of this establishment at the extremity of the globe in a country where
the astonished settler sees nothing, not even the grass under his feet, which
is not different to whatever had before met his eye, could not but present
objects of great interest to the European reader" (Flinders 1814: 1.xcv). Using
negatives like planks to bridge the distance between the surprise of the settler
and the curiosity of the reader is a technique vindicated by the observations
of voyagers to Australia before and after him, who represented novelties as
horrid instances of the absence of what might have been expected. William
Dampier, the first Englishman to describe Australia, made an irritable list of
things that were not to be found on the coast of Shark Bay—water, trees,
birds, beasts, even ostrich eggs, were all absent, he complained; and it was
not much better on his second visit: "There were no Trees, Shrubs, or Grass
to be seen.... I saw there was no Harbour here ... a place where there was
no shelter.... We searched for Water but could find none, nor any Houses,
nor People, for they were all gone" (Dampier 1729: 2.118; 1939: 82, 100).
Flinders himself made the same judgment of Keppel Bay when he sailed
north from Port Jackson: "No good anchorage was found, nor was there
wood or water upon the island worth the attention of a ship" (Flinders 1814:
2.25). Twenty-five years later, Dumont D'Urville saw Raffles Bay in the same

mood, reporting, "There is nothing, not even a coconut palm" (Dumont D'Urville 1987: 2.411). On his arrival at the Australian coast almost two hundred years after Dampier, Anthony Trollope found that nothing, exactly nothing, had changed since Dampier's day: "There were no animals giving meat, no trees giving fruit, no yams, no bread trees, no cocoa-nuts, no bananas" (cited in Walker and Roberts 1988: 11). It was not just that flora and fauna were different—trees so hard they broke axes, animals that looked like giant rats, mammals that laid eggs, foliage with which one "could hold no manner of fellowship" (Barron Field in Carter 1987: 43)—they were utterly untranslatable into the kind of history or narrative of settlement that Flinders believed ought to have been inaugurated when he first set out on his discoveries. Fifty years later, John Mitchel, the Irish political prisoner, confirmed this, "No indigenous plant in all Van Diemen's Land is identical with any European plant: even the grass is altogether different" (Mitchel 1868: 262). So from a narrative point of view, to recollect the experience of negotiating such a space was to tell the story of something so exotic it was comparable with nothing, destitute of resemblance or analogy within the system of knowledge. From a dietetic angle, a place "so utterly destitute of the means of affording subsistence to either man or beast" (Oxley 1820: 54) provided nothing but a story of starvation and the inevitable onset of scurvy.

When John Oxley found during his second expedition beyond the Blue Mountains that his chronometer had stopped working and, shortly afterwards, that his compass had reversed itself, he says simply, "We knew not which way to turn ourselves." But when he climbed a hill and saw the ocean his desolation came to an end, and "every difficulty vanished" (252–71). If the empty narrative of a land tedious in its singular barrenness were to be replaced by a history, then something like Oxley's ocean needed to be manifest: something, that is to say, capable of framing and coordinating one's position in a waste whose undistinguishable extent embarrassed the resources of brain and tongue. The sheer contingency of immeasurable time and space has to become subject to the will if it is to be known. Paul Carter puts it most succinctly when he says of imperial narrative, "This kind of history ... reduces space to a stage [and] pays attention to events unfolding in time alone.... The governor erects a tent here rather than there; the soldier blazes a trail in that direction rather than this.... For these [facts] ... unlike the material uncertainties of lived time and space, are durable objects which can be treated as typical, as further evidence of a universal historical process" (Carter 1987: xvi). Flinders's Australian grass was a phenomenon experienced in the lived time and space of the present moment, for it had no relation to any other; but some kind of necessary connection with other data was needed if a bare fact was to have a meaning and a narrative, so that the unknown might find its place in the scheme of knowledge. A narrative of unrelated facts was not a history but a romance, a cluster of contingent things and events that might just as well not have happened. As we have heard Henry James point out, "The real represents ... the things we cannot possibly *not* know, sooner or later, in one way or another.... The romantic stands, on the other hand, for the things that ... we never can directly know" (James 1947: 31). The spectator of history in the making verifies the public disposition, sequence, and importance of immanently intelligible events, teaching the reader an appreciation of the epic and imperial necessity of what is being seen and done. When Alejandro Malaspina arrived in 1793 at Port Jackson this was his function, as David Collins pointed out on his behalf: "It was a pleasure to follow [in the track of Captain Cook], as it left him nothing to attend to, but to remark the accuracy of his observations" (Collins 1798: 1.273). Flinders was claiming something similar for himself when he said that Cook had "reaped the harvest of discovery," leaving some grains for a gleaner such as himself (Flinders 1814: 1.lxxxiii). The job of the witness was to confirm the truth of an earlier discovery of land entirely new by retracing the sequence of its being known.

At the beginning of his first narrative of the Australian colony, Watkin Tench fancied just such an observer witnessing the early scenes of disembarkation at Sydney Cove: "The scene to an indifferent spectator, at leisure to contemplate it, would have been highly picturesque and amusing" (Tench 1789: 38). That is to say, it would have been representable to a virtual witness, just like an experiment at the Royal Society; but along with all the other missing things noted by Dampier, Dumont d'Urville, Oxley, and Trollope, there was no one occupying that position. The rising structure of historical facts was swamped by misfortunes, generally expressed as the absence of all necessary adjuncts for a worthwhile scene. As Paul Carter affirms, "*There was no spectator*, no gallery, no surveyor-like comprehension" (Carter 1987: xvi). When Tench finally turned round to observe the lonely predicament of the settlement, haphazardly placed at the margin of trackless forests and impassable stone outcrops, he wrote, "I record with regret that this extended view presented not a single gleam of change, which could encourage hope, or stimulate industry, to attempt its culture" (Tench 1793: 118). All he could see was just more of the same material chaos, "piles of mis-shapen desolation ... whose unvarying appearance renders them incapable of affording either novelty or gratification" (26). It was a particular grief to him and all the other settlers who had read Captain Cook's account of his voyage to Botany Bay that none of the alleged meadows, greenswards, and parcels of fertile soil he mentioned in his journal was found to exist in Botany Bay. The one explorer of the shore who might have been relied upon to provide the organizing principle of settlement—he who first identified the place as agreeable to European sensibilities and answerable to their needs—had in effect absconded from the scene, either because what he saw was no longer visible or because it never had been there in the first place. Were it not for Cook's accurate charts of the three bays, it was Tench's opinion that "there would exist the utmost reason to believe, that those who have described the contiguous country, had never seen it" (30); or (like Dampier and the others) that all he had seen was nothing.

David Collins began his *Account of the English Colony in New South Wales* (1798) by telling his reader in the third person, "It was not a romance that he had to give to the world; nor has he gone out of the track that actual circumstances prepared for him, to furnish food for sickly minds, by fictitious relation of adventures that never happened, but which are by a certain description of readers perused with avidity, and not unfrequently considered as the only passages worthy of notice" (1.vii). The whole enterprise of the Australian colony had begun as a search for knowledge, with Banks collecting specimens and carefully drying them on sheets of paper while his artists drew faithful pictures of them. Natural history was enshrined in the name, Botany Bay, by which Port Jackson was known in England for many years after settlement. The pictures brought home by John White, John Hunter, George Raper, and others showing the strange reptiles, insects, animals, and birds that abounded in the colony, were offered to the public as a treasure trove of visible experimental knowledge. What was more, the region named New South Wales by Cook was soon chosen as the site of a radical departure from previous modes of penal transportation, for instead of convicts being sent to a colony already settled and economically viable, such as Virginia, Bermuda, or the Cape, they were themselves to found it, explore it, till it, build its roads, and make it grow. So two experimental projects were launched simultaneously in Botany Bay, one concerning the ordonnance of species and the other a utopian exercise in the management and reformation of criminals as citizens of a new commonwealth. The Linnaean system of nature and the principles of civil society were to be jointly set upon a new foundation.

But it was evident from the beginning that for an ill-assorted collection of people, bewildered by the strangeness of all the living things they saw and dogged at every step by famine, disease, severe punishment, displacement, and the uncertain susceptibilities of the scorbutic temperament, no inventory of its natural history was likely to be a crystalline addition to taxonomy. Simply in terms of binomial designation there were difficulties. Harold Carter has pointed out that the kangaroo's genus *macropus* was settled straight away, but that its species adjective is still in doubt. The ring-tailed possum acquired seven different binomial titles (Carter 1988: 90). Although partly responsible for this confusion, and notwithstanding his disputable estimate of the vegetable resources of Botany Bay, Banks was an inveterate Linnaean and utterly hostile to naturalists such as Buffon and Louis-Jean-Marie Daubenton whose attitudes to events such as extinction and speciation were much more fluid than his own (Lamb 2009). So he encouraged his correspondents to mock this careless thinking as romance. In a letter to Banks, Georg Forster refers to Buffon's works as "the Romance of Natural-History" (Banks 2007: 1.203), and so does T. A. Mann, "Buffon's System must be only a romance in Natural Philosophy" (1.175). What they have in mind is his skeptical approach to hard and fast categorizations of species that had been anticipated by Locke and more recently Philibert Commerson, the botanist on Bougainville's voyage to the Pacific. In his *Histoire Naturelle* (1758) for instance, Buffon writes, "La Nature marche par des gradations inconnues, & par conséquent elle ne peut pas se préter totalement à ces divisions, puisqu'elle passé d'une éspèce à une autre éspèce, & souvent d'un genre à une autre genre, par des nuances imperceptibles; de sorte qu'il se trouve un grand nombre d'éspèces moyennes & des objets mi-partis ... qui dérangent necessairement le projet du systeme générale" [Nature advances by invisible gradations, consequently she cannot adhere totally to these divisions because she is moving from one species, and often from one genus, to another by imperceptible degrees, and in such a manner as to exhibit many transitional species and hybrid objects] (Buffon and Daubenton 1758: 1.13). A species for Buffon was like a kangaroo for Banks, always potentially verging on the nondescript. But such descriptive fluidity was no consolation to those who felt they were themselves losing a grip on their coordinates and identities.

As for the land itself, it was no more clearly perceived than any of the other landfalls in the South Seas and the Indian Ocean made by people starved, dehydrated, and sick. There was seldom any unanimity there about what lay in front of the scorbutic eyes of visitors to Pacific islands, so when John Byron discovered giants in Patagonia, or when Jacob Roggeveen exposed what he believed were the lies told by Woodes Rogers and William Dampier about the island of Juan Fernandez, or when Kerguelen not once but twice discovered the edge of the Terra Incognita, the disagreements of experimental observers and the unsteadiness of scorbutic sensoria joined hands in what was generally referred to as romance. And so they did off the coast of Australia. Francois Peron, Baudin's zoologist, brought this disparity of views up to date when challenging Dampier's description of Shark Bay, but he was about to make his own contribution to the history of such enigmas in his account of the happy circumstances of Port Jackson, where (he alleged) murder and robbery were unheard of, corporal punishment seldom administered, marriages between the convicts blissful and long-lasting, and their offspring "raised with the greatest care by their parents and ... dressed with remarkable cleanliness" (Peron 2014: 169, 171–73). He alleged he was impressed by the results of the whole penological experiment: "Never, perhaps, has there been so conspicuous a specimen of the effect of good laws upon a criminal people, the result of which has been to reform the most abandoned vagabonds, and transmute the most ignominious robbers of Great Britain into honest and peaceable subjects" (idem 1810: vii).

When Peron went on to claim the land from the western part of Bass Strait up to the southern tip of New South Wales as a French discovery, naming it Terre Napoleon, Flinders was incredulous: "How came M. Peron to advance what was so contrary to Truth? Was he a man destitute of all principle?" (Flinders 1814: 1.193). But Flinders himself had reported how strangely that coast had reacted with his eye, so that the looming of a distinct headland would be absorbed into the featureless line of the shore as soon one came abreast of it, and no bearing could be got of it a second time (1.92). Baudin's journal is replete with examples of his hallucinations, of whales looking like rocks awash, of a fabulous wooded promontory extending far into the sea, of a vast rock that appeared first like a barn and then like a perfectly symmetrical upturned bowl, and of a cape he had named but failed to recognize because it turned out he had never seen it (Baudin 2004: 206, 169, 243, 410). With the exception of Peron, who lied from policy as well as inclination, it was not that these sailors intended to be deceived, or to deceive others, but internal pressures of malnutrition and depression conspired with the sheer novelty of what was witnessed to make accuracy impossible and to render the testimony of discoverers tendentious, impulsive, and extravagant. The colony after all was founded not on a systematic and civic ideal, but on a mistake jointly made by Banks, an eminent natural historian with vast political clout, and Cook, perhaps the most celebrated cartographer of the age.

The question Flinders addressed rhetorically to Peron was one Baudin often had reason to put to the man himself, for although Peron advertised himself as a volatile sentimentalist, prone to freaks and obsessions whose effects on others he found impossible to gauge, it is clear that there was (eventually at least) a strong vein of political calculation in his self-publicity that was ultimately very damaging to the reputation of his own commander. So he provides an important test-case for estimating the degree to which scurvy may overwhelm the precision of science with romance, even without the added pressure of an unknown landscape, and also for measuring how much calculation may enter into the stream of contingent events, either at the moment they occur or later.

In the overture to his own narrative of the voyage, Peron sets out his Linnaean credentials with great éclat. His booty is 100,000 specimens, many important genera, and 2,500 new species, all collected according to "a constant and regular plan ... in order to form a general history of the social condition and varied particulars of [New Holland]" (Peron 1810: iv). His narrative however consists of rapid alternations between alarming bulletins of the progress of scurvy among the crew ("The number of our sick increased every hour.... The sick below made the vessel re-echo with their painful cries.... Not one of us is free from it" [260, 281]) and excited observations about a giant squid together with a theory concerning the whiteness of species in the high latitudes, interspersed with hymns to the charming groves of Melaleuca, Thesium, Conchyum, and Evodia he finds growing in Tasmania (182). Baudin noticed that Peron's behaviour became increasingly deranged, especially in his frequent and potentially fatal disappearances into uncharted places, prompted by his rising passion for collecting shells. Peron came to believe they would reveal the secret of New Holland's origins as a shallow sea, thrown up into land by some primordial geological event. The astonishing number of his specimens, and the bulging outline of the history they were supposed to tell, may be explained, according to Baudin, by the suddenness of his zoologist's conversion from the natural history of animals to that of bivalves, mollusks, and gastropods. "From the beginning of our stay in Timor this scientist had plunged headlong, furiously even, into collecting shells, although he had no knowledge in that field. A periwinkle, a nerita, etc., was a treasure to him" (Baudin 2004: 277). Perhaps Peron's immoderate enthusiasm was a variant of calenture blended with ambition, for by the time the scurvy was at its worst, Baudin reported that his mode of collection had become indiscriminate. Shells were to him what green tints were to the masthead dreamer. Baudin observed, "He has one or two cases of broken shells, for in several places along the shore one can shovel them up" (494). Henri de Freycinet told Flinders that if the French had not wasted their time picking up shells they might have had a legitimate claim to some of Australia's Southwest Coast (Flinders 1814: 1.193).

Peron's ambitions for his reputation as a naturalist began, then, in what looks like compulsive compensatory activity, pursued because he felt himself to be dwelling in the shadow of scorbutic death. It is Peron to whom Baudin alludes when in May 1802 he writes in his journal of the first serious visitation of scurvy, and adds, "One of our scientists ... was struck down by a most unusual illness. He was seized with a fear of dying and was convinced his career had ended" (Baudin 2004: 399). In this respect, Peron's delirium was not too remote from the deplorable physical and psychological state of affairs in the colony, which he congratulated so warmly on the penological brilliance of its conception and execution.

In previous chapters, I have been disturbing the cartographical and historical certainties generally associated with the word "situation," so it might be best to begin the story of scurvy in Australia by outlining the strains the concept and experience of situation come under during the first years of the Port Jackson colony. Its topography is summarized by Tench after he has lost all faith in the charm and reliability of mediated viewpoints:

> Let the reader now cast his eye on the relative situation of Port Jackson. He will see it cut off from communication with the northward by Broken Bay, and with the southward by Botany Bay; and what is worse, the whole space of intervening country yet explored (except a narrow strip called the Kangaroo ground) in both directions is so bad as to preclude cultivation. (Tench 1793: 161)

As for relations between the new arrivals and the local population, he found them far from good. From the outset, the convicts were in violent competition with local people for available resources and were fiercely resented on that account. Summing up this miserable state of affairs, Tench says, "I have impartially stated the situation of matters as they stand while I write between the natives and us" (idem 1789: 92).

The situation thus stated is an unhappy alignment of roughly a thousand Europeans, destitute of the means of subsistence except for what they have carried there, with an inaccessible terrain—the south, north, and west were closed off by serious physical obstacles to further exploration and rendered even more impenetrable by a hostile population. Not only was the colony boxed up in this rocky, soggy, and infertile set of bays, each terminating in mangrove swamps, it was dependent for the visible future on imported food and equipment that had to be shipped from the other side of the world, a line of supply that could not have been lengthier or more prone to accidents, as events were to prove. The only real advantage was the anchorage at Port Jackson, a deepwater harbor of considerable strategic value. Governor Arthur Phillip made the best of things in his published account of the foundation of the settlement, but in a letter written to Lord Sydney six months after landing, he confessed, "No country offers less assistance to the first settlers than this does; nor do I think any country could be more disadvantageously placed with respect to support from the mother country, on which for a few years we must entirely depend" (HRA 1914: I, xx, 51).

Since no supply ships reached New South Wales for the next two years, the situation rapidly became what David Collins calls "peculiar." When the *Sirius* left to get emergency rations from the Cape, no ships were left in Port Jackson, for the transports had all gone home and the armed tender, *Supply*, was in Norfolk Island. The colony was in effect stranded, "a circumstance ... which forcibly drew our attention to the peculiarity of our situation." "Most peculiarly situated" were those convicts whose terms of transportation were up but whose papers had not arrived from London, leaving them cast away in a very strange place on an isthmus between bondage and liberty. Meanwhile the food ration had to be cut to a point where people were so weak with hunger and scurvy they could no longer work, a "peculiar situation" (Collins 1798: 1.77, 74, 109). Perhaps Collins was thinking of Lind's use of the same word when talking of "the peculiar situation of such places ... where [scurvy] is found to be a constant endemic disease" (Lind 1753: 188).

Lodged inside these peculiarities, it was difficult to give any account of them. When the *Sirius* found itself embayed on the Tasmanian coast, the commander John Hunter wrote, "Our situation was such that not a man could have escaped to have told where the rest suffered" (Hunter 1968: 83). Even with a hypothetical survivor the peculiarity of such situations was not readily communicable, for "our situation was now become extremely irksome: we had been oppressed by feelings more distressing than I can find words to express" (131). With all channels of communication blocked, rations successively cut, and mortality levels rising, many in the colony must have felt that their situation was that of Hunter's crew, only on a larger scale: without a messenger or a message. The most dramatic symptom of their suffocating sense of isolation was madness. Daniel Gordon, charged with theft of clothes and provisions, was the first convict deemed unfit to plead on grounds of insanity, "a wretch who either had not, or affected not to have, a sufficient sense of his situation" (Collins 1798: 1.79). Many more were to follow him into distraction, including officers such as Lieutenant George Maxwell who tried to kill himself in a rowing boat on the headland of the bay, and who planted his guineas in the hospital garden in hopes of a heavy crop the next season (Nagle 1988: 111). Convicts still in their wits nevertheless made desperate plans for escape, sailing in impossibly small craft for destinations they called Timor and Tahiti. In November 1791, twenty-one convicts absconded "with the chimerical idea of walking to China," prey to the common delusion that northern Australia was joined to the bottom of East Asia (Collins 1798: 1.82). That situations like theirs were not susceptible to reason was evident in the reply given by Henri de Freycinet to Nicolas Baudin when he demanded of his lieutenant why he had beached a tender and left it in a filthy condition: "It was an inevitable consequence of the situation she was in" (Baudin 2004: 473). It recalls what Rousseau says in the *Confessions* (1782) of shames and misfortunes happening of their own accord: the only explanation of the situation was the situation itself; all other reasons were irrelevant. The painful circularity of situational thinking overwhelmed the crew of the *Sirius* when they found themselves wrecked on Norfolk Island: "Day after day we talked to one another respecting our situation, as no other subject seemed to occupy the mind of any one of us: We were situated upon an island of only five miles long, three in breadth, three hundred leagues from the nearest part of the coast of New South Wales, deprived of any hope of finding any relief by a change of situation" (Hunter 1968: 124). The magnetic force of a single idea anticipates the obsessive conversations about food among the members of Scott's expeditions to the South Pole.

Whether they were being forced to labor in the colony or impelled to navigate the unexplored portions of its coastline, everyone caught in the restrictive sameness of this unrecognizable landscape, where one anomaly kept desolate company with another, was acquainted with scurvy and preoccupied with food. Flinders and his crew suffered badly from the disease while exploring the Gulf of Carpentaria, and then succumbed to a worse outbreak while navigating the Great Australian Bight on the way back to Port Jackson. Baudin and his men endured the same torment as they made their way from Shark Bay to Tasmania, and thence to Port Jackson. The First Fleet arrived in 1788 in good health, but scurvy grew into a problem almost as soon as disembarkation was complete. In Botany Bay and Port Jackson, there was a dearth of the antiscorbutic plants that had restored Anson's sick crews on Juan Fernandez and Tinian. Constitutions already weakened by months at sea with nothing to eat but salt meat, pulse, and biscuit found little or nothing to prevent the onset of scorbutic symptoms once they were on dry land. Supplies of fish were occasionally lavish, but never dependable as a regular supplement for the diet of the whole population. Game, chiefly in the form of kangaroo meat, was much scarcer. Ralph Clark had anticipated a good supply of greens at Botany, but "Botany Bay Greens" was slang for a vile concoction of boiled seaweed. The natural pharmacopoeia of New South Wales was exiguous, consisting of a native currant, *leptomeria acida*, so scarce it was of no nutritional significance; a creeper, *smilax glycophylla*, taken as an infusion and said to taste like liquorice; together with a few other edible herbs and greens (Walker and Roberts 1988: 3; Bradley 1969: 135). "We sometimes met with a little wild spinach, parsley, and sorrel," John Hunter remembered,

Fig. 10. Port Jackson Painter, *Aborigines Attacking a Sailor*. Port Jackson Coll. 44.
© The Trustees of the Natural History Museum, London.

"but in too small quantities to expect it to be of advantage to the seamen" (Hunter 1968: 49). Among later explorers of the colony, *trigonella suavissima*, a clover-like plant, was reckoned to be a useful remedy, but scurvy remained a problem (Mitchell 1835: 1.378). Tench thought that "the list of esculent vegetables, and wild fruits, is too contemptible to deserve notice" (Tench 1793: 164). William Bradley reported that parties were sent out to collect vegetable food for the sick, but were attacked by Aborigines (Bradley 1969: 114, 118), a scene commemorated by the Port Jackson Painter (Fig. 10). It shows a terrified sailor fleeing from the spears of local tribesman; his hat is on the ground, but his bunch of greens remains tightly clasped under his arm.

As for prepared antiscorbutics, there were scarcely any. The two Navy ships were carrying essence of malt, portable soup, and dried cabbage, while small quantities of acidic supplements such as oil of vitriol and lime juice were reserved as a cure in extreme cases. On the transports, some care had been taken to deal with scurvy. John White dosed his convicts with oil of tar, sauerkraut, and malt; but once they were ashore, it was impossible to treat the growing number of sick with anything but what the land or sea afforded, which was nowhere near enough. It was ironic that the terrible outbreaks of scurvy in Australia should have occurred in the decade when the disease was being tamed aboard British naval ships by more or less regular doses of lime juice, although it is doubtful that the circumstances of preparation and storage would have ensured sufficient amounts of the right quality, either on the transports or in the colony. Even if they had, there were surgeons on the convict ships who had no faith in citrus as a remedy: One of them wrote, "I have come to the conclusion … that Lemon Juice or any other acid never was, and never can be, either a preventive or cure for Scurvy;—and is a perfectly useless expense to the Service" (Henderson 1835).

There were signs that the regimen of the later transports tried to keep pace with developments in antiscorbutic medicine since Cook, with some atrocious exceptions such as the *Neptune* in the Second Fleet. Charles Cameron, the surgeon of the *Fergusson*, sailing from Ireland to Port Jackson in 1828, dealt with a bad outbreak of scurvy by using nitre from gunpowder mixed with lime juice, an innovation pioneered by a surgeon called David Paterson and mentioned by Gilbert Blane with approval (Cameron 1829: 6; Blane 1799: 496; Foxhall 2012: 134). The Select Committee Report on Transportation (RSCHC 1837: Appendix 20, 352) shows that convicts by then were expected to be supplied with the same ration of lime juice that was by then nominally standard in navy vessels: an ounce a day for each person. There was, however, no guarantee of its being shipped or prescribed, as we shall see; and since its efficacy after a long time at sea in extreme weather and temperatures was not to be relied upon, its failure would justify the skepticism of surgeons such as Andrew Henderson. In his surgeon's report, John Washington Price of the *Minerva*, a transport badly attacked by scurvy, lists remedies of portable soup and spruce, but doesn't mention lime juice at all (Price 2000: 140). The only reliable treatment for incipient scurvy on the way to Australia was to take in a good stock of vegetables and fruit at Bermuda, Brazil, or the Cape, but it cost cash or barter: "During our stay we got a little fresh meat, some oranges, limes, pine apples etc.," wrote Edward Laing, surgeon of the *Pitt*, "but you may suppose few or none of these fell to the share of the convicts" (Laing in Trotter 1795: 72). No provision was made for the sequel of settlement because never before had it been necessary to supplement the diet of scorbutic sailors and voyagers who had succeeded in making land. There had always been scurvy-grass, sea-celery, bountiful catches of fish, or a decent supply of animal flesh to ensure a speedy return to health. Blane believed that merely putting foot on land was enough, "insomuch as I have known men … in unfrequented islands, recover with very little change of diet" (Blane 1799: 497). But that remedy didn't work in Australia.

There is evidence that surgeons themselves shared the difficulties of the other visitors and settlers in identifying facts. Although he was punctual in listing the incidence of scurvy during the voyage of the *Pitt*, Laing seems to have been blind to what forcibly struck Phillip and Tench as a serious outbreak of the disease in the colony. He was apparently convinced the endemic sickness there was dysentery, accompanied by an obstinate rheumatism. In 1852, W. McCrea, surgeon superintendent of the *Anna Maria*, was so determined to supplant scurvy with dysentery he actually referred to it as a nutritional disease (Staniforth 1996: 119–22). Aching joints were often chosen as the scorbutic symptom easiest to confound with a less opprobrious sickness, such as rheumatism. Bligh had opted for it on the *Bounty*, and so did John Price on the *Minerva* (Price 2000: 132). Price agreed with Laing that dysentery was the chief disease ashore, "almost the only species of sickness they know here" (151), despite the fact that a bloody flux was common in the later stages of scurvy, too. As late as 1853, the *Phoebe Dunbar* arrived in Fremantle with a cargo of convicts all sick of scurvy, sixteen dead and thirty-five dangerously ill, whom the surgeon, John Bowler, had confidently diagnosed as suffering from cholera, perhaps because he didn't want to be blamed for not having taken on fresh fruit and vegetables in Rio de Janeiro (Weaver 1993: 235–36).

Once in the colony, reports about the availability of food varied wildly. Laing was sure that great relief would be afforded sufferers by the creeper *smilax glycophylla* and the native currant, despite supplies of both being so meager and perilous to harvest (Trotter 1795: 102). John Turnbull, who visited Port Jackson shortly afterward, wrote, "The colonists have not as yet found any species of vegetables which they could apply to culinary purposes. Nor have the colonists found that the natives were acquainted with any thing of this kind, excepting the fern root" (Turnbull 1805: 3.151). Regarding produce and fruit, Price says they are very plentiful in Hobart, if a little dear, but the potential for farms and gardens on Tasmania was generally judged to be poor, with a soil generally sour and wet. Turnbull reports a blight in the crops around Port Jackson and a terrible flood in the Hawkesbury, resulting in distress and ruin for those raising food (Price 2000: 161; Turnbull 1805: 3.144). Peron mentions the flood, but he adds that the production of food has fulfilled all expectations: "Fruit, potatoes, and vegetables of all kinds are in abundance" (Peron 2014: 347).

So Botany Bay and its environs were exceptional in the annals of scurvy in providing a landfall effectually destitute of antiscorbutics and proving, if any proof were really needed, the truth of Cullen's claim that land-scurvy and sea-scurvy were the same disease. There were those who construed such a state of affairs as an advantage. Marcus Clarke's Mr. Pounce thanks God for Tasmania: "This island seems specially adapted by Providence for a convict settlement; for with an admirable climate it carries little indigenous vegetation which will support human life" (Clarke 1899: 242). So from the prisoners' point of view, things could only get worse. "Soon after landing," Phillip reported, "scurvy began to rage with a virulence which kept the hospital tents supplied with patients" (Phillip 1790: 58–59). John White, the chief surgeon of the settlement, agreed that the situation was critical: "The sick have increased since our landing to such a degree, that a spot for a general hospital has been marked out" (White 1790: 127). Collins noted that by June 1788 there were 154 people listed sick, three out of every twenty, mostly with scurvy. The scorbutic diathesis (or chronic shortage of ascorbate) of the rest must have hovered close to the critical point of depletion, at which signs begin to show themselves on the body. It was no respecter of rank or station, although the convicts still on their feet, subject to shrinking rations and an unremitting regime of hard labor, would have borne the brunt. A journal entry for September 1790 notes a limited distribution of antiscorbutics, "The Governor has ordered All the free People to be served Oil and Vinager" (Cobley 1963: 289). Tench wrote, "Had it not been for a stray kangaroo … we should have been utter strangers to fresh food. Thus situated, the scurvy began its usual ravages, and extended its baneful influence, more or less, through all descriptions of persons" (Tench 1789: 107). The situation was certainly peculiar in this regard. The colony was nutritionally speaking no better than a becalmed large ship: its inhabitants hemmed in by the sea, bush, mountains, and angry indigenes; its crew and passengers reliant for protein on salt meat, for carbohydrate on ship's biscuit and a little flour, and for fat on a tiny ration of butter. Vitamins A, B, and C were absent from their diet, apart from what they could pick up from game, fish, and the sparse supplies of greens and berries, pieced out by boiled seaweed (Hughes 1988: 125).

In the Port Jackson Painter's two pictures of the wounding of Governor Phillip, the amphitheater of trees beyond the beach is represented as an impenetrable barrier to the Europeans, but as the porous habitat of those who menace them (Fig. 11 and Fig. 12). Phillip blamed the diet of salt meat and the damp conditions for the early outbreak of scurvy, yet neither the diet nor the cramped conditions of the colony were going to change. The three supply ships with the fleet landed salt meat, pulse, butter and carbohydrate. The vessels eagerly awaited after the failure of the *Guardian*, so severely damaged by an iceberg south of the Cape it was left a drifting hulk, were fetching identical cargoes which, on a full ration, worked out at a weekly issue of 7 lb. salt beef or pork, 1 lb. flour, 7 lb. biscuit, 3 pints of peas, and 6 oz. butter. The only reliable method of keeping scurvy at bay was to prepare gardens, plant seeds, and hope that they grew rapidly into edible plants, but the soil available to the first settlers was largely infertile. Yet it is common to meet with very positive descriptions of the available produce in the early colony, of which Francois Peron's is typically the most fulsome and mendacious. There is no

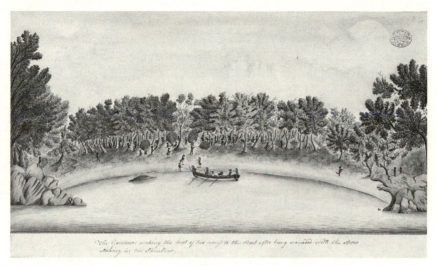

Fig. 11. Port Jackson Painter, *Mr. Waterhouse Endeavouring to Break the Spear after Govr Phillips [sic] Was Wounded.* Watling Coll. 22. © The Trustees of the Natural History Museum, London.

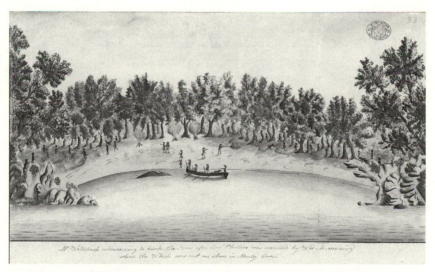

Fig. 12. Port Jackson Painter, *The Governor Making the Best of His Way to the Boat after Being Wounded with the Spear.* Watling Coll. 23. © The Trustees of the Natural History Museum, London.

doubt that little or none of this abundance reached convicts working in gangs and that it played no part in the diet of those working in penal settlements. Edward Laing wrote to Thomas Trotter that the newly landed convict "must boil his allowance of greens (if he can get any) with a scrap of pork or beef, which has been four or five years in salt" (Trotter 1795: 89). Writing from Van Diemen's Land during what was a widespread outbreak of scurvy in the penal settlements more than thirty years later, Dr. Charles Turner begged to state "that the deprivation of nourishment in the supplies of the Colony has operated equally to cause the principal disease, Scurvy, as to retard and in some instances to prevent the cure of it" (HRA 1927: III, vi, 650).

## Part II

> You have read of Captain Cook our late worthy commander,
> The great Sir Joseph Banks, and Doctor Solander,
> They sailed round the world, were perplexed and teiz'd too,
> To find out a place where the King might send thieves to.
> —Botany Bay Song

Joseph Banks was a guiltier party to this desperate state of affairs than Cook, for he had seen from the start that Botany Bay offered very little in the way of immediate sustenance. Recollecting his experiences there, he wrote in his journal:

> A soil so barren and at the same time intirely void of the helps derived from cultivation could not be supposd to yeild much towards the support of man. We had been so long at sea with but a scanty supply of fresh provisions that we had long usd to eat every thing we could lay our hands upon, fish, flesh, or vegetable ... yet we could but now and then procure a dish of bad greens for our own table and never but in the place where the ship was careend [Endeavour River] met with a sufficient quantity to supply the ship. (Banks 1962: 2.113–14)

The trees were so hard, he added, that they spoilt the tools used on them. But when some years later he gave evidence to a Parliamentary Select Committee enquiring into alternatives to Virginia, Bermuda, and the Cape as destinations for convicted felons, his views had grown much more positive: "The Proportion of rich Soil was ... sufficient to support a very large Number of People.... The Grass was long and luxuriant, and there were some eatable Vegetables, particularly a Sort of Wild spinage.... There was abundance of

Timber and Fuel, sufficient for any Number of Buildings" (RSCHC, xxxvii, 1799: 311; cited by Beaglehole in Banks 1962: 2.113 n. 2). The only possible basis for such a view of the resources of Botany Bay lay in his belief (for it could be nothing more than that) that a good supply of seeds and cuttings would thrive in ground quickly tilled and cultivated by forced labor. Exactly how soon that might occur was for Banks a speculation (he predicted that a year after settlement a rapidly increasingly quantity of vegetables would be available, and he thought it to no purpose to send out large supplies of essence of malt, his preferred antiscorbutic [HRNSW I, part 2, 1783–92, 1892: 232]); but for those actually inhabiting the peculiar situation, it was a desperate gamble that was not paying off. With soil scanty and poor, laborers enfeebled by exhaustion and disease, and with crops meagre and vulnerable to assaults from vermin, drought, tempest, and thieves, there was no chance of supplying everyone with a sufficient amount of fresh food. Two years into the life of the colony, famine was killing those already weakened by scurvy. "It was not hard labour that destroyed them," Collins noticed, "it was an entire want of strength in the constitution to receive nourishment" (Collins 1798: 1.209)—a judgment supported by Edward Laing who noticed, "food passes through the intestinal canal without any part of it being assimilated" (Laing in Trotter 1795: 1.89).

Even when the Nepean and Hawkesbury Valleys started to come into production in the early 1800s, supplemented by the relatively fertile ground of Rose Hill (later Parramatta), scurvy was still common. In April 1792, Collins reported a dreadful sick list with convicts running mad for no apparent reason. The death toll for that year was 482 men, women, and children (Collins 1798: 204, 246). Though mortality decreased in Port Jackson over the next two years, Collins warned Hobart in 1804 that scurvy had broken out in Tasmania "to an alarming degree" (HRA 1921: III, i, 286). He reported disastrous harvests for New South Wales in 1799 (Collins 1802: 2.142). For the next three decades, scurvy was endemic in Tasmania—at Macquarie Harbour and Port Arthur—and in settlements on the northern coasts—Moreton Bay, Melville Island, and Port Raffles, as well as in rural districts remote from coastal settlements. A brief exchange before the Select Committee of 1837 indicates what a low place it took at that time among the priorities of the British Government compared with the other travails of the colony, and probably always had. Sir Charles Lemon (*nomen sine omen*) was putting questions to John Russell, formerly the surgeon at Port Arthur:

What is the diet of the convicts at Port Arthur?—It is plentiful in quantity, but it is salt rations. Do they not get fresh meat?—Up to the time I

left (1833) they got no fresh meat. Had they plenty of vegetables?—
Not up to that period.... The ground was very barren, and there was
great difficulty in working it. What was the effect of this sort of diet
on the health of the convicts?—It produced a great deal of scurvy
among them. (RSCHC 1837: xix, 2.50)

There were no further questions and no recommendations for improvement.
It had transpired in the course of questioning that the only fresh meat available at Port Arthur was reserved for guard dogs, to keep them alert and fierce.
Saxby Pridmore estimates that scurvy was the greatest cause of sickness in
Tasmania for the first four decades of settlement (Pridmore 1983: 277). Detailed descriptions of the outbreaks on Melville Island were given by Dr. John
Gold in a letter of advice to the Commandant at Port Raffles, where things
were so bad the settlement was abandoned two years after its foundation in
1827 (Gold 1827: 79–80). In 1825 Dr. Charles Turner summed up the causes
of the deplorable condition of the people in Tasmania as follows: "Exhaustion from labour in a Tropical Climate and exposure to damp during the rainy
season have been of much injury.... No supplies of fresh meat, fish, turtle or
kangaroo have been procured, and the growth of vegetables has been unpro-
ductive.... The supply of Lime juice was soon exhausted, as also the less efficacious remedies of mineral acids and Cinchona Bark" (HRA 1927: III, vi).
Fifty years after the founding of New South Wales, serious droughts could
ruin harvests even in the most fertile areas. "It was estimated that one third
of the wheat and flour consumed in the colony was imported in the years
1839 to 1842" (HRA 1924: I, xx, x). This particular event resulted in a collapse of credit, and people desperate for food found it impossible to obtain
any without ready cash, a resource limited to the troops and upper echelons
of the settlement.

In June 1790, five ships arrived at Port Jackson, including the long-awaited
store ship *Justinian*, which had narrowly avoided disaster on the voyage. The
other four transports were in a bad way, especially the *Neptune*. Out of its
human cargo of 530 persons, 163 were dead and 269 sick. The master, Denis
Trail, was subsequently to be investigated in England for criminal neglect,
but the effect on a colony already reeling from scurvy and starvation was
devastating. Out of a total of 1038 new convicts, 273 had perished and
almost 500 were too sick to move, the bulk of them scorbutic. For the next
three months, the average sick bill at the hospital stood at 400 (Tench 1793:
45; Collins 1798: 1. 119–24; Pearn and O'Carrigan 1983: 15). Although
never again was a shipment of convicts to arrive in quite such a pitiable state,
it turned out that for propitious weather and good health, the voyage of the

First Fleet had been exceptional. After that, the colony was destined to be a net importer, as well as a local producer, of scorbutic personnel. In May 1792, the *Queen* arrived from Ireland with only 50 out of her cargo of 122 convicts surviving; in October of the same year, 80 convicts were landed sick from the *Royal Admiral* (Collins 1798: 210, 237). Baudin, commanding the French expedition off the southwestern coast, brought in two ships paralyzed with scurvy in 1802. On his own vessel, the *Geographe*, an original complement of 126 was reduced to 75, of whom only 6 (some say 4) could keep the deck when he made Port Jackson, where he remained for six months in order to reestablish his crews (Peron 1810: 260–81; Cornell 1965: 48). When Flinders got back to Port Jackson in the *Investigator*, nine of his crew had died of scurvy, and eleven were put straight into the hospital (Mack 1972: 145–46, 151–55; McCalman 2014: 47–49).

These influxes of scorbutic victims did not stop. There were regular arrivals of ships with high mortality and surviving convicts in a deplorable condition. Of the 1040 male convicts carried to the colony in 1794, 175 died, a ratio of 1 to 6 (Bateson 1959: 43). The *Hillsborough* sailed from Gravesend in 1798 and lost 95 people by the time she reached Port Jackson, chiefly of jail fever. The *Albemarle* had 32 deaths, the *Admiral Baring* 36. Governor King reported "great debility" on the *Royal Admiral* with 12 dead and 83 sick (Shaw 1966: 113). Although a proportion of these deaths was owing to diseases other than scurvy, this was not the case with ships from Ireland where prisoners, often starving and sick when embarked, were cheated of food and clean air during the voyage and succumbed in large numbers to scurvy. The most scandalous example of brutal treatment was the *Britannia*, sailing from Cork in 1796 with a sadistic master and a spineless surgeon who declined to object to the cruelties he saw inflicted on the people for whom he was responsible. Already held in appalling conditions in the bowels of the ship, the convicts were suspected of a mutiny and its alleged ringleaders were in effect tortured to death with the lash. Of the 144 male convicts, 10 were dead on arrival, and the rest very ill. After the uprising in 1798, the treatment of Irish convicts was particularly bad. Those on the *Hercules* and the *Atlas II* were convinced that that they had been deliberately starved, with 44 deaths on the former and 65 on the latter when they arrived in 1802 (ibid.; Bateson 1959: 142–47; Hughes 1988: 190). The *Three Bees* landed scorbutic Irish prisoners in 1813, having lost seven dead; the *Chapman* arrived in 1817 from Cork with twelve dead and a large sick list.

Although deaths from scurvy on the transports gradually decreased, there were some notable spikes. In 1835, the *George III* sighted the Tasmanian coast having lost twelve prisoners to scurvy, and with a further fifty prostrate.

Attempting to make an urgent run to Hobart, she struck a rock near the Actaeon Reefs and sank, with 133 deaths. In 1839, Sir George Gipps reported to Lord Glenelg that of the 330 convicts landed from the *Lord Lynedoch*, only 89 had escaped the scurvy: 28 were dead, 114 were instantly hospitalized, 23 were permanent invalids, and many of the rest would "feel the effects of the disease for the rest of their lives" (HRA 1914: I, xx, 57; Bateson 1959: 249). Katherine Foxhall's careful scrutiny of convict transportation reveals that even though the death rate was flatlining, scurvy was actually on the rise in convict ships throughout the 1830s (Foxhall 2012: 122). Even in the 1850s, ships from Ireland bore fatal evidence of neglect and cruelty. The *Robert Small* and the *Phoebe Dunbar* reported six and ten fatalities, respectively, but the figures were as unreliable as the diagnoses. It was Bowler, the *Phoebe Dunbar*'s surgeon, who tried to persuade an enquiry that his patients had died of cholera, but it is clear that scurvy, caused by a combination of a debilitating prison diet before the voyage and short-rationing throughout it, was to blame. An ugly irony attended this vicious lie, for official instructions for an outbreak of ship-borne cholera required that all fruit and vegetables be thrown overboard (Reece 1993: 8–9; Weaver in ibid.: 231–55; Bateson 1959: 341).

Besides its homegrown scurvy, then, Australia had been importing new additions to its stock with every shipment of convicts, aside from scorbutic mariners such as those arriving with Baudin and Flinders, and later with Dumont D'Urville. From the start, it was also an exporter of the disease. The ships of the First Fleet that turned for home in June 1788, having disembarked their passengers, were crewed by sailors who had lived on salt rations for more than a year. Unless they had spent their time in Port Jackson culling the few antiscorbutic plants the land afforded, they had no chance of avoiding scurvy. When the *Sirius* set off for the Cape to obtain some food, she was carrying "a dozen heartless cabbages & as many young Brocoli plants & those the produce of the Governor's Garden" (Bradley 1969: 135). Her supply of essence of malt for what it was worth was too small to share among the whole crew, so it was stopped. When they got to the Cape, 40 men were incapable of work, and there was no one on board not scorbutic to some degree (158). John Fowell reported a similar disaster aboard the *Supply*, sailing back to England via Cape Horn (Fowell 1789: 3). Phillip gave vivid accounts of the return voyages of the transports. On the *Alexander*, scurvy was rife, the only remedies being porter, spruce beer, and wine. She was sailing in company with the *Friendship* where the situation was no better. Only six hands were fit for work on the one, and five on the other, so they merged crews and scuttled the

*Friendship*. When they drifted into Batavia no one was capable of handling the vessel. Only four of the original two crews shipped home from there, the rest being dead or too ill. The *Scarborough* landed fifteen scorbutic hands at Tinian, Anson's old refuge from scurvy, while the *Lady Penrhyn*, when she got to Tahiti, had only five men who could go on deck, "the rest of the crew were in a truly deplorable state" (Phillip 1790: 206–69). Hunter heard that the *Prince of Wales* and the *Borrowdale* arrived at their first ports of call in as helpless a condition (Hunter 1968: 77).

Conceive of New South Wales if you like as a pump for scurvy—sucking it in, seasoning it, and sending it out again—its taint recirculated and strengthened as if it were the sinister obverse of the seeds received, restored, and dispatched from Edmund Spenser's Garden of Adonis. I have talked before of the leakage of scurvy through various boundaries and membranes: from the blood vessels to the bone or the skin, from the body to the mind, and the mind to the body; then from the body to the ship, from the ship to the air, air to the sea, and so on. Here the same principle is at work but figured as the circulation of a putrid current between the center of the world and its outermost limit, and (as Tench would say) casting its baneful influence as far as dimensions of the globe allowed. Just such a scorbutic eddy was evident in the difficulties of reconciling Banks's and Cook's accounts of their Australian landfall with the reality of Botany Bay endured by the First Fleet. On the *Endeavour*, they were short of greens and low on ascorbate when they saw it, and Banks was going to develop mouth ulcers and scorbutic nostalgia in the Arafura Sea a few weeks later. Whether this was responsible for the mistakes they made about the fertility of Botany Bay, indulging wistful projections of what they needed rather than making accurate descriptions of what they saw, the outcome was a situation in which incipient scurvy bred the conditions for the production of more when the convicts were landed. The image of a corrupt influence pulsing back and forth through the sea-lanes of the globe is active in the minds of the settlers of the Port Jackson colony, as if they were encircled by an evil tide, piling up its filthy jetsam with each successive fleet. Collins thought scurvy might not have been such a scourge had better food been supplied: then it "would not have met so powerful an ally in its ravages among the emaciated and debilitated objects which the gaols had crowded into transports, and the transports had landed in these settlements" (Collins 1798: 1.206). He was right about the jails, nurseries of scurvy, which was sometimes referred to as the "Millbank disease" after the London prison where diets appeared calculated to produce it; and of which, without careful superintendence, the transports completed the culture. Except for the fresh produce purchased at Bermuda, Rio de Janeiro, or the Cape, the food was always going to be the wrong food, and other remedies for the most part useless.

If the perpetual reinvigoration of a cause within the circle of its own effects is, as Trotter might have said, the genius of scurvy, then its pattern is traceable in many of the ungovernable peculiarities of the situation in Australia. These often suggested a vortex of morbific human atoms, as the drift of debilitated men and women from jail to hulk, hulk to transport, transport to hard labor, and finally from labor to the hospital made up the tale of Collins's dismal record of settlement. For example, when Sir William Molesworth, chairman of the Select Committee convened in 1837 to assess the first fifty years of the colony, decided to supply Watkin Tench's desideratum by taking a long view of its short history, this is what he saw: "It would appear ... as if the main business of us all were the commission of crime and the punishment of it, as if the whole colony were continually in motion towards several courts of justice" (RSCHC 1837: 1.71). Ralph Clark saw it more as the perpetual motion of crime toward fresh outrages against property: "I think we are in a fine stat we brought nothing but thefs out with use [us] to find nothing but thefs" (Clark 1981: 100). Even the labor of construction wore a troubling appearance of ceaseless blind ferment. The huts of the road gangs, according to one witness, "were like bee-hives, the inhabitants busily pouring in and out, but with this difference, the one works by day, the other by night" (RSCHC 1837, 1.72).

Night work in the colony was generally devoted to other activities. The convict culture of theft was preternaturally strong and getting stronger: "It is impossible for any body to attemp[t] to raise any Gardin stuff for before it comes to perfection the[y] will steal it," Clark complained (Clark 1981: 11)—a fact Collins gloomily confirmed. At first he thought of devising a plan to prevent it, but then acknowledged that it wouldn't work. "While there was a vegetable to steal, there were those who would steal it, wholly regardless as to the injustice done" (Collins 1798: 1.111). Trying for some useful critical angle on a situation that was becoming daily more peculiar, with the punishment of theft (reduced rations being a standard sentence) seemingly designed to precipitate more crime and yet more punishment, he turned to the shortage of food as the fundamental cause of the collapse of what was left of the social fabric of the settlement. Had there been enough to eat, he reasoned with himself, "the plea of hunger could not have been advanced as a motive and excuse for thefts" (206), and something might have been done to ease the cascade of prisoners from one form of crime and punishment to another. But Collins knew that shortage of bulk in the convict diet was not the first problem encountered in the early days of the colony, any more than it was at Port Arthur later. Though hunger was certainly a serious issue in Port Jackson after the loss of the *Guardian*, the chief problem in the beginning was scurvy. People stole not because they didn't have sufficient food, but because they were craving the right kind, which was why gardens were the main target and why vegetables were the prize. This was the motive for people already being punished for stealing produce to steal more: not from greed or inveterate criminality but from dreams of greens. Their cravings were impelling them toward the nutritive juice, or *succus nutritius*, of live plants. Even before the great famine of 1845–48, many of the Irish had been transported for this very offence in the famine years of 1800, 1817, and 1822, and they were now locked in a cycle of iterations: starvation, scurvy, theft, punishment, starvation, and so on.

Those who raided gardens were looking for the same relief as Bradley when he collected wild spinach, acid berries, and local samphire. It is even possible to trace a faint correlation between the sort of theft committed and the kind of nutritional deficit it was supplying. A sixth of the colony's crop of Indian corn was stolen in the summer 1792. The germ of the corn was the only source of vitamin B until the *Atlantic* landed rice (presumably unpolished) from India in June. The *Pitt* had arrived in February with nothing but salt meat. Some time between these two dates, there was an outbreak of insanity among the female convicts who were getting sufficient protein and were not forced to labor, so "it was difficult to assign a cause for this disorder" (Collins 1798: 1.204). It was most likely a symptom of pellagra, often called black scurvy for the scabs and scale it raised on the legs and forearms; later to be known in Australia as "barcoo rot" (Furphy 1981: 161). It was caused by depletion of niacin that the thieves, exhibiting the somatic intuition so typical of victims of nutritional disease, knew the germ of the corn would contain. Certainly it was pellagra that afflicted the sail-maker Peter White who had been four days starving in the bush in the winter of 1788 and was suffering the loss of mental focus and motor control typical of a serious niacin deficit. He seemed to Collins to be intoxicated, and to Hunter too: "He ... appeared to us to be stupid and ... he staggered like a man drunk" (Hunter 1968: 108; Collins 1798: 1.73). He was behaving exactly like the *Muselmaenner* in Primo Levi's account of Auschwitz, who also swayed when they sat, plunged when they walked, and became incoherent when they talked, all for want of the same vitamin.

## PART III

> I'll be judge, I'll be jury,
> Cried wicked old Fury,
> I'll try the whole cause,
> And condemn you to death.
> —Lewis Carroll, *Alice in Wonderland* (1865)

Of the many contingencies listed by Collins as responsible for the dire state of the colony—the shipwrecks of the *Sirius* and the *Guardian*, the failure and theft of crops, mortality levels that for a while were almost outstripping the numbers of new arrivals, the difficulty of communicating the emergency to officials in London, the exhaustion of the convicts, and the general loss among them of hope and of any principle social cohesion—I mean to place scurvy at the head of them as the first accident that no one had anticipated, and the one whose effects were felt the longest in Australia, long after food was plentiful and the constitutions of most inhabitants were restored to health and robustness.

I will begin by rehearsing those traits of the disease that are going to achieve salience in the colony, although not always identified as specifically scorbutic. By the middle of 1789, Collins was amazed by the frequency of thefts, particularly the robbing of gardens, which he and Clark agreed was a singular injustice done to the gardeners (1.77). Yet Governor Phillip had enough candor to tell Dundas, "I can recollect very few crimes during the last three years but what have been committed to procure the necessaries of life" (HRA 1914: I, i, 373). Collins was able to be more specific. He knew that theft of produce, often immature, while it struck at the one resource that was critical to the colony's future, was not taken just to satisfy hunger; but he was not quite clear about the motives and passions attending the crimes. He assumed that everyone, convicts too, knew they were in a battle to survive, and that they would understand, like him, how nothing but a common effort of self-preservation would ensure the survival of the maximum number of people. Gradually, he assembled in his journal a dossier of convict depravity that looked to him like a suicidal festival of dishonesty and excess. Undoubtedly, anarchy was strengthened by famine; but the blind and importunate yearning for what might satisfy the craving of a scorbutic body was not in his list. The suspension of moral will that was to lead to subsequent disorder in the community, and what many (including Collins) attributed to an inveterate depravity of mind among its inhabitants, began with the pursuit of greens. Clark himself experienced its first stages on the voyage out, vainly expecting it would be satisfied when he got to Botany. In the report of the Admiralty Committee charged with investigating the outbreaks of scurvy on the ships of George Nares's expedition to the Arctic, Dr. Buzzard was asked about scorbutic dreams of food: "Would you consider these phenomena as a mental expression, urging the dreamer to the choice of food best suited to his condition?—yes, I should certainly" (RCA 1877: 198). Buzzard had already reported that scorbutic slumber is "accompanied by dreams in which the luxuries of fruits and vegetables are vividly pictured" (196). Bradley, Hunter, and the Port Jackson Painter's sailor risked their lives trying to lay their hands on acid fruit and esculent plants. Collins wanted to interpret the same impulsive behaviour among the convicts as one more missing thing in the desolate moral landscape of New South Wales, namely "the want of that ingredient, so necessary to society" (Collins 1798: 1.501). He means self-restraint, or the *succus socialis*, and it is interesting he should identify it as an item deficient in the moral economy of these individuals, just as "vegetable effluvia" are the deficient element in a body fed too long with a sea diet: for one deficiency ushers in the other (Melville ML MS Q 36, 250).

Collins also noted the effects of hunger and forced labor on newly arrived convicts, who "wore in general a most miserable and emaciated appearance, and numbers died daily.... Of those who could handle the hoe or spade by far the greater part carried hunger in their countenances" (Collins 1798: 1.201, 204). He observed in the faces of Norfolk Island prisoners the distinctive sunken eyes and horrid pallor of scurvy, "a very unhealthy and cadaverous appearance" (1.233). It was a unique symptom, that leaden hue and hollow eye, that makes the scorbutic victim so frightful to look at. Bligh saw it in the faces of his companions at Coupang, and they in his. The crews of his two ships were so disfigured by scurvy, Dumont D'Urville recalled, that one of his officers, "seeing the victims being carried up from below ... could not recognize his own men, and had to ask them their names." Buzzard had told the Nares enquiry, confirming Bligh's account of the phenomenon, "The change of aspect ... will be noticed by them of each other, whilst the observer is unconscious he is presenting the same appearance" (Bligh 1937: 227; Dumont D'Urville 1987: 2.347; Blane 1799: 483; RCA 1877: 196). This misrecognition of a general metamorphosis, akin to the "horrid sympathy" experienced by Milton's Satan when (himself now transformed into a serpent) he watches his legions going through the same change, was typical of scorbutic depressions and excitements alike. The extreme isolation of the scorbutic patient was demonstrated in the faces and emotions of these wasted prisoners: while they ceased to be recognizable as the persons they had formerly been, each individual supposed this to be true of everyone else but him- or herself, removing the basis of any sympathetic and certainly any social community. Their dreams and depressions were as peculiar to each of them as their new situation and the skeletal masks they wore in it, unshareable though common to all. This was "the intense concentration of self in the middle of a heartless immensity," to borrow a phrase from *Moby-Dick* (Melville 1972: 525).

In its more exuberant aspect, this isolation manifested itself as Saturnalian energy. Alcohol came to Sydney Cove in 1792—porter was licensed for sale, and 7,597 gallons of spirits were unloaded from an American vessel later the same year. Liquor became "the parent of every irregularity" (Collins 1798: 1.240, 255). Soon it was the preferred form of payment, better than provisions, clothing, or coin; it would buy anything, so people would observe no restraint in trying to get hold of it: "It operated like a mania, there being nothing which they would not risk to obtain it" (327; RSCHC 1812: 5). Once to hand, it promoted every form of excess, including the destruction of all traces of prudence: "Breaking out from the restraint to which they had been subject, [the settlers] conducted themselves with the greatest impropriety, beating their wives, destroying their stock, trampling on and injuring their crops in the ground, and destroying each other's property" (Collins 1798: 1.240). There is a strict reverse symmetry in the anarchy of these people when viewed as the repudiation of Collins's visionary scheme for social order. They behaved as if they were intent on eradicating the last atom of "that ingredient, so necessary to society" (consensus, deferred gratification, common purpose, hard work, altruism, etc.) because it appeared to them to have turned to poison. There was nothing in this fantastic land of pain that was desirable except its opposite, supplied in this case by the delirium of drunkenness. Convicts figured the odds until what was rationally eligible became of all things least worthy of choice. Collins was aghast, "They now committed thefts as if they stole from principle" (idem 1802: 2.146). "It is not from me," averred the soi-disant settler Edmund Gibbon Wakefield, "that you have to learn why a people governed from afar by an absolute power, and continually increasing their territory, must have a continually increasing tendency to rebellion" (Wakefield 1829: 66). It was the lesson already exemplified in the "swing" of Kenelm Digby's bean plant and in the history of Neville's *The Isle of Pines*: "When the inferior Members that should study nothing but obedience, have gotten the power into their hands ... then every one of them following their impetuous inclinations, the whole is brought into confusion" (Digby 1669: 211).

Convicts knew of no reason not to risk the life that was already risked, and often lost, in gratifying an urgent and immediate need for food or clothing; and having got the habit of acting impulsively to obtain the things the body demanded, there was no fancy, no matter how improbable or obscene, that was not admitted and indulged when regular holidays from privation and depression were supplied by rum. After all, they were the children of contingency, as Collins had partly recognized. Georg Steller, chronicler of Bering's expedition, noted that many who recovered from scurvy became addicted to gambling (Steller 1988: 143). The long odds that Collins quoted for the establishment of civil society read to convicts like a catalogue of fatal miseries, so why not welcome chance and risk as your only friends, and be as heedless of the law and its consequences as if there were no one but oneself to witness the folly of what was imagined or the madness of what was done? Collins reported that by 1794 gambling had become a general obsession of the colony (Collins 1798: 359). Locked inside Newgate while pretending to be in Australia, Wakefield imagined with some shrewdness what life was like in Sydney before he ever got there. Among convicts, he said, everyone was involved in the perpetual calculation of odds, "and in this pursuit they show such a degree of acuteness and arrive at such just conclusions, as would be surprising, if one did not consider the excitement of their minds" (quoted in Scheckter 1998: 82). This excitement was the subterfuge of thought and desire, as James calls it, which sustained the contingent romance of Botany Bay.

However, one form of excess was met by another. On Norfolk Island, a convict called Castleton was flogged to death for refusing to work when presumably he did not have a choice (Shaw 1966: 208). At Port Arthur, a surgeon called Scott acknowledged that making scorbutic prisoners work was torture, and Charles Sturt described what it was like to suffer it: "The back muscles of my legs are all contracted, and refuse their office and to attempt to straighten them gives me most intolerable pain" (TSA/CSO1 641–14418, 11 January 1835; Sturt 2002: 313). Edward Evans remembered what it was like to go over snow with scurvy: "I suffered absolute agonies in forcing my way along.... I somehow waddled on ski until one day I fainted to striving to start a march" (Evans 1926: 199). Yet the numbers reduced to strict immobility made it more imperative that those who could still walk, work harder. Marcus Clarke's hero Rufus Dawes says, "They made me work when I couldn't stand.... It is wonderful what spirit the cat gives a man. There's nothing like work to get rid of aching muscles—so they used to tell me" (Clarke 1899: 168). Their labor was necessary to meet an objective Collins saw as "of the first magnitude and importance," which was "to endeavour speedily, and by every possible exertion, to place [the colony's] inhabitants in a situation that accident and delay might not affect." He wanted an end to the devastating plight of being "the sport of contingency" and of dwelling in a peculiar situation: he believed he was planning an intelligible and morally equitable line of conduct that any impartial spectator would understand and approve because the alternative was unthinkable (209, 204, 357).

Collins's only line of action was to pursue regimes of punishment unparalleled in their cruelty. If thieves were so compulsively criminal as to seem to steal by principle, then they must be flogged according to a stronger principle; "not dreading anything that was not immediately present to their own feelings," they should endure correction at their nerve endings (Collins 1798: 1.196). Even restrictions such as a bread and water diet (sometimes imposed for as long as three years) or banishment to solitary confinement on a nearby island were, in these circumstances, a death sentence; a fact apparently not understood until they were ended in 1832 (Walker and Roberts 1988: 5). But the use of the lash on scorbutic bodies, already stressed with loss of collagen and pathologically susceptible to all sense impressions, was torture upon torture. White recalls, "Two men ... sentenced to receive five hundred lashes each ... could not undergo the whole of that punishment, as, like most of the persons in the colony, they were much afflicted with scurvy" (White 1790: 159). Punishments as excessive as these would have been regarded as the equivalent of flogging round the fleet in the navy (generally set at 300 lashes), an ordeal from which the victim was not expected to recover, consequently it was a sentence usually remitted (Rodger 2004: 494). In Australia, the tally went up and up. In October 1788, a man was sentenced to 500 lashes for stealing soap valued at eightpence (White 1790: 216); John Hudson was given 500 for selling his clothes (Clark 1981: 215); a thief who wounded the commissary's assistant was sentenced to 800 (Collins 1798: 1.473). In 1790, a convict was given 1,000 lashes for stealing 3 lb. of potatoes (Hughes 1988: 101). Joseph Menbury was flogged so often that he lost much of the flesh on his back and shoulders; his collar bones were described as standing out like two horns (114). People were hanged for offences that were plainly not capital—James Bennett, a seventeen-year-old, for stealing goods worth five shillings; James Collington, for stealing food. It was he who said on the gallows something that stuck in Collins's mind and maybe in his conscience: namely that he stole because he was famished and would rather die than live like that (Collins 1798: 1.200).

This hideous cruelty made no difference, as everyone agreed. George Worgan noted "plenty of floggings, but I believe the Devil's in them, and can't be flogged out" (Worgan 2009: 99). Collins admits at the start of his second volume of memoirs that relentless cruelty has brought no improvement in the social sense of the population. A member of the 1837 Report made a hideous joke out of it when he asked, "Would it not be better to burn them alive all at once?" (RSCHC 1837: 88). That joke reverberated with another, told by Tench as he picked out the infernally heroic dimensions of the struggle to settle Australia while narrating his trip to the Hawkesbury River with Phillip, where they made their way up to the top of yet another shapeless stone outcrop: "I must not forget to relate, that to this pile of desolation, on which, like the fallen angel on the top of Niphates, we stood contemplating our nether Eden, his excellency was pleased to give the name of Tench's Prospect Mount" (Tench 1793: 1.119). If Tench takes Satan's part in the Australian version of *Paradise Lost*, the governor God's, and the convicts Adam's and Eve's, then the parallel suggests that the retribution exacted is quite as disproportionate and capricious as God's punishment for eating the fruit of the Tree of Knowledge, for the new settlers in this upside down Eden were only following the example of their first parents, seizing the food their nature craved in defiance of the law. The three great questions surrounding original sin were applicable here, too, namely whether exile was a just penalty for their disobedience; whether the defence of property-as-food in a state of nature was a viable restriction; and whether the colonization of the earth would ever result in redemption.

The trouble with Collins's project for social salvation was that it meant dying, as Smollett pointed out in a parallel situation in his novel, *Roderick Random* (1748). To force people suffering from scurvy into violent exertion was to hasten their death. When Charles Sturt decided to send James Poole, desperately sick with scurvy, back to Adelaide from their base camp at Depot Glen during his expedition of 1844, the man was dead of a burst artery within twenty-four hours (Sturt 2002: 203). Striving to keep the trajectory of the imperial enterprise intact, Collins evolved a plan whose cost was the very outcome it was intended to abridge; and in the end, he had to confess that his was a "visionary speculation," no different from that which drove the convicts to emigrate to a fantastic China in order to avoid the chain gang. Until quite late, he was playing an involuntary part in the scorbutic push and pull of colonial life, whose outcome was not an equitable division of labor leading to reformation, but the perpetual circuit of misery, sickness, theft, punishment, more misery, sickness, and pain, ending only in death. John Mitchel was not joking when he got to Tasmania and saw the full effects of this systematic degradation, "What a blessing to these creatures, if they had been hanged" (Mitchel 1868: 256). A convict was cited in the 1837 Report as saying, before being sentenced to punishment, "Let a man be what he will when he comes here, he is soon as bad as the rest; a man's heart is taken from him, and there is given to him the heart of a beast" (71).

Such a transformation was doubtless partly the result of the savage excesses indulged by the convict multitude, but much of it was caused by a system of convict management that deliberately imposed corporal punishments beyond the limits of the law. Tench had been astonished at the powers invested in the governor by the Royal Commission, which embraced cases of property, life, and death effectually unrestrained by precedent and common law. Gubernatorial power was extended by the Marine Mutiny Act to include civilian offenders within the regime of martial law (Tench 1789: 66–68; RSCHC 1812: 7–11). "Colonial Regulations," as they were called, allowed the governor to order lashes up to 500, but often punishments exceeded this limit (RSCHC 1812: 42, 54). No officer was supposed to order a flogging without recourse to a magistrate or the criminal court, but they seldom did so, especially in the penal settlements. The extension of this arbitrary power explains why corporal punishment not only exceeded the limits of the common law but even the regimen of naval ships insofar as disproportionate sentences were imposed for minor offences such as disobedience, insolence, and pilfering, well outside the scope of mutiny. The governor could at his own discretion expand the criminal court of judicature into a military tribunal in order to make inequitable laws more terrible and punishment more wanton. A boy called John Hudson, for instance, was given fifty lashes for being out of his hut after nine at night (Clark 1981: 183). The returns quoted in the 1837 Report show that the average flogging was by then fifty lashes, but the protocol of each return required it be carefully noted whether the skin broke, how much blood flowed, and what cries came from the victim: "John Orr, Hercules, neglect of duty, 12 lashes. (A boy) cried out very much" (RSCHC 1837: Appendix 1, 68). As for greater penalties, a state of exception was easy to invoke by means of colonial regulation, and of all categories of punishment, that was the one the prisoners most feared.

Jeremy Bentham and Sir Samuel Romilly were appalled when they read Collins's book. Bentham called it "a disgusting narrative of atrocious crimes and most severe and cruel punishments." Bentham produced a powerful genealogy of the descent of so-called justice in the colony into "so much lawless violence," tracing it from its origin "in a country … of itself yielding nothing in the way of sustenance" to the "Nullity of Legislation in New South Wales for want of an Assembly to consent," and to the "Nullity of Governor's Ordinances, for want of a Court to try Offences" (Bentham 1803: 39, 24, 20). Nothing, it was evident, comes of nothing. He wrote a memoir to Lord Pelham on the breaches of civil rights committed by the New South Wales government and began to frame an alternative concept of penal servitude based on a system of silence and perpetual surveillance, which he believed would put an end to the excesses of convicts and jailers alike (Currey 2000: 140; RSCHC 1799). However, when it was put into practice at Port Arthur, this innovation proved to be a refinement only of cruelty, transferring the infliction of violence from the body to the mind. Soon after the innovation of solitary confinement in silence was established there, an insane asylum was deemed necessary, too. The part played by the theory and practice of Benthamite penology in the evolution of punishment in Australia would doubtless have shocked Bentham himself, but it pointed to an insidious use of disease both mental and physical in the calculated infliction of physical and psychological trauma.

George Arthur took over from William Sorell as Lt. Governor of Macquarie Harbour on Sarah Island, off the coast of Tasmania, in 1822. James Backhouse describes it as a violent and miserable place of flogging and overwork in wet conditions. Nutritionally, it was a nightmare: the ground peaty and soaked, yielding no grain and few vegetables, with scarcely any livestock, for even goats found it impossible to survive on such terrain. Arthur oversaw the removal of the prison from there to Port Arthur in 1833. Backhouse doesn't say much about scurvy at Port Macquarie, but he cites a prisoner who said the convicts were worked so hard on short rations that they could not "appease the cravings of exhausted nature," and a surgeon who reported "that when the men became ill, the tone of their constitution was so low that they were difficult to recover [and] some were affected with scurvy long after leaving the settlement" (Backhouse 1843: 54–55). His visit coincided with the removal to the new location, and when he arrived at Port Arthur, he noted that "cases of scurvy have of late increased," adding, "no prisoners are now allowed private gardens," and only those manning boats were allowed to fish; nobody was allowed to hunt (167–68). This suggests that prisoners were allowed garden plots at Macquarie Island, but even had that been the case, it was unlikely that they were productive. Ever since Collins's day, convicts there had been allowed to scour Sarah Island for "Botany Bay Greens" and for the dried-up fragments of blubber (Herman Melville calls them "fritters"; the convicts called them "crap" [21]) washed ashore from whaleships. Arthur's reason for a decision to halt the supply of fresh food is hard to fathom since the daily allowance of food—at 1.5 lbs. of flour, 1 lb. of salt beef, or 10 oz. of salt pork, with tea and sugar (MacFie and Bonet 1985: 5), a ration that was halved for those accused of shirking or sentenced to hard labor—was bound to cause scurvy. No surprise then that the disease was instantly observable at Port Arthur, since an embargo had effectually been placed on all forms of its prevention. In November 1834, Backhouse summed up the sequel: "In consequence of the prisoners living on salt meat, and being defectively supplied with vegetables, a large number were suffering from scurvy" (Backhouse 1843: 167, 226). The diet, by this time lacking in both tea and vegetables, was not enough for men on forced labor and explains the rising number of scorbutic patients in the hospital. Backhouse saw nineteen at one time, "as appalling a picture of human wretchedness, as I recollect ever to have witnessed" (226). Yet Arthur claimed in his testimony to the 1837 Select Committee that since 1834 the convicts "have almost wholly supplied themselves with vegetables," adding a furher untruth: "It is found that the soil produces potatoes and vegetables of all kinds remarkably well" (RSCHC 1937: Appendix 16, 309).

The statistics for 1830–35 paint a picture more dire than this. From seven cases of hospitalized scorbutics in 1830, there were 321 in 1833 and 412 the following year, dropping to 26 after a supply of vegetables was reestablished in 1835, although that seems to have been restricted (at least at first) to those in the hospital. The official rate of mortality for scurvy in 1834 was a single death, but many more were ascribed to a long list of various genera of disease, obviously arranged according to Cullen's nosology, such as apoplexia, asthma, cerebritis, debilitas, morbus cordis, rheumatismus, and spasma—any of which might plausibly have been cited as a fatal symptom of scurvy (Ross 1995: 41). The technique of widening the spread of scurvy over a multitude of ills was already being specialized on the transports, where diarrhea, dysentery, erysipelas, and, as we have seen, even cholera took the blame for scurvy (Staniforth 1996: 122). Peter MacFie and Marissa Bonet are right to warn readers of this material that there is often a difference between official correspondence and what was actually happening. If out of a population of 887 prisoners in 1834 (Backhouse 1843: 225), there were 412 cases of scurvy diagnosed the same year (and it is plain from Backhouse's descriptions of the ones he saw in the hospital that they had to be seriously ill to get lodged there), and if the death rate for those on hard labor stood at an average of 48 per 1,000 (Maxwell-Stewart and Bradley 2006: 10), how is it conceivable that only one person died of scurvy in 1834? The official figure for convict mortality as a whole that year was 25 (Kamphuis 2007: 26), but given the carelessness with which deaths were listed and sometimes only conjectured (Backhouse 1843: 49), it seems likely the figure was higher and that fatal cases of scurvy were misreported, either by accident or deliberately. In the official correspondence, everyone, with one exception, seems ready to recognize an outbreak of scurvy and equally willing to explain that a rapid application of remedies such as vinegar and vegetables is putting an end to it. Charles Booth, the captain commandant, was enthusiastic about vinegar, perhaps because he had calculated that it would take six tons of vegetables a month "to supply the prisoners agreeably to the new scale" and that in the present state of the gardens it would be impossible to do that any time soon (TSA/CSO1–706–15459, 15 April 1834). Eight months later, the surgeon Gavin Casey, another fan of vinegar, wrote that "three fourths of the men exempted [from work] are cases of 'scurvy' which is at present very prevalent here, it being with great difficulty that vegetables can be obtained for the use of the Hospital" (TSA/CSO1 735–15912, 9 December 1834). He added, "The scorbutic diathesis [borderline scurvy] is so general among the men that the slightest wound or abrasion assumes this character, and proves extremely tedious." A few days earlier the colonial secretary wrote disingenuously, "Mr Blackburn, who has lately returned from Port Arthur, informs me that the Scurvy prevails there to a very principal extent! Has this been reported? I have no recollection of having the circumstances brought under my special attention in any way for some months past" (ibid.).

Surgeons were in a difficult position because one of their more important tasks was to curb rates of malingering (Maxwell-Stewart and Bradley 2006: 23). But with no indication that Arthur retreated from his veto of private gardens before 1835, or even then took any effective steps to increase the supply of vegetables, the number of the sick was rising and genuine cures were in short supply. Casey said that in the latter part of 1834 there were 50–70 people coming to the dispensary with scurvy, and he wants to say that an emergency issue of vegetables in the most urgent cases, together with a general application of vinegar, "has subdued in a very considerable degree this tendency" (CSO1 735–15912, 1 January 1835). But it is evident from what he says about scorbutic diathesis that almost all the prisoners, with the exception of the boys at Point Puer, were scorbutic or close to it. J. Scott, another surgeon joining in this improbable conspiracy to declare a temporary state of alleviation, said that labor could now "be extorted from the Prisoners without fear of inflicting torture" (CSO1 641–14418, 12 November 1834). John Burnett put it rather differently when, much earlier in the crisis, he wrote, "There is about a third or a fourth, more or less subject to the scurvy. Some of them seriously ill but who continue to exert themselves" (CSO1 484–10750, 10 November 1831). Exertion was well known to be a torment for those with scurvy, especially once the tendons in the legs were contracted, and no doubt the surgeons were aware of this, no matter how much pressure was placed on them to pretend that they had made these men fit for work. So one is left with Arthur himself, who was perfectly well aware of the consequence of his decisions on the health of the colony, but who evidently felt that extra pain and suffering was not just a risk but a necessary addition to the suite of devices needed to cure a worse disease, for as Backhouse said, "Penal discipline my [*sic*] be regarded as a medicine for the remedy and removal of moral evil" (Backhouse 1843: Appendix D, xlvii). So in order to expel a moral evil, a physical one was permitted to flourish just below the horizon of official notice or public indignation.

There is a strong possibility that Arthur was beginning to institute a change in the nature of the purgative of moral evil, which up until now had consisted largely of flogging and capital punishment. Occasionally serious re-offending would be punished with isolation and a reduced diet in some inaccessible place, usually an island, but by the time Backhouse visited Port Arthur a shift in emphasis had already taken place. In 1826–28 at Macquarie Harbour, 188 prisoners sentenced to punishment had shared 6,280 lashes, but solitary confinement and the treadmill were beginning to replace flogging as preferred instruments of reformation and were already in use at the Hobart penitentiary. It would be a mistake to think that Arthur had any interest in reformation as such; terror, or what he called "dread," was the motor of his version of the vicious circle of the convict system, designed as "a natural and unceasing process of classification." So he operated seven circles in his particular hell, in which every prisoner would find his class. At the bottom were those condemned to scanty rations and work of "the most incessant and galling description the settlement can produce," working and sleeping in irons (RSCHC 1837: Appendix 1, 55, 60). Arthur calculated the convicts in this class throughout the whole colony numbered more than a quarter of the total, meaning that under his control at Port Arthur, he had roughly three hundred prisoners condemned to arduous labor fed with the slenderest of diets "upon whom I think no punishment appears to have any effect whatsoever" (Appendix 15, 290).

In the next thirty years, the architecture of Port Arthur was to bear witness to the improvements in discipline and surveillance already recommended by Bentham: a block of solitary cells where total silence was enforced; a chapel with pews like stalls occluding vision anywhere but forward to the pulpit; and a mental asylum for those driven mad. For the emphasis lay now on sensory deprivation—dark cells and total silence—reinforced by the older punishment of short rations. Those in solitary confinement received nothing but 12 oz. of flour and a little salt (Backhouse 1843: 51; *Rules and Regulations* 1868: 65). For his part, Arthur thought this a very cushy number: "They take their food and sleep; and it is astonishing the health they keep in, quite remarkable." He believed solitary confinement of use only when accompanied by solitary labor, such as the treadmill (RSCHC 1837: Appendix 15, 304).

It has become evident to several commentators on scurvy in prisons that the withdrawal of vegetables from the inmates' diet was deliberate policy (Foxhall 2012: 133; Harrison 2013: 7–25). The allowance of potatoes had been curtailed at Millbank Prison in London not because of any shortage of the vegetable but because—officially at least—the diet of prisoners was not to be more "full" than that of the poor. So in 1822, a diet of meat and potatoes was exchanged for one of bread, broth, and gruel (Carpenter 1986: 99; Harrison 2013: 15). When almost half the prison's population started to exhibit scorbutic symptoms, a Parliamentary select committee ordered that the new diet be changed. However, the pressures that had induced the original change remained, for as Mark Harrison points out, "Prison doctors ... were charged with managing scorbutic symptoms by dietary manipulation, allowing rations to be kept to the bare minimum compatible with health" (Harrison 2013: 17). That is to say, scorbutic diathesis was the rule. Each prisoner was made to hover on the border between occult scurvy and its manifestation, probably on the assumption that this would have the same dampening effect on the spirits of prisoners as corporal punishment, but without the moral odium. In making this calculation in a system of heavy labor, often undertaken by ironed men working in water and heavy mud, and still subject to corporal punishment, Arthur could not have been less ignorant than his surgeons of the excruciating torture inseparable from even ordinary levels of physical exertion that accompanied cases of incipient scurvy. His experiment in replacing a system of flogging with one of calculated starvation and heavy labor was simply part of the lawless violence of a delinquent governing class, already destitute of any active principle of human community and social good. It was less flamboyant than convict anarchy, but no less excessive.

In naming scurvy "the Millbank Disease" in 1828, Charles Cameron was being more accurate than he knew, for in 1832, the Admiralty copied the Millbank system by reducing rations for convict transports, including lime juice—the same date fresh food was eliminated from the convict diet at Port Arthur and that solitary confinement with restricted rations was integrated into the system of discipline in Van Diemen's Land and New South Wales (Foxhall 2012: 134; Walter and Roberts 1988: 5). There was nothing accidental about the disappearance of vegetables, potatoes in particular. It was innovation in penological practice, and Arthur was acting in concert with people such as Dr. Hutchinson, the senior physician at Millbank, and Sir James MacGrigor, director of the army medical department, both of whom had denied that there was any serious dietary problem in the prison after potatoes and meat were removed from meals. The Select Committee on Transportation of 1837 restored the ration of lime juice to convict transports, and it was sufficiently alert to what had been happening during Arthur's tenure to justify his recall in March of the same year.

That committee had been asked to assess the efficacy of punishment in Australia, and to estimate its "Influence on the Moral State of Society in the Penal Colonies" (1.iii). After using the telling image of a flux of perpetual crime and punishment, Sir William Molesworth, said, "The most painful reflection of all must be, that so many capital sentences and the execution of them have not had the effect of preventing crime by way of example" (1.71).

The headlong chase after momentary gratification—noted by Collins as being responsible for the spread of sexual license, drunkenness, and addictive gambling—was noted again now, but on a much vaster scale, embracing free settlers and the government of the colony as well as convicts, currencies (creole children), and emancipists (former convicts). Prostitution, concubinage, pederasty, sodomy, and bestiality were rife (although "cases with animals are not common in Sydney" [1.67]); alcohol consumption was six gallons of spirits a year for every man, woman, and child, aside from what was drunk from illicit stills. This addictive culture led to violent extremes of wealth and destitution. Lucky emancipists, now owners of their own labor, were making vast fortunes in property speculation, boasting incomes (by 1837 at least) of between £20,000 and £30,000 (1.14). The officers of the Seventy-third Regiment of Foot (the New South Wales Corps) held the monopoly on all goods entering the colony: food, manufactures, and most important of all, liquor. In an economy operating almost entirely by barter, this meant that in effect they owned all labor in the colony outside the slave economy of government works since it was paid for exclusively in truck (RSCHC 1812: 53). D. D. Heath, a resident magistrate, concluded that New South Wales was not strictly a social union at all because it was so fraught with moral evil. The best that could be hoped from it as a functioning unit, he thought, was that it might become a Tunis of the South Seas, a kind of pirate state (1.261–71). Wakefield had earlier, on the basis of his reading, predicted the same thing, namely that Australia was "peculiarly adapted to become, and at a very early period too, the abode of a Tartar people" (Wakefield 1829: 69). Bligh was not exaggerating when he said that with the mutiny that put him out of his governorship in 1808, organized by the Seventy-third Regiment of Foot to protect their profits of 800%, "the civil power was annihilated" (Bligh 2011: 55).

## Part IV

> The habit of giving indirect answers, which I certainly found to
> be prevalent in Ireland, is not peculiar to the Irish, but may be
> induced by certain treatment in any country, or any climate.
> —Richard Lovell Edgeworth and Maria Edgeworth, *Essay on Irish Bulls*
> (1802)

For those oppressed, like Collins, by the anarchic turbulence of the colony, the last resource was to insist desperately on one last inviolable difference: namely the superlative degeneracy of the Irish. The grounds of such a difference lay in economic, political, temperamental, and medical factors, some of which were known to the administrators of the colony and emphasized by them, and some of which they cared not to know, or could not know. Of those they did, the criminal history of the Irish they were aware of fell into two related parts. The bulk of Irish convicts came from rural districts, not urban centers like the Scots and the English. Their numbers increased directly in proportion to the shortage of food, so the famine years of 1800, 1817, 1822, and especially 1846–48, saw upsurges in the kind of rural crime that Collins now believed to be the worst of all: the theft of produce, livestock, and crops. After 1798, the number of rural offenders from Ireland was augmented by those convicted for sedition and rebellion in the wake of Wolfe Tone's uprising. Thereafter, a steady stream of Irish political prisoners flowed to Australia, ending only with the cessation of transportation itself in 1865—United Irishmen, Young Irelanders, and lastly Fenians, dovetailing with agrarian insurgent groups such as the White Boys and Ribbon Men. There is no doubt that the atrocious treatment meted out to Irish convicts in the *Anne* (1800), *Atlas II* (1801), and *Hercules* (1801) was owing to the extreme prejudice of their guards, sharpened by the events of 1798, and that the attempted mutinies on the *Anne* and the *Hercules* were at least partly owing to the exasperation provoked by abuse and the suffocating conditions in which the prisoners were held. There were several attempted mutinies on transports out of Ireland—Price suspected one was being planned on the *Minerva*, bringing the first cargo of Irish insurgents. Even before the rebellion, the mere rumour of mutiny led to one of the worst voyages from Ireland to Australia, the *Britannia* in 1796, with eleven dead of torture and the remainder arriving in an appalling state of health. The first Irish transport, the *Queen* (1791), arrived with starving prisoners—seven had died and many were sick—complaining that they had been cheated of their rations and living space by a venal crew cramming the hold full of trade goods. Once in New South Wales, Irish convicts, nutritionally speaking, were placed exactly in their former predicament, needing to steal food to survive. They rose up at Castle Hill in 1804. Those who escaped hanging were sent to work coal at Newcastle on starvation diets (Hughes 1988: 195). Peron was convinced the Irish would readily have supported a French invasion of the colony—for him a rare intersection with the truth.

It was to people like these that David Collins was alluding when he spoke of "natural vicious propensities" characteristic of creatures less than human, compared with whom naked savages were enlightened, and now rendered so turbulent and refractory by hunger and disappointment that only the severest punishments could get them to work (Collins 1802: 2.130; Costello 1987: 23). Governor King called the Irish convicts in the *Anne* "the most desperate and diabolical characters that could be selected" (cited in Bateson 1959: 158). William Sorell said they were among "the most depraved and unprincipled people in the universe" (Shaw 1966: 187). Henderson, surgeon of the *Royal Admiral*, said they were addicted to melancholy and incurably indolent, a judgment endorsed by Sir William Denison, to which he added that they were unapt for labor and were wanting in intelligence and demeanour (338).

Centuries of prejudice and tyranny had gone into fashioning these stereotypes, but other factors were involved. The stupor and lethargy of scurvy victims is evident here, together with an obsessive tendency to nurture a sensation and to specify its every nuance, which Trotter found so typical of his scorbutics; also a desperate carelessness of consequences—the result of too little hope and too much pain—that could manifest itself either in sullen obduracy or in irrational fits of mental and imaginative energy. Why would the Irish be more likely to succumb to these symptoms than other prisoners? And even if they were, why should their symptoms be regarded as so despicable?

James Poole was a red-headed Irishman and a free man, recruited as Charles Sturt's lieutenant on his 1844 trek into the Simpson Desert, who soon was immobilized by scurvy. His symptoms were so much more severe than anyone else's that Sturt thought Poole's was a more virulent strain of the disease (Sturt 2002: 185). He was one of thousands of Irish in Australia whose sufferings were probably accelerated by hemochromatosis, a condition common among Irish males. A mutation in the C282Y gene is responsible for storage of surplus iron in the blood that overtaxes the body's supplies of vitamin C. In women carrying the gene, the excess is significantly less damaging because it is shed during menstruation. The effects of hemochromatosis are made worse among the men by the haptoglobin phenotype, which increases the oxidative stress of dealing with surplus iron. With roughly 14% of Irish males carrying the mutated gene, and the phenotype occurring in 41% of the population, their vulnerability to fatal scurvy in a famine like that of 1846–48, where 55% of their source of vitamin C disappeared, was a hundred times greater than in France and fifty times greater than in Germany (Delanghe et al. 2013: 3583–84). During the Great Famine, J. O. Curran used the difference between the numbers of men and women presenting symptoms of scurvy in order to emphasize the unique vulnerability of Irish males to the disease that was inevitably to afflict them in colonial Australia much more severely than any other nationality. He said, "The very small number of females amongst those seen by me in this city is remarkable.... Dr Christison treated 32 males and three females, and this is nearly the ratio of the sexes in the cases sent me from the country. Some practitioners ... have never seen scurvy in a female.... The comparative exemption ... is hard to explain" (Curran 1847:

116; Quigley forthcoming). It is likely that James Poole was carrying the gene and the phenotype. The melancholy and nostalgia to which the Irish were most prone, including their powerful belief in a China to the north and a nation of white people to the southwest, were exhibitions of the same dramatic and passionate register in which the longings of all victims of scurvy are cast. However, the lack of sympathy that Trotter's scorbutic sailors were quick to recognize in others was endemic in Australia. Poole's symptoms attracted only exasperation from his leader. As he was dying, Sturt wrote in his journal, "The whole case is embarrassing as it is lamentable ... most painful and embarrassing" (Sturt 2002: 188, 190). The speed at which the Irish succumbed to scurvy was always regarded as dubious—a trick of malingering, a ruse to further a plot, a sign of savagery or stupidity, or immoveable obstinacy.

There was one more characteristic—or scorbutic symptom—for which the Irish were singular, and it was an extravagant or awkward use of words, often derided as Irish "bulls" ("I was a beautiful baby but they changed me"). The impossibility of communicating scorbutic impressions, whether painful or rapturous, has been mentioned before: cravings so powerful they are comparable with no other sensation and therefore unimaginable to those who have never suffered them. When Phillip first encountered "this dreadful scourge" in the colony, he declared, its severity "cannot easily be conceived even by those who have been placed in similar scenes" (Phillip 1790: 206). Many of the Irish coming from rural districts spoke only Erse, a monoglottism the governors and guards alike found to be one more proof of an ignorant and headstrong nation; but like the rest of their unpleasant characteristics, this linguistic isolation was a drawback they could not help or that was compounded by their situation. There were, of course, manifold alterations in the language of the new colony for which the Irish were not responsible, such as rhyming slang, but they could conveniently stand as exemplars of the strains language was under. Under the explosive pressure of contingency, peculiarity, and violence, words could not sustain their freight of a common meaning nor a general currency.

This twist in the tongue was cannily to be improved into a private language by the Irish as they laced what English they had with Erse and tinker's cant, making their sentiments even more impenetrable to official reason. From the start, convict language had been notable for its vivid vernacular: Clark heard a woman called Elizabeth Barbus invite Meredith, the captain of marines, "to come and kiss her C ... for he was nothing but a Lousy Rascall" (Clark 1981: 28). Thomas Reid complained that exposure to such depraved expressions licensed conduct that would contaminate even the most innocent

female, leaving them "so absolutely vitiated as scarcely to retain the consciousness of a single virtuous thought" (Reid 1822: 296). On the other hand, Joy
Damousi explains that women used this sort of language not just abusively,
or even provocatively, but rather as a tonic, to keep their spirits up (Damousi
1997: 20–21; Foxhall 2012: 128). As well as Billingsgate, convicts talked
thieves' cant, "the *flash* or *kiddy* language," of which James Hardy Vaux
supplied a vocabulary at the end of his *Memoirs* (1819). On Norfolk Island,
it was noticed that the meanings of words began to be inverted: a bad man
was called a good fellow, a pretty woman an ugly whore, and so on (Tench
1793: 207; RSCHC 1837: 1.xvi; Backhouse 1843: 278). This was a linguistic feature of the most ancient language of Irish itinerants called Shelta, or
*garu cainnt* (bad speech), in which a syllable or whole word of Irish would
be turned backwards, so that *gabhail* (taking) became *bagail* and *gap* (kiss)
turned into *pog* (Macalister 1937: 172–73, 182).

The refusal to talk intelligible English was everywhere greeted with indignation and outrage by those with an interest in correct forms of speech. A
priest called Ullathorne was at one with Thomas Reid when he said convicts
perverted language in order "to adapt themselves to the complete subversion
of the human heart" (quoted in Shaw 1966: 206). What had begun as instinc-
tive reactions to intolerable conditions had by now become "grafted upon
the colony"—not out of consensus, habit, or mutinous intent, but as heterogeneous manifestations of displacement, passion, dissidence, imagination, and
(finally) taste (RSCHC 1837: 1.28). Wakefield complained that the news-
paper reports of crime and punishment in were given in "low-lived slang and
flash": "The base language of English thieves is becoming the established language of the colony. Terms of slang and flash are used, as a matter of course,
everywhere, from the gaols to the Viceroy's palace" (Wakefield 1829: 106;
Hughes 1988: 258). Dumont D'Urville noticed the same thing in Hobart
where the public prints were filled with scurrilous epithets (Dumont D'Urville
1987: 503). Reid singled out the superintendent of the Female Factory at
Parramatta, Mr. Hutchinson, as someone whose profanities and expletives
formed an expert descant to the noisy obscenities his charges: "His own expressions outstripped and completely eclipsed theirs in wickedness and revolting filthiness" (Reid 1822: 273).

I mean to argue that scurvy and its satellite formations provided an edge
of creative friction between the evolution of the convict system and what was
to end up as the culture of the new colony. That, of course, was not how it
was experienced or understood by either side, at first. Collins's despair as he
watched his epic narrative dwindling into an accidental and squalid romance
is fully equal in terms of imbecility to the desperation of convicts destitute of

any reason for living. As Rufus Dawes is being flogged in order to make him cry out in Marcus Clarke's convict epic, it is not the tyranny of his torturers to which he yields but the radical peculiarity of his situation, falling into an ecstasy of exasperation with the incoherence of his life, cursing and blaspheming noisily and indiscriminately until his words are nothing more than an agonized glossolalia (Clarke 1899: 284–97). I want to advance the idea that the victim of such a situation might find a way out of it by means of something like the principle of corroboration discussed in the previous chapter, where the spontaneous manifestation of incommunicable passion is answered or echoed by an event or a response that endows it with a degree of Coleridge's "outness," what used to be called an objective correlative.

That which Sylvia in *For the Term of His Natural Life* (1899) calls "the hideous Freemasonry of crime and suffering" (260) found an early home on the stage, popular from the first days of the colony. Almost as soon as they were settled in Sydney Cove, the convicts performed George Farquhar's *The Recruiting Officer* (1706), and later Edward Young's *The Revenge* (1721), cheering up a despondent pseudocommunity with a little make-believe. A theater was soon built in Sydney, but it is evident from the 1837 Report that the repertoire had coarsened:

> Is there a theatre at Sydney?—Yes. Is it well attended?—By a certain class. What class?—The emancipists and convicts. The respectable people do not go there?—Very rarely.... They are quite disgusted, the way the theatre is conducted, in allowing all sorts of people free access to what they term the dress boxes.... What kind of dramatic exhibitions are there?—Inferior to the very lowest description that you meet with in provincial towns in Britain. You do not think that the theatre at Sydney is an exhibition at all calculated to refine or improve the people?—The very reverse. (RSCHC 1837: 1.108)

An audience of prostitutes and emancipists, already proof against barbarous corporal punishment that had been originally justified as exemplary, but by now had been supplanted by more occult forms of torture, is at least consistent in expelling from their amusements any exemplary theme or purpose. Presumably the witness in this exchange means that they were enjoying the illegitimate theater of pantomimes, harlequinades, melodramas, and music hall that had already worn Australian dress on the English stage with performances of William Moncrieff's *Van Diemen's Land! Or, Settlers and Natives*, first shown in 1830: "an intirely new Serio-Comical, Operatical, Melo Dramatical, Pantomimical, Characteristical, Satirical, Tasmanian, Australian

Extravaganza" (Worrall, 2007: 175–78). Two characters in it, Howe and Wildgoose, use flash language: "Wait till the old man is safe in his dab … draw his barking irons unbetty the lock … [and] snap the swag" (187). Talking flash, cant, or Gypsy French had already been common in harlequinades such as John Thurmond's *Harlequin Sheppard*, where the great prison-breaker's feats are commemorated: "He broke thro' all Rubbs in the Whitt, / And chiv'd his Darbies in twain; / But fileing of a Rumbo Ken, / My Boman is snabbled again" (Thurmond 1774: 20). The plot and the language look forward to the celebrated elegy of a bushranger, "Waltzing Matilda," where the colloquial vigour of convict speech—a blend of Aboriginal names for natural phenomena and flash—shames the strained pastoral of Robert Southey's *Botany Bay Eclogues* (1794) and of Therese Huber's early novel, *Adventures on a Journey to New Holland* (1801).

The welcome given to a dissident jargon of cant and rhyming slang, whether in conversation or on the stage, was in the first place a refusal of the decencies of life in whose name and defense the convicts were being made to suffer. When they started hiding or inverting the meanings of words, they demonstrated that there was no principle of difference left, whether moral, lexical, legal, or existential. They demanded from all modes of representation, whether a performance or a sentence, the very opposite of what Collins expected from an imperial history of settlement. If he relied on the necessary and inevitable articulation of facts within settled structures of space and time—the staged epic of the history of settlement—they called for the recognition of "disconnected and uncontrolled experience … disengaged, disembroiled, disencumbered, exempt from the conditions that usually attach to it … [namely] the inconvenience of a related, a measurable state, a state subject to all our vulgar communities" (James 1947: 33–34). They wanted either to probe the extent of the unintelligibility of their misery or to enjoy the exuberance of an upside down world where Harlequin drinks tea from a cauldron and keeps his coals in an eggcup, and where Punch, having been transported to the Antipodes, sings in nothing but scrambled syllables: "Riberi, biberi, bino, / Faddledy, diddledy, faddledy, daddledy, / Robbery, bobbery, ribery, bibery, / Faddledy, daddledy, dino"—invoking the same world as theirs, the antipodes of reason, language, and law, a fantastic and peculiar situation where the inventive thief escapes from his enemies with his greens under his arm and lives happily ever after (Dibdin 1779: 10; Heller-Roazen 2013: 37). Even if the outcome should end less comically in death, as it does in "Waltzing Matilda" and the history of Ned Kelly, it is evident that the medium of fiction and nonsense is satisfying in itself. According to the terms of the Licensing Act of 1741, harlequinades shown in Britain had been forbidden to use articulate dialogue, the lowest class of drama being condemned by law to nonsense only. But now the tables were turned, and what had been forbidden is transformed as if by magic into the freest of free expression, linguistic kindred of the obscenities of Billingsgate and the back-to-front formations of Shelta.

When James Backhouse visited Norfolk Island and contemplated the damage done to the law by the unlimited mendacity of those who were supposed to be brought to acknowledge both its force and utility, he wrote, "Where the standard [of truth] is properly maintained, [perjured oaths] are useless, yea being yea, and nay, nay" (Backhouse 1843: 272). But that standard had already been harmed by its guardians—not just by Mr. Hutchinson with his bad language, but by all the people serving Arthur's system of applied nutritional disease and calling it something else, such as dysentery or rheumatism. They might have thought there was a floor or limit to this sort of deception, but the convicts knew better. So many boundaries had been transgressed in human relations that a new standard of values had been established. Backhouse himself admitted that the last sanction of the law had been rendered useless by transportation, which, "under its present system of penal discipline, is, in our opinion, more to be dreaded than death itself" (Appendix F, lxi). On Norfolk Island, it was now true that good men were the bad ones and that "this Island, beautiful by nature and comparable to the Garden of Eden, is rendered, not only a moral wilderness, but a place of torment to these men … by their conduct to one another" (267). Of course he meant the convicts themselves, but there was a level beneath that including many more than them.

Although the bulk of convicts were transported for felonies, a good proportion of educated convicts were forgers. Reid, surgeon on the *Neptune* in 1817, said that 41 of 121 convicts being transported were guilty of forgery. Dr. Halloran, who was the first person to teach classics in the colony, had been sent there for forgery; so had Captain Fitch, formerly of the Royal Navy; likewise Thomas Watling, one of the most skilled of the early artists, who arrived in 1792 on the *Royal Admiral* (RSCHC 1837: 1.3, 21). Manifold, a character in *Van Diemen's Land!*, "was sent here for being a little too literary … the bankers don't like plagiarism" (Worrall 2007: 181). Joseph Lycett, whose *Views in Australia* (1824) provided the first successful pictorial publicity of the colony (or what he called without any trace of irony the "ocular demonstration of the benign effects of a paternal Government upon those rude and distant regions" [Lycett 1824: Dedication]), was transported for forgery in 1814. Like Watling, who turned again to forgery as soon as he got back to England, Lycett couldn't resist the temptation of committing the crime even before the end of his term, passing forged government paper in Sydney in 1815 and bills of exchange when once more he was back in England in 1822. Tench complained that men came out with letters of recommendation that they had forged themselves, or had paid others to forge for them (Tench 1793: 207). Edmund Gibbon Wakefield, who laid out the city of Adelaide before heading for New Zealand to found two further cities, entered the spirit of things when he expounded his theory of Australian settlement in *A Letter from Sydney* (1829), penned in Newgate Prison. "I am standing with my head downwards, as it were, almost under your feet," he lied, at the same time as proving his eligibility for the *monde renversee* of the new colony, where truth and lies had effectually changed places (Wakefield 1829: 30). Even before the first ships had landed, Thomas Barret had assembled a group of counterfeiters on the *Alexander* who minted Brazilian quarter dollars so cleverly John White was compelled to admire their "ingenuity, cunning, caution and address" (White 1790: 45). White also mentions a man called Daily who filed some genuine gold into a mixture of other ores and earth, and swore it was dust from a mine he had discovered (211).

Forgeries and counterfeits play a part in the peculiar romance of Botany Bay because of the special obstacles encountered there by people who made a living by passing illicit copies of coins and financial instruments. At the outset, everything in the settlement had to be substantially what it was, intact, unrepresentable in any other medium, resembling nothing that had been known before, and therefore impossible or at least difficult to represent. The convicts' exclusive susceptibility to impressions immediately present to their feelings, reacting instantly to the radical novelty of the natural history of the land in which they found themselves, ensured a degree of originality in their sensations that was curiously sharpened by official policy. There were strong reasons why everything had to be as it was, and not another thing. Food, equipment, and clothing were distributed without reference to the amount of labor performed in order to wear it, or to any standard of desert or exchange. Until you were free, you were "on the store" and entitled to the same ration as everyone else unless it was forfeited as a punishment. Barter was forbidden among convicts; and even if it hadn't been, the illicit exchange of blankets and food was restricted by necessity and hunger. Until sex became the chief commodity and liquor the medium by which it was obtained, nothing masqueraded as the sign of goods or value. Flinders pointed out that literally nothing was the measure of the natural novelties of the place; likewise food and labor, the two prime articles generated by the colony according to Collins's scheme of development, were for several years incommensurable and had no price, owing to the fact that the government set the price of wheat and had the monopoly on labor. To prevent them from becoming commodities was part of the government plan, but it was distorted by events beyond official control. The food ration was subject to sudden contractions in efforts to stave off mass starvation; meanwhile convicts, not working for wages, did all they could not to work productively unless they were to be paid in goods and liquor on the private and very unequal market that was being developed by the New South Wales Corps (Wakefield 1829: 24, 37). So faith in these two essential articles of existence was shaken, and there was no substitute to make up the shortfall because there was no productive economy that was not dominated by the military monopoly. The great paradox of the colony even after it solved these problems was the absence of money. Ten thousand pounds' worth of silver sent to New South Wales disappeared instantly. The wealthy could assemble a sufficient number of government receipts to exchange for Treasury notes, but even these were paid out not in cash but goods and stock; meanwhile, the poor went to market with their genitals or their muscles because copper money could buy nothing (RSCHC 1812: 43, 33, 57). "You may roll in plenty, without possessing anything of exchangeable value," Wakefield pointed out (Wakefield 1829: 19). These were the contradictions William Bligh was trying to iron out, with widespread prostitution on the one hand and profits of 800% from manufactured goods on the other, when a conspiracy of rich men and wealthy officers organized the mutiny that cost him his command, the second of his career (Bligh 2011: 72).

There was a strict correlation between the peculiar situation of property and the decay, or at least the innovations, in language. If there was no settled measure of value with regard to things, there was none with respect to the meanings attached to words. They could mean whatever you wanted them to; or, as Humpty Dumpty stated in his own antipodal situation, you could mean whatever you said; but, since there was no consensus about what words stood for, you could not say what you meant. That is to say, there was no conventional system of representation in either the circulation of goods or in the communication of sentiments. As theft supplanted exchange, and monopoly a fair market, so nonsense supplanted speech. For a few critical years, there was nothing therefore that might be copied or represented as a public or consensual sign of value. Tench came to hear of the counterfeit letters of introduction because there was no culture of polite exchange to disguise the presentation of false ones. The man who pretended to have discovered gold was exposed because there was as yet no use for the metal, never mind a market in it; the fake Brazilian quarter dollars belonged to a cash economy that was still to come about. Nevertheless, the inhabitants were preparing for a time when the products of imagination might begin to circulate. Tench "found the convicts particularly happy in fertility of invention and exaggerated descriptions" (Tench 1793: 6). George Arthur mentioned in his testimony to the Committee on Transportation that many of the letters sent home by the convicts consisted of falsehoods. He said a convict on a road gang wrote to his wife in England that he was earning £60 a year and was in comfortable circumstances, when really he was in irons on a road gang (RSCHC 1837: Appendix 15, 287). In the colony, everything was tinged with romance to the degree that there was no generally acknowledged principle of authenticity to guide the hand of a correspondent or a forger who wished to copy something and pass it on as genuine. There was no identity, no reputation or character to usurp; there was not much to remember and no reason for hope. As everything was subject to contingency, phenomena were disconnected and disengaged and, as James puts it, measureless. People tried pretending that circumstances in Sydney were parallel with those in London, such as Quintus Servinton, the hero of one of the earliest colonial novels and a convicted forger, but to attempt a representation of anything not colonial was inevitably a mockery, like the grotesque mass marriages encouraged by Governor Phillip and so enthusiastically witnessed by Peron, most of which were soon renounced by the newlyweds.

Watling and Lycett were in an odd position. Not only was there no object worth imitating, whatever they produced was not their own property and could not be sold. Among the early convict community, a signature was valueless. It was thought a great effrontery in Watling that he wanted to put his name to his own pictures, and John White, the assigned owner of his labor, forbade him to do so. In any case, he had portrayed things so strange and unparalleled that no one, either in the colony or back in Britain where they were sent, was in a position to make a judgment about their fidelity. Creatures such as the potoroo, a marsupial cross between a kangaroo and a rat, or the two-headed lizard (leaf-tailed gecko), or the red-necked avocet with its long wry bill, might just as well have been imagined. There was no standard or norm by which to measure—in one sense, a golden opportunity for a forger insofar as he could paint or invent what he pleased, and equally a disaster inasmuch as his product had no relation to any other and therefore was not current. The artist was driven (or liberated) into absolute originality, his skill no longer measurable by likeness but solely by the force of the image acting as an image. Whether they wanted to or not, these artists were obliged to explore a new aesthetic of primitivism where, as Banks said of Maori carving, the design is like nothing but itself (Banks 1962: 2.24).

Paul Carter is right to suggest that the closest a European genre comes to this kind of exercise is mid-seventeenth century still life, where attention is commanded not by imitation or allegorical significance but by the powerful and immediate illusion of presence from which the human is excluded. In the art of the First Fleet, a curious kind of confusion is the result: ocean views show sculpted waves; rocks shimmer like water; human figures coil like tree-branches; and tree-bark looks like skin. The French artists on Baudin's expedition were not convict labor. Apart from commemorating Peron's fantastic assemblage of shells, Charles-Alexandre Lesueur and Nicolas-Martin Petit could pursue their own interests. Neither had shipped as artists—they were rated as assistant gunners on the *Geographe*—but with the defections of the official artists in Mauritius, they advanced to that position, producing remarkable watercolors, pastels, and drawings of indigenes. Lesueur's work has the delicacy of tone and the intricacy of detail that Paul Carter compares with the best Dutch artists; while Petit's ethnographic work, such as his pastel and charcoal sketch of the Aborigine Morore, is astonishing for its physiognomic mobility, oscillating in this portrait between symptoms of bewilderment, anxiety, and truculence. Petit, was an ingenious mimic and prestidigitator, capable of inveigling his subjects into the poses and expressions he found most dramatic, spending much less time than his British counterparts on ethnographic details such as tools, scarification, and body painting, and evidently enjoying what he was doing, unlike Watling. Some of his Tasmanian profiles, perhaps at his own instigation, seem to emerge from an Antipodean *commedia dell'arte*—the figure often amused and grinning, sometimes overflowing into caricature (Bonnemains, Forsyth, and Smith 1988: 147).

Common to the British and French pictorial records of Australia is a fascination with novelty, an involuntary originality that incorporates effects I have referred to in other contexts as ellipsis or dazzle. While rather less evident in the work of the French, this fluent alternation between opposite qualities and moods seems to lodge not just in the objects depicted but also in the sensibility of the artist. Having at some length suggested how the convict system contributed to the inversion of hierarchies and the rapid alternation between opposite impulses and emotions, I want briefly to generalize this tendency in the whole experience of the early experience of Australia before returning to its exemplification in the art of the First Fleet. In *The Road to Botany Bay* (1987), Paul Carter identified an ecstatic state of mind frequent among the early surveyors of inland Australia, whose fatigue of body and eye could erupt into delight as they identified features of the landscape that were simply not there—hills or lakes—or that certainly existed, but for them alone (Carter 1987: 82). Empiricist commitment to rendering an exact report of things consorted with an attachment to details that either had no scientific value or whose importance was reoriented by the emphatic force of the sensations accompanying them. For instance, Thomas Mitchell begins the account of his three expeditions by announcing that "the sole merit [the author] claims is that of having faithfully described what he attentively observed." When circumstances disturb the author and his eye, the scene shifts from proofs of experimental accuracy to the throes of private agitation. Shortly before he reports that scurvy has broken out among his party, Mitchell begins, "The scene which followed I cannot satisfactorily describe, or represent, although I shall never forget it." It consists of a group of Aborigines, dancing in a ring with furious precision while moving backward, all by way of demonstrating their anger with these white men who fired a gun to frighten them. Mitchell's description of this dance of defiance-by-retreat is a heap of broken phrases: "crouching and leaping to a war-song ... their hideous, crouching, measured gestures, and low jumps, all to the tune of a wild song, and the fiendish glare of their countenances ... all eyes and teeth ... thus these savages slowly retired along the river bank, all the while dancing in a circle" (Mitchell 1835: 1.iii; 245–46). When John Oxley explored the Lachlan River some years earlier he was discomfitted not by the natives but by the unreadable features of the land, "so extremely singular, that a conjecture on the subject is hardly hazarded before it is overturned; and every thing seems to run counter to the ordinary course of nature in other countries" (Oxley 1820: 81). Charles Sturt was badly scurvied during his expedition to the Simpson Desert, where he was longing for rain so that some vegetation (sow thistles or *rhagodia*) might spring up and vary their diet. The symptoms of his shattered condition are evident in the obsessive and repetitive entries he made in his journal, dramatic illustrations of the egotistical particularity of the victim Trotter noted in the later stages of the disease: "There is no hope of rain.... I am beyond measure provoked at being unable to ride.... As I have said before there is no hope of rain.... There is not at the present moment the slightest indication of Rain.... As I believe I have already observed yesterday was unbearably hot.... I am sorry to say that I have all but lost the use of my limbs.... Today has been insufferably hot.... I regret to say that I have all but lost the use of my limbs" (Sturt 2002: 304–11).

Now I shall try to link these scorbutic disturbances of the eye, tongue, and temperament with aspects of the work of the anonymous artist (or artists [Rosenthal, forthcoming]), known only as the Port Jackson Painter, for he more than any other offers a sublime reconciliation of the consternation of Mitchell's failed description of eyes and teeth on the one hand, and the purely hallucinatory glow of things and human figures occupying space whose lack of coordinates has rendered it peculiar and incommunicable. Bernard Smith believes that the Port Jackson Painter was a naval man and not a convict. His most likely candidate is Henry Brewer, an older seaman who came out with Phillip as his steward, was listed as a midshipman on the *Sirius*, and ended up as acting provost marshall until shortly before he died in 1796 (Smith and Wheeler 1988: 220–24). He appears as the partner of Duckling, a female convict, in Wertenbaker's play, *Our Country's Good* (2015). Trained as an architect and as a naval draughtsman, Brewer had a framework for visual notation that can remain primitive and perfunctory when the scene is populated by multiple figures, but which grows considerably in subtlety and power when he embarks on isolated individuals or portraits. Whoever he was (or they were) the Port Jackson Painter worked closely with the forger Thomas Watling and George Raper (likewise a midshipman on the *Sirius*), copying some of their pictures and perhaps learning new skills in drawing, particularly from Watling, who had run an art school in Dumfries. From the start, the Port Jackson Painter's work was distinctive in its treatment of Australian subject matter; for besides pictures of animals and birds (which Bernard Smith regards as his best work), he was drawn to a kind of ethnographic sketching that differed in key respects from the work of Watling and, as it turned out, Petit, although it had something important in common with Lycett's and Westall's.

Both Watling and Petit were trained artists, and they posed their subjects as if for a studio portrait—Watling's often in three-quarter view, with the eyes seldom directly meeting the viewer's. Petit's are either in three-quarter view or profile. Some small attention is paid to body painting and scarring, but the emphasis falls on the face as specimen (Watling) or as an opportunity for sympathetic humor (Petit). The question of context and local custom (fishing, healing, cookery, revenge) is ignored, and even in the portraits where the emotions of the Aboriginal sitters are on display, as in the blend of contrary emotions in Morore (Petit's finest portrait) or in Watling's picture of the Aboriginal chief Wearrung in an aggressive mood, attention is concentrated on the exactness of the unfamiliar shape of the head and the unusual angles of the lips and jaw. Watling wrote of another attempt at Wearrung's head, "The likeness is very strong indeed ... [a] pretty good resemblance" (Smith and Wheeler 1988: 65). But it wasn't one upon which Watling piqued himself, for he thought his subjects "extremely ill-featured," destitute of moral virtue and prey to the worst emotions: "Irascibility, ferocity, cunning, treachery, revenge, filth and immodesty are strikingly their dark characteristics," he wrote (Watling 1793: 11; cited in Clendinnen 2005: 249–50). Possibly Watling's attitude to the people was influenced by the humiliating position in which he felt placed, as a convict artist assigned to John White, told what to draw and with no financial interest in the result. His performances, he told his aunt, "are, in consequence, such as may be expected from a genius in bondage to a very mercenary and sordid person" (Watling 1793: 20). But it is also possible that he found his skill as a copier/forger worth nothing not simply because he couldn't sell what he produced, but because he found the original so unlovely and farouche.

Petit exhibits none of this impatience, but evidently he felt Aboriginal humanity had to be called out by means of amusing tricks, otherwise it would remain invisible. He is free from this constraint only when he is drawing things, such as boats and huts. His masterpiece, highly valued by himself judging by the numbers of copies he made, is of a bark windbreak entitled *Aboriginal Dwelling* (Bonnemains, Forsyth, and Smith 1988: 123; Fig. 13 and Plate 4). A structure apparently dilapidated from its conception, its purpose so uncertain and casual, might bear comparison with Dutch genre pieces of tumbledown wayside inns on sandy roads were it not that it asserts itself so strongly as what it is and nothing else. Lesueur's attempt at the same thing ends up with strangely humanoid forms of tree stumps, one appearing to warm its branches at a fire and another in what Rabelais would call a "divaricating posture," as if he were determined to find some kind of resemblance to domesticity and sexuality (ibid.; Fig. 14). Flinders described an enigma similar to Petit's in the north, made of stones and bark: "Of the intention of setting up these stones under a shed, no person could form a reasonable conjecture; the first idea was, that it had some relation to the dead, and we dug underneath to satisfy our curiosity; but nothing was found" (Flinders 1814: 2.172). Flinders's artist, William Westall, drew a picture of it, not as fine as Petit's; but on the other hand, his portrait entitled *Port Jackson: A Native* comes close to the mixture of pathos and menace in the Port Jackson painter's pictures of Bolladeree and Nga-na-nga-na (Westall 1962: illus. 36; Fig. 15).

Lycett's work is divided between landscape and ethnographic illustration. Although he advertises his *Views* as "absolute fac-similes of scenes and places," they are intended to attract new settlers to a terrain that is at least potentially recognizable, for it is accompanied by a map of the Port Jackson hinterland distinguished by labels such as "rich land," "good valleys," "fine open country" (Lycett 1824: ii). The commentary on views of houses such as Captain Piper's Naval Villa read like a real estate advertisement; and the landmarks such as the Nepean River, Beckett's Falls, and the Wingeecarrabee River could be anywhere; as John Crowley says, "perfectly conventional in the repertoire of European landscape appreciation" (Crowley 2011: 225). But occasionally—in his picture of Cape Pillar, for instance—he produces a dazzle-picture, in this case of a vast beehive shape sliced away at the side to reveal narrow vertical pillars arranged on such a different axis from the lateral bands of the

Fig. 13. Nicolas-Martin Petit, *Aboriginal Dwelling*. No. 18013. Courtesy of the Muséum d'histoire naturell, Le Havre. See also Plate 4.

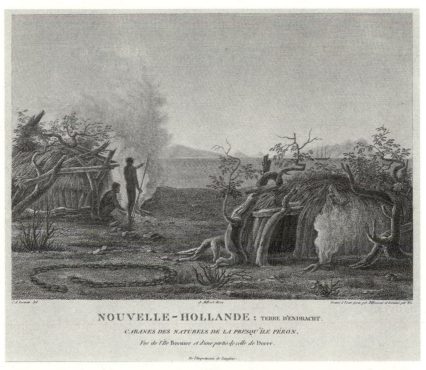

NOUVELLE ~ HOLLANDE : TERRE D'ENDRACHT.

CABANES DES NATURELS DE LA PRESQU'ÎLE PÉRON.

Vue de l'Île Bernier et d'une partie de celle de Dorre.

Fig. 14. Charles-Alexandre Lesueur, *Aborigines' Huts on Peron Peninsula*. 1807, hand-colored etching. Courtesy of the National Gallery of Australia, Canberra. The Wordsworth Collection, purchased 2010.

Fig. 15. William Westall, *Port Jackson, A Native*. No. 2098456. Courtesy of the National Library of Australia.

beehive that the eye is excited by the elliptical tension (Fig. 16). His Aboriginal figures, lithe silhouettes, are engaged in the landscape: fishing, hunting, or fighting Europeans (Fig. 17 and Plate 5). They have the same athletic thin black bodies we see in the work of the Port Jackson Painter, exhibiting the shocking vividness of shape and tonality for which Banks blamed Dampier when he wrote in his journal, "We stood in with the land near enough to discern 5 people who appeard through our glasses to be enormously black: so far did the prejudices which we had built on Dampiers account influence us that we fancied we could see their Colour when we could scarce distinguish whether or not they were men" (Banks 1962: 2.50). Dampier was not to blame. Such collisions between black and other colors, especially white, recall some of Hodges's finest work among icebergs: the unearthly sheen of *Ice Island* emphasized by the inky sea and most astonishing of all the explosive shards

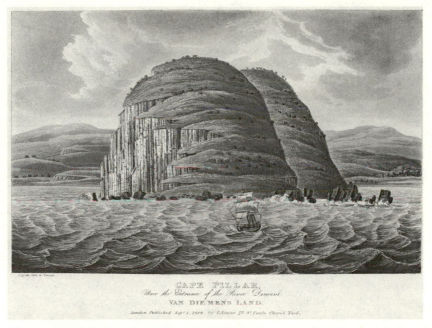

CAPE PILLAR,
*Near the Entrance of the River Derwent*
VAN DIEMENS LAND.

*London, Published Sep.t 1, 1824, by I. Souter 73 St Pauls Church Yard.*

Fig. 16. Joseph Lycett, *Cape Pillar*. No. 563837. Courtesy of the National Gallery of Australia.

of black and white comprising *The Island of South Georgia* (Fig. 18 and Fig. 19). The Port Jackson Painter's black silhouettes are as sharply outlined; like Hodges's iceberg, they look as if they had been cut out of different fabric, or appliquéd. An Irish bull of the Edgeworths catches the dazzle effect when for want of more adequate language pure white is cited as a "beautiful black" (Edgeworth and Edgeworth 1802: 93): Here is something like beautiful white.

With one notable exception, the Port Jackson Painter is not interested in objects that have no use, nor in the limitations of representational decorum defined by studio techniques. His Aboriginal groups are always situated in a landscape that establishes some intimate relation between land, water, and sky. By means of spears, throwing sticks, fizzgigs, boomerangs, canoes, paddles, threads, and axes, human beings negotiate the various elements around them in order to eat and travel, to maintain health and make war. That these people are unrecognizable silhouettes mainly in profile, bearing some faint resemblance to ancient Greek or Etruscan figures on kraters and urns, emphasizes their distance from the viewer and their proximity to a terrain intractably foreign. The diagrammatic arrangement of these pieces leaves no room for a vanishing point or its correlate, a spectator whose eye makes a property

Fig. 17. Joseph Lycett, *Aborigines Spearing Fish*. No. 2962715. Courtesy of the
National Gallery of Australia. See also Plate 5.

of the scene. Here are three men climbing trees, curvilinear shapes on a light
ground; here is a woman curing her child of a stomachache (Fig. 20, Fig. 21,
and Plate 6). Any attempt at lifelike detailing would be out of keeping with
the internal éclat of each composition, which is the Port Jackson Painter's
homage to an environment and a social organization that is evidently active
and productive, but not in a way he understands.

How deeply he is engaged with what he doesn't understand is evident in
the morphology of the rocks in the remarkable painting, *Grotto Point in the
Entrance of Port Jackson* (Fig. 22 and Plate 7). Here, there are no natives in
the picture, so the strangeness of the natural forms has free rein. He gives full
value to the opulent scallopings of the rock in the foreground and to the
coils, twists, and laminations of the sea-carved sandstone above. The kind of
painful detailing with which Raper represents the waves of the ocean, crisp-
ing them into solid shapes, is transferred by the Port Jackson Painter to the
sandstone, but with the opposite effect. Instead of immobilizing liquid, he
liquefies what is solid. The striations of the more distant cliffs, like the lavish
shapes of those closer in, are reflected in the ripples of the sea below and
mimicked by the ropes of vegetation above. The compositional brilliance of
*Grotto Point* accomplishes likewise the very opposite of Watling's careful notation of strange countenances because it dazzles the viewer's eye instead of
concentrating it on a definite object. In this respect, it is very different from

Fig. 18. William Hodges, *Ice Island*. SAFE/PXD 11, f. 27a. Courtesy of the Mitchell Library, State Library of New South Wales.

Fig. 19. William Hodges, *Island of South Georgia*. SAFE/PXD 11, f. 33. Courtesy of the Mitchell Library, State Library of New South Wales.

Fig. 20. Port Jackson Painter, *Method of Climbing Trees*. Watling Coll. 75. © The Trustees of the Natural History Museum, London.

Petit's bark windbreak, insofar as the Frenchman's contemplative rendition of what Felix Vallotton called the décor of ruin is here supplanted by the activity of the rock and vegetation; they vibrate with energy and won't stand still.

When the Port Jackson Painter places Europeans alongside the Aborigines, the dazzle is reduced to a diagrammatic opposition between unaligned movements or shapes. In the two pictures showing the wounding of Governor Phillip, I have already pointed out how the stick figures of the Aborigines move freely in and out of the fringe of bush, while the Europeans find safety

Fig. 21. Port Jackson Painter, *Aboriginal Woman Curing Her Child*. Watling Coll. 62. © The Trustees of the Natural History Museum, London. See also Plate 6.

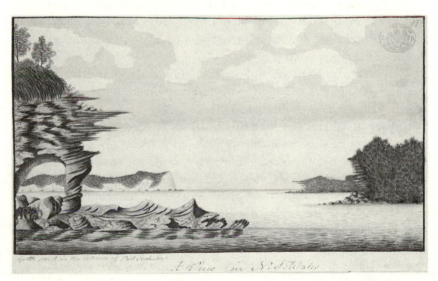

Fig. 22. Port Jackson Painter, *Grotto Point in the Entrance of Port Jackson*. Watling Coll. 7. © The Trustees of the Natural History Museum, London. See also Plate 7.

in retreating to the much more narrow confines of a longboat. In *Aborigines Attacking a Sailor,* the dramatic opposition between the European running for his life and the natives holding their ground is reinforced by eye-play, with the sailor's glance cast longingly from the left hand of the picture at his hat, which has fallen off. On the right, three bright eyes and three missiles are focused entirely on him. The viewer meets the sailor's eyes along the same axis as the hat, a piece of property now as alien as the hats, caps, and shoes washed up on Crusoe's island. The center of urgency is the satchel of greens underneath his arm, attracting and impelling all the movement of the piece: they are what he has risked his life for, and what will save it. *Mssrs. White, Harris, and Laing* is a painting partly commemorating Edward Laing, surgeon of the New South Wales Corps who wrote the highly esteemed letter to Thomas Trotter, printed in the second edition of his *Observations on Scurvy*, concerning scurvy in the new colony. He is one of three Europeans at the center of the picture, each carrying fowling pieces, with a platoon of marines away to the right, standing stiffly like toy soldiers. Apart from the sole Aborigine who is sitting meditatively on a log in the right foreground, all the vitality of the scene is concentrated in the Aboriginal bodies lying underneath the tree at the left, disposed in groups of twos and threes, each entwined with or folded into another. The easy blend of figures seems partly sexual and partly maternal, or familial, and stands in sharp contrast with the attitudes and the business of the men in red (Fig. 23).

In the field of portraiture, the Port Jackson Painter is divided between two compositional modes, the anthropological and something harder to characterize and much more elliptical. In 1791, a series of skirmishes occurred after Bolladeree, an enterprising local fisherman who traded with the settlement, found his new canoe stove in, for which he took revenge by spearing a convict. The issue was not resolved until Bolladeree died of a fever later that year, mourned by Baneelon and Colebee (Clendinnen 2005: 212–16). The Port Jackson Painter did a number of pictures of Aborigines covered in funeral paint ("moobee") in attendance at Bolladeree's burial in Governor Phillip's garden, including Colebee and Barangaroo, Baneelon's wife. With the exception of Barangaroo's, these are profiles exhibiting red and white ochre over the brows and cheeks, and a mosaic of colors across the shoulders and breast. The faces are clumsily or at least perfunctorily executed, especially that of Barangaroo, who was reckoned to be an extraordinarily attractive woman (224)— they are merely frames for the "moobee" mourning. Like the landscape pictures of fishing, stabbing, hunting, cooking, and embracing, the human figure is less important, or certainly no more important, than the situation in which

Fig. 23. Port Jackson Painter, *Mssrs. White, Harris, and Laing.* Watling Coll. 25.
© The Trustees of the Natural History Museum, London.

he or she is found, whether it is actively traversing a scene or bearing the cer-
emonial signatures of grief.

All that has changed in the profile of Gna-na-gna-na, tentatively identified as Baneelon's brother-in-law Gnunga-Gnunga (Smith and Wheeler 1988:
20; Collins 1798: 1.299; Fig. 24). The astonishing confidence of the brush
that fetches in the musculature of the neck and left shoulder is consistent
with the assured emphasis of the turn of the nose and the swell of the lips,
reinforced by the rivetting glance sent from the orb of the single visible eye.
The artist catches an expression whose power arises from a coalition of pure
surprise and undistracted concentration, as if the subject were simultaneously reacting to and radiating some sort of threat. This is a mixture already
evident in the earlier picture, *Woman Curing a Child* (Fig. 21), where the
woman's staring eyes and spread arms and thighs express concentration and
distraction in equal measure as she wields the string that connects her gums
to the child's stomach in order that she might bleed from her own body the
pain afflicting her infant. The formal signatures of emotion diagrammatically
represented in "moobee" ornaments have here been transferred to the angles
and features of actual bodies and faces, all alive to the call of a phenomenon
in front of them. A transitional example is entitled, *The Manner in which the
Natives of New South Wales Ornament Themselves,* a profile of a chief called
Goo-roo-dar (Fig. 25). The angle of the mouth, lips, and eye focus a glance as

Fig. 24. Port Jackson Painter, *Portrait of Gna-na-gna-na*. Watling Coll. 60. © The Trustees of the Natural History Museum, London.

pointed as the barbed spear in his hand. The same alertness of the glance is evident in *A Native Carrying a Water-basket*, so sharp it prevents the head from serving as any other purpose, anthropological or pictorial.

The Port Jackson Painter drops a hint about his second mode of portraiture in the title, *Ben-nel-long When Angry after Botany Bay Colebee Was Wounded*. Collins documented the circumstances of Baneelong's anger in a reference to his presence at a scene of ritual vengeance taken on Colebee. After gazing at the unequal and shocking violence, apparently impassively, he did what he was often prone to do, and ran wild: "On a sudden, he chose to be in a rage at something or other, and threw a spear among the soldiers, which, dreadfully took effect on one of them, entering at his back and coming out at the belly, close to the navel.... Mr Smith the provost-marshal ... brought him away, boiling with the most savage rage" (Collins 1802: 2.68). Sturt was to notice the same coexistence of aloofness and violent excitement in his favorite guide Toonda, who had the eye of a Roman patrician but

Fig. 25. Port Jackson Painter, *The Manner in which the Natives of New South Wales Ornament Themselves: Goo-roo-da*. Watling Coll. 57. © The Trustees of the Natural History Museum, London.

would chew a blanket in his rage (Sturt 2002: 42). Techniques of coloring are being tried out by the Port Jackson Painter in the Ben-nel-long (Baneelong) picture that are to be perfected in the portraits of Gna-na-gna-na and Balloderree (Boladeree). The hair is carefully built up as a framework for the eyes, with a dark pigment providing the underlay of strands, curls, and tiny ornaments of bone, while a rich brown on top gives it body. The eyes are emphasized by the white hairband above and the stripe of white ochre over the bridge of the nose; and emphasized again by the brightness of the whites. The lips, faintly outlined in pink, carry the affective freight of the glance, which is not beamed to the right, as in the profiles, but straight at the viewer. The anger is directed there, nowhere else.

These techniques arrive at a most startling conjunction in the picture of Balloderree, who is also staring out of the picture at the viewer with a concentrated force not dissipated by the scale, attitude, or ornaments of the figure

Fig. 26. Port Jackson Painter, *Portrait of Balloderree*. Watling Coll. 58.
© The Trustees of the Natural History Museum, London. See also Plate 8.

(Fig. 26 and Plate 8). It is as if Gna-na-gna-na had turned to face us, for the muscles of the neck are rendered with the same confidence and the face is animated by two emotions, comprising something like surprise or apprehension together held in the sort of pause that precedes sudden and aggressive movement. The medusa-like halo of hair, made of black flourishes on a brown ground, combines with the fine hatching of the beard to emphasize the key features of the eyes, nose, and lips. The depth of the eyes is emphasized by the black circles around the carefully mottled irises, counterbalanced by the protrusion of the lips, delicately insisted upon by the lightened color of the lower one and by the deep crimson line where it meets the upper.

By the time the Port Jackson Painter finished the portrait of Balloderree, he would have known that the young man had witnessed a savage attack upon his own wife by Baneelon, who justified his rage by saying she was the child of his enemy. Balloderree had not intervened, any more than Baneelon did when witnessing the terrible beating of Colebee, although very troubled by an event none of the Europeans could understand. However, when Balloderree's canoe was wrecked by convicts, he was enraged and announced to Govenor Phillip his intention to avenge his loss, even if it meant killing the governor himself. Phillip soothed him by having the perpetrators flogged in front of him, but Balloderree still speared a convict, although not fatally, as payback for the canoe; whereupon he became a kind of outlaw until he fell sick. He was cared for by Baneelon, at whose request Balloderree was lodged in John White's hospital (at that point filled with over four hundred scorbutic invalids). And at his death, Baneelon was chief mourner. So it is a tangled tale that fits none of the models of native romance that were later to be produced by authors such as George Barrington in Australia and Alfred Domett in New Zealand. It is a collection of contingent events provoked by passions ranging from insensate rage to the tenderest concern, with no intelligible link between any of them except perhaps the consequences of the vengeance taken for the damaged canoe. The title of the narrative of either of the two portraits could be, "Anything Might Happen."

So what did the Port Jackson Painter see in those eyes that searched his own? Well, rising emotions with no visible cause, unless the Painter saw himself as providing one. Perhaps the reason he wanted to give shape and coloring to the enigma of Balloderree, the man so active in pursuit of revenge and yet so tolerant of a nearly fatal attack on his wife, lay adjacent to the dying hero in White's hospital. They were the scorbutics, some of whom would recover to form the stem of what Collins called "a people regardless of the future and not dreading anything that was not immediately present to their own feelings," a judgment that might be applied equally to the convicts and the Aborigines they were displacing (Collins 1798: 1.196). The scorbutic romance that shaped the colony was notable for experiences that were disconnected and uncontrolled, especially after alcohol drove the settlers to beat their womenfolk and destroy their own stock and crops for no reason at all (240). However, the Port Jackson Painter is not making an allegory out of the countenance of Bolladerree and using it to represent what it was like to be a settler in Botany Bay. His absorption is a strangely tense equilibrium of opposite impulses that is owing to no likeness or referent at all, but solely to the present fact of what life in that place happens to be: standing still to be hurt, or to witness hurt; or running wild in sudden defiance of such violence, or yielding to a reverie that is inexpressible because there is no messenger and no message, just the extraordinary peculiarity of the situation.

In his section on the metamorphosis commodities, Marx wrote of the form needed to reconcile the contradiction of their movements, when "one body [is] constantly falling towards another and at the same time constantly flying away from it." He concluded that "the ellipse is a form of motion within which this contradiction is both realised and resolved" (Marx 1978: 198). The same contradiction is evident in these two pictures of Bolladerree and Gna-na-gna-na, where the figure, held in the pause of what could be apprehension, surprise, or anger, is suspended between two contrary movements— forward to touch or wound, or backward to avoid a threat or a puzzle. The Port Jackson Painter has dealt with this paradox in the form of a pictorial ellipse, where both options are present and buoyant, but not yet taken up. Mitchell observed the same ellipse in the terrifying, circling dance of aggressive retreat performed by the Aborigines he had tried to frighten with a gunshot. Collins observed one of the two foci of the Botany Bay ellipse in the conduct of prisoners whose social germ had been extirpated. Bentham and Bligh located its counterpart in the lawless violence of a delinquent governing class. It is heard in the guarded nonsense of flash talk and bulls, generally turning on the co-presence of identity and its opposite: "When I first saw you I thought it was you, but now I see it is your brother" (Edgeworth and Edgeworth 1802: 227). It was inaugurated in Ballance's upside-down advice to Plume in *The Recruiting Officer*, regarding the right way to treat Sylvia: "Be modishly ungrateful, because she has been unfashionably kind, and use her worse than you would any Body else, because you can't use her so well as she deserves" (Farquhar 1721: 1.85). The Port Jackson Painter recognized the same ellipse even in the landscape as it vibrated into shapes and reflections that placed the maximum strain on volume, line, consistency, and figure, generating the kind of dazzle that bursts out of Georg Forster's *Ice Islands with Ice Blink* and is carefully secreted beneath the sunlit pastoral of William Hodges's *Pickersgill Harbour*. Between these visual ellipses and the efforts of Tench, Flinders, and others to write of Australian somethings by reference to nothings, a formal resemblance exists that I shall try to explore further in the next chapter.

# CHAPTER FIVE

## Genera Mixta

The Scurvy owns not one universal cause, but is the Bastard of
many Parents.
—Everard Maynwaringe, *Morbus Polyrhizos et Polymorphaeus: A Treatise
of the Scurvy* (1666)

Fiction has shadowed and skirted the discussion of scurvy through the
four previous chapters. Insofar as the disease has been associated with
remote and unknown places, difficult to comprehend or to imagine from a
distance and outrageous in its impact on the nerves of the sufferer, it is part
of the blur that makes experience in such places hard to absorb, and the reports of it difficult to authenticate or even to understand. Novelty combined
with extreme mood swings makes for a volatile and improbable mixture, with
the result that uncorroborated accounts of events so far from the ordinary
are received as fiction. Readers decide to understand them as having been
adjusted by fancy to advance self-interest, justify diffidence, or indulge a taste
for the marvellous. Thus William Dampier's caution in a sea fight, George
Anson's lordly treatment of the Chinese, John Byron's encounters with the
Patagonian giants, and George Shelvocke's strangely muted tale of taking
Spanish treasure ships, are all greeted back home with the same derisive cry
of "romance."

Voyagers themselves are keenly aware that in very alien places events lack
the symmetry of form and the order of succession upon which any claim for
probability must rest. When situations become (in David Collins's lexicon)
"peculiar," such as John Oxley's when he finds his chronometer stopped and
his compass reversed in the middle of the Australian desert, or Bligh when
he lands in a dream at Coupang, the experience is unpleasantly unreal and
overwhelming, and the place itself usurps the authority of those who came
to measure it. Under these circumstances, Collins himself reaches for the dis-
reputable category of romance as the only one available. Just before he died
exploring central Australia, W. J. Wills bravely fetched a comic parallel from

a novel, writing in his journal, "Nothing now but the greatest good luck can save any of us.... I can only look out, like Mr Micawber, for something to turn up" (Scheckter 1998: 37). So in this chapter, I want to see how fiction, and the genre of romance in particular, supplies the scorbutic imagination with models suitable for "peculiarity" and, having suggested in the previous chapter that there may be something like an aesthetics of the dazzled eye, to explore the possibility of a connection between that and the genres of scorbutic literature.

The histories of scurvy and fiction arrive at a curious junction in the work of Trotter and Beddoes when, in their later work on the nervous temperament, they cite novels not so much as an epistemological or aesthetic blemish on modern culture but as a significant challenge to public health. Earlier in the century, George Cheyne had estimated nervous diseases as comprising a third of all maladies; but by the beginning of the nineteenth century, they had grown and diversified so rapidly that, according to Trotter, they represented two-thirds of the whole catalogue. Explaining why nervous diseases had displaced fevers as the leading sickness of the age, he wrote: "The passion of novel reading is intitled to a place here. In the present age it is one of the great causes of nervous disorders. The mind that can amuse itself with the love-sick trash of most modern compositions of this kind, seeks enjoyment beneath the level of a rational being. It creates for itself an ideal world, on the loose descriptions of romantic love, that leave passion without any moral guide in the real occurrences of life" (Trotter 1807: 89).

The "deluded and vivid imaginations" of novel readers enslave them, he declared, to "all that is extravagant ... or absurd in fiction" (218), seducing them into an involuntary state of excitement typical of reverie. In Trotter's opinion, this addiction aggravates a general nervous weakness that is now endemic among the middle class, owing chiefly to the incalculable effects of chance on life in a market-driven economy. The vagaries of credit, the heavy investment in shipborne trade, and the undulations of stock prices and interest rates all conspire to wind the nerves up to an unhealthy pitch. Similarly, innovations in the conduct of fashionable life and warfare are liable to expose people to extremes of boredom and excitement: the longueurs of country life interspersed with sudden charms of the London Season, or the explosive interludes of battle between the tedious duty of naval blockades (idem 1812: 165). The alternating spikes and troughs of the passions provoked by books, speculation, fashion, trade, and war then invade even the meditative intervals of interiority, as castles built in air collapse and torpor (or something worse) supervenes, and people "sit for hours in the same posture, without paying the least attention to what is going on around them," or suddenly burst into pointless laughter (192, 175). The odd fantasies about the depletion of the body noticed by Darwin, such as the lady who could see that her head had rolled on the floor and was presently in the corner being chewed by the dog (Darwin 1801: 2.362), are evident too: patients who feel their insides are nothing but a vacuum, a man who was sure his lower part was connected to his upper by a waist as narrow as a wasp's, a disease now known as Cotard's Syndrome (Trotter 1812: 205). Trotter was beholding in the neuroses of the most privileged enclaves of metropolitan society oddities of perception and affect he had witnessed among the stressed crews of naval vessels and the tormented human cargo of slave ships. The difference between them, in his opinion, was that the symptoms of the one were induced by a culpable indulgence in absurd lies, while those of the other were involuntary reactions to the extreme privations of service or servitude. Yet they had much in common, not only in the exhibition of mental distress but also in the circumstances causing it, which one way or other flowed from an expanding global market economy that included a large trade in ephemeral literature.

Beddoes wrote a survey of health called *Hygeia* (1802) with the same intention as Trotter, exploring "the Causes affecting the personal state of our middling and affluent classes." He culled his examples from the same sources: fashion, print culture, public opinion, the hurry and fatigue of business, all necessary to keep up the whirl of artificial pleasures and ambitions that displaced attention from what Trotter called "the real occurrences of life." News had become a signal of excitement, not of information: "Did you see the papers today? Have you read the new play?—the new poem—the new pamphlet—the last novel?" Equipping oneself with the opinions necessary to answer these questions encourages a hubbub in the brain that Beddoes compares with the madness of Don Quixote (Beddoes 1802: 3.163–64). He finds several examples of Quixotic reader-errantry: a woman "so affected by the reading of a romance, having for its subject a disappointment in love, as to be deprived of her senses and fall into convulsions," and a lunatic who immersed himself in a novel, changing the names of the characters into those of his own acquaintance, so that the pathos of the story he had half-invented overwhelmed him, and he went on transposing and reading in floods of tears (3.164, 55). Pathologies like these provide the material for imitations of Cervantes in the eighteenth century, such as Charlotte Lennox's *The Female Quixote* (1752) and Jane Austen's *Northanger Abbey* (1817). Among the theorists of reverie most directly concerned with the confusions that exercised Trotter and Beddoes, Lord Kames stands out as one who had linked it directly to the experience of reading fiction. He had concluded (as, with some reservations, Coleridge was to conclude later) that absorption in a fiction was productive of a trance he called "ideal presence." He compared it to a dream or reverie in being in an unreflective and immediate engagement with images instantly presented to the reader's mind as an event. In such a receptive state the will is suspended, time is compressed into an instant, and space into the compass of what is imagined; for "if reflection be laid aside," Kames wrote, "history stands upon the same footing with fable: what effect either may have to raise our sympathy, depends on the vivacity of the ideas they raise; and with respect to that circumstance, fable is generally more successful than history" (Kames 2005: 1.71; Vickers 2004: 71; Dames 2007: 13–16).

Like many of his contemporaries, Kames was intrigued by the more lawless feats of the imagination since they were often pleasurable, placing (as Kames says) "every object in our sight," turning the patient into a spectator and the mind into a theater (1.70). Nor were they limited to the middle class, being observable (as Erasmus Darwin and William Wordsworth had shown, and Trotter, too, at length) in the lives of ordinary seamen and rural laborers. Calenture, somnambulism, sensory hallucinations, and second sight were all worthy of interest. Samuel Johnson's tour of the Highlands gave him an opportunity to authenticate evidence of the last, and Darwin provides a specimen in his *Zoonomia* of a woman who went upstairs to find her sister in her bedroom at the very instant the same sister died twenty miles away. Darwin lists it along with these other pathologies of perception under the genus of "Increased Actions of the Organs of Sense" (Darwin 1801: 2.358). After his theory of septon had been dispatched, Samuel Mitchill went on to write an analysis of fifteen distinct genera of the class of reveries he called *somnia*, including second sight (Mitchill 1815: 124–40).

None of these commentators were quite as eager as Trotter and Beddoes to blame fiction for weakening the nervous temperament; and given their collaboration on a pneumatic theory of scurvy, as well as the extensive investigation of factitious airs pursued at the Pneumatic Institute under Beddoes's supervision, especially Davy's self-experiments with nitrous oxide, it was surprising that novels should have borne the brunt of their disapproval. After all, Beddoes was not only a witness of the effects of nitrous oxide, he enjoyed its magic, along with Coleridge, Southey, and all the others, in a public setting. There on a stage the impulses of pointless pleasure and incipient discomfort were vociferated, uncontrollable impulses set free, personalities split in two, and the unreality of a parallel world penetrated (Jay 2009: 172–73). Next to this incandescent display of sensory and imaginative extravagance, the emotional disturbances incident to the most gullible reading of the worst novel were faint and anodyne. This seems like a whale Trotter and Beddoes could not get out of—a vast gas-inflated volitional malady produced like an evil djinn out of a bag, releasing a catalogue of psychotic symptoms of which scurvy was the enigmatic prototype, and for which nitrous oxide acted as the portable artificial stimulant. Trotter's inclination to historicize symptoms of nervous disease could never get further than to dream, like Beddoes, that there might be a place in time and space where mental and physical health could remain inviolate.

Since these were matters that preoccupied authors as well as physicians, it might be as well briefly to explore the genres of fiction to which Trotter and Beddoes are alluding, since they seem to identify the popular novel as a major culprit ("a variety of prevalent indispositions ... may be caught from a circulating library" [Beddoes 1802: 3.165]). Strongly implied in their judgments therefore is a distinction between the fiction of overexcitement, such as the Gothic romances of Anne Radcliffe, whose popularity was then at its height, and those novels that provide both rational enjoyment and a guide for life by depicting the real occurrences of it—the criterion set out by Samuel Johnson in his fourth *Rambler* paper (1750) and by Clara Reeve in *The Progress of Romance* (1785). The division between romance and the novel had been magisterially laid down by William Congreve a hundred years earlier when, in the preface to *Incognita: A Novel*, he distinguished between modern fictions that "come near us, and represent ... Accidents and odd Events, but not such as are wholly unusual or unpresidented" and "Romances [that] are generally composed of ... lofty Language, miraculous Contingencies and impossible Performances [which] elevate and surprize the Reader into a giddy Delight" (Congreve 1923: 1.111).

So when Trotter refers to fiction as an irrational enjoyment, he means that it appeals to the imagination and the passions by representing what is not real as if it were. When he says that it conjures up an ideal world, he means that it is like a reverie to the extent that it shuts out the world we commonly cognize, rendering us obedient solely to internal stimuli. When he adds that the relationships shared in this world are romantic and loosely described, he means that they are erotic and conducted under the anarchic licence of what Charlotte Lennox's Arabella, heroine of *The Female Quixote*, calls "the Empire of Love." And when he tells us that the story serves as no guide for conduct, he means that the law of moral consequence plays no part in it, and the events described are governed not by the will and intelligence of the characters nor by the moral tendency of the narrative, but merely by chance and caprice in league with an aroused imagination.

Johnson drew an important division not, as might have been expected, between realist fiction and romance, but between two different types of modern realism. Romance was so improbable that he believed, unlike Hobbes, it had no function except as a trivial amusement, for no one could possibly mistake flying horses and iron men for an imitation of real life. But in some novels, we are given an accurate but indiscriminate picture of the world as it is, where good and bad elements are combined in a single character, and there is no guarantee that a preponderance of virtue will be rewarded with happiness, or of vice with its opposite. The bad example set by such a character, he says, can "take possession of the memory by a kind of violence, and produce effects almost without the intervention of the will" (*Rambler* 4). He argues that it is not a mirror of the world that fiction ought to supply, but a selective imitation, for "if the world be promiscuously described, I cannot see of what use it can be to read the account." Mimetic realism on the other hand engages the reader's mind not as a photograph or moving picture of life as it is lived from moment to moment, but as a set of hypotheses, "mock encounters in the art of necessary defence." This is how fiction works to defend social norms, Johnson suggests: in presenting a speculative account of risks that will likely be faced in the reader's future and inviting imaginative participation in figuring how they might best be managed, the author prepares the reader's imagination as a training ground for real experience. So the most dangerous fiction young people are likely to read does not concern castles, dragons, enchanters, and ladies pent in durance vile; it is a probable reflection of real life from which the conjectural element of self-defense has been removed.

In a recent important essay on fictionality, Catherine Gallagher has shown how intimately Johnson's brief program for the novel intersected with the psychological developments accompanying the Financial Revolution. Johnson expected novelists not only to honor a contract with the reader regarding the outcome of invented stories (virtue rewarded, vice ending in ignominy) but also to use the form of contract for these mock encounters with danger—what, if I were in that position, would I feel, say, do, and what upshot would I expect?—so that a credible estimate might be made of how a speculative situation might develop over time. Gallagher endorses this judgment: "Readers were invited to make suppositional predictions ... to speculate upon the action, entertaining various hypotheses [until it becomes clear that] the reality of the story itself is a kind of suppositional speculation" (Gallagher 2006: 1.346). Author and reader alike were entering into the same imagined relation to the future that governed all transactions in a commercial society and every kind of contractual social engagement, from dancing to marriage. The ur-conjecture out of which all these representations, bargains, properties, identities, experiments, and histories are made is simply this: "What if what I do not know, I did?" Once the bargain is struck or the contract made, life itself is shaped, not merely reflected, in the working out of that hypothesis, and reference to the real is displaced by conjecture—the reader's own working fiction—as the primary device of edification and social cohesion. The licence given to the reader's imagination by the novelist is cognate with that empowering every party to a contract in the world of global commerce, "merchants and insurers calculating risks, investors extending credit," a place where "no enterprise could prosper without some degree of imaginative play" (ibid.).

Despite the limits placed on this sort of contractual reading, it is continuous with the world of calculations and promissory exchanges that Trotter and Beddoes blame for the rise in mental illness. That which makes the novel dangerous in their opinion—the substitution of imagination for cognition—is what makes it useful in Johnson's and Gallagher's, with one important proviso—that the will of the reader not be compromised. With the exception of Kames, those discussing the reading of fiction from Johnson to Gallagher refer the suspension of disbelief and the play of imagination to the will. This "temporary half-faith" is what prevents illusion from becoming delusion and keeps reading from becoming what Darwin calls a disease of volition (Coleridge 1930: 200). It preserves an ironic credulity that keeps all that is imagined within the zone of a provisional hypothesis, "What if what I do not know, I did?"

The opposite of this speculation is the radical form of skepticism adopted by Descartes in order to secure himself in the inviolable certainty that he thinks, and that because he thinks, he exists: namely, "What if what I do know, I didn't?" There is nothing provisional or conditional about such an exercise, because it requires that all forms of empirical cognition be entirely abandoned in favor of an indisputable action of the mind. As Bernard Williams points out, this is a project of pure enquiry whose prize is not knowledge but truth (Williams 1978: 35), and it bears an astonishing similarity to the strategy adopted by Don Quixote on his chivalric quest when he chooses to assign all evidence of an empirical and nonheroic reality to the malice of enchanters, leaving his imagination as the sole warrant of all existent things: a barber's basin is truly the helmet of Mambrino, a peasant girl on an ass is truly Dulcinea del Toboso, Alonso Quesada is truly Don Quixote de la Mancha, and so on. The exclusion of empirical actuality typical of reverie in its most extreme form is typical also of romance, and this is why Trotter and Beddoes, unlike Johnson, associate the worst kinds of fiction with that genre, so the task ahead is to isolate those features of romance that will loom large in what I shall be identifying as scorbutic fiction.

Up until now, I have distinguished scorbutic reveries as arising from situations that are, as Collins calls them, peculiar in the sense that they are unrelated to any other, uniquely horrible and repellent, or sometimes singularly attractive. Henry James's definition of romance as a "fantasy of unrelatedness" has frequently been cited, and his characterization of the experience of such dislocation as unknowable has also been useful. The sense of inhabiting a fiction whose plot is unfathomable, so often presented in Gothic romance as the predicament of the heroine ("Her present life appeared like the dream of a distempered imagination, or like one of those frightful fictions, in which the wild genius of the poets sometimes delighted" [Radcliffe 1989: 296–97]), distinguishes the temporality both of romance and scorbutic isolation, where there is no orderly succession of events ("reflection brought only regret, and anticipation terror" [ibid.]), and no community is standing by to offer comfort, advice, or models of behavior.

Nevertheless, romance is remarkable for scenes of colossal confidence, such as Quixote's grand catalogue of the heroes of the rival armies he conjures out of the dust clouds raised by two flocks of sheep, or his explanation to the Toledan merchants that Dulcinea's excellence is not to be vouched for by sight or hearsay, but simply by their unlimited faith in the truth of the words in which he praises it. He exemplifies this implicit faith, for while he presents his mistress as the creature of his imagination and his will ("I fancy her to be just such as I would have her"), he worships this idol of his own brain as completely as if it were an existent thing and an extrinsic force, irresistible: "She fights and overcomes in me; I live and breathe in her, holding Life and Being from her" (Cervantes 1991: 1.229, 292). When Lennox's Arabella lists the duties of constancy governing the knight in love, they soar above standard responsibilities to friends and nation; for she declares it is sometimes proof of the most exalted passion to embrace the cause of one's enemy, or one's rival in love, instances of what she, too, calls peculiarity: "It is in that peculiarity ... that his generosity consists, for certainly there is nothing extraordinary in fighting for one's father and one's country; but when a man has arrived to such pitch of greatness of soul as to neglect those mean and selfish considerations, and, loving virtue in the persons of his enemies, can prefer their glory before his own particular interest, he is then a perfect hero indeed" (Lennox 1973: 229). It is clear that the peculiarity of passionate isolation can sometimes elicit the most shocking acts of independence, all of them familiar to a student of the Empire of Love, where "Love requires an Obedience which is circumscribed by no Laws whatever, and dependent upon nothing but itself" (321).

The words repeated most frequently in this world of confident disorder are "questionless" and "doubtless." They underwrite a conviction of boundless exceptionality that is summed up in the exclamation of Mme de Lafayette's heroine in *The Princess of Cleves* (1678): "In the whole world there is not another case like mine" (Lafayette 1994: 77). It is not an embarrassment but a triumph to be so peculiar and to lack the analogies suitable for such a situation, for the truth is there are none, nor any means of communicating its glory. As Eliza Haywood avers, "There is no greater proof of a vast and elegant passion, than the being incapable of expressing it" (Haywood 2000: 101). The singular and tautologous locutions of romance are heard again when the miseries and ecstasies of voyaging in the South Seas fail to find language adequate to their peculiarity, belonging as they do to a Terra Incognita where may be found, as Peter Heylyn ironically observed of the newly discovered parts of the South Seas, "the Isle of Adamants in Sir Huon of Bordeaux; the Firm Island in the History Amadis de Gaul; the hidden island, and that of the Sage Aliart in Sir Palmerin of England" (Heylyn 1667: 1094). However, the absolute peculiarity of a single and incommunicable situation is not always in itself eligible: some analogy, contradiction, or corroboration is needed if its advantages are to be known.

As compendiously as I can, I shall try to arrange those features of romance it shares with reverie, before embarking on a discussion of fictions associated with, or arising from, an isolation that may be termed scorbutic. First, there is the enclosed space of the imagination in a state of reverie, so well secured from outside impressions that it can figure to itself another world that is entire, self-comprehended, and yielding, as Haywood says, "one great unutterable comprehensive meaning" (Haywood 2000: 122, 89). That is to say, it is not an entity either able or inclined to enter into engagements with others. The topographical equivalent of a reverie is, as Heylyn wittily suggests, an island, like those proliferating over an ocean that Epeli Hau'ofa, inverting the properties of sea and land, has named the "sea of islands." If all one starts out with is a peninsula, then it behooves aspiring Utopians to cut a breach in the isthmus and make it an island, as Thomas More's have done. The architectural structure that bears analogy to an island is a prison, which no doubt is why islands were so often chosen as ideal sites for the incarceration of those guilty of heinous crimes or afflicted with the most contagious diseases, such as Norfolk Island and Tasmania in the new convict colony of Australia, and Molokai, where the lepers of Hawai'i endured permanent exile and civil death. Port Arthur is the perfect antitype of Utopia, an island born of an island by having the isthmus sealed, not by water but by a line of savage dogs. So the places most closely allied to reverie are those that insulate corruption or protect innocence. On old maps, the Terra Incognita was sometimes marked with the fancied location of the terrestrial paradise, and navigators arriving at islands in the South Seas were prone to think that they had arrived at it. In Vanuatu, Pedro Fernandez de Quiros held a chivalric and liturgical celebration of the New Jerusalem; and at Tahiti, Cook cautiously noted that men and women went naked without shame and obtained their bread from trees, not the sweat of their brows. Sometimes the symptoms of innocence and corruption were confounded in the one place, for at Tahiti, the *arioi*, an aristocratic sect of itinerant men and women devoted to theatrical displays and free love, behaved like troubadours in pursuing their own brand of *amour courtois*, having "enter'd into a resolution of injoying free liberty in love without being troubled or disturbed by its consequences" (Cook 1955: 127–28). The consequences were either infanticide or illegitimacy on a large scale, a common phenomenon in romance where "bastardy was in peculiar reputation" (Dunlop 1816: 1.174).

Illegitimacy is perhaps the most dramatic illustration of a lack of concern about consequences, penetrating not only the action but also the structure and temporality of romance. Doubts about origin and posterity are scarcely ever raised, for it is almost as if birth out of wedlock provides a nondynastic title to positions of greatest power and esteem, a custom honored in the accession of bastards to principal vacancies that are especially numerous after whole fellowships, such as Arthur's and Charlemagne's, are obliterated. Galahad, the issue of Lancelot's enchanted encounter with Elaine le Blank, succeeds in the quest for the Sangraal while his father doesn't; Isaie le Triste, the illegitimate son of Iseult and Tristan, restores the chivalry of Britain after the deaths of Arthur and Lancelot (1.276). There is a corresponding hierarchy of adultery. Tristan and Iseult, the wife of King Mark, elope to Joyous Garde where they are hosted by Lancelot, vindicator of the pseudoinnocence of Guinevere. When Isaie sets about to reform the abuses of the Round Table, he begins by seducing the niece of King Irion and having an illegitimate son called Mark (of all names) who does the same with the daughter of a Saracen admiral, forming as it were a lineage of fornication and bends sinister (1.286–88). Genealogy and posterity, the past and the future, are both rendered uncertain, allowing the present time to be the medium of an undistracted attention to pure action as knights roam "in quest of adventures for the mere pleasure of achieving them" (1.182). The receptacle of memory is no longer a representation of lived experience but a fog of miraculous contingencies, and the future is a cloud of incalculable events and impossible performances. Those experiences that might have formed the basis of common sense or national history are set aside.

So on the island of time formed out of an everexpanding romantic present moment, the senses are liberated from any duty of comparison or reflection. Each sensation is exquisite in its pain or pleasure because there is nothing like it. "Love creates intolerable torments! Unspeakable joys! Raises us to the highest heaven of happiness or sinks us to the lowest hell of misery" (Haywood 2000: 165). The only way to manage such extremes is by formalizing the moves appropriate to each adventure, turning the phases of seduction into a fantastic but powerfully erotic protocol, or the chaos of war into elaborate repetitions of violence required by the code of the tournament. Every now and then, contingency breaks through these ceremonies of passionate action and returns a scene to its original peculiarity, such as Guinevere's bed linen so strangely soaked with the blood from Lancelot's wounded hand, or the arrow shot by sheer accident into his buttock from the bow of a female huntress while he rested by a brook en route for a joust (Malory 1969: 2.438, 418).

The eruption of startling physical actuality into the sealed system of romance signals an alliance between the imagination and the senses that is always important. They are like the scenes of corroboration that prevent scorbutic nostalgia from attenuating, or like the moment of surprise in an experiment that Boyle (from the other side of the spectrum) compared to a passage in romance. In their more extreme form, they produce what seems like a breach in the system, as when it is discovered that the baby Merlin was a demonic implant in the virgin womb of his mother, requiring emergency baptism if the promise of Christian salvation is to remain operative (Dunlop 1816: 1.203). The symmetry of the rituals of love and war must be shadowed and sometimes assaulted by an opposite principle if the energy of the stylized system of seduction or jousting is to be sustained. This is achieved to some degree by the unpredictable outcomes of actions taken regardless of consequences, but more so by what Niklas Luhmann calls the "paradoxicalisation" of romance (Luhmann 1986: 45). This occurs when a contradiction surfaces between the regulations of the code and an irruptive contingency, or between the code of chivalry and the code of love, or between the Christian ideal of the Sangraal and the demonic malice of magicians and enchantresses. When seduction occurs, for instance, there is on both sides of the gender divide a contradiction between the "passivity" of lovers overwhelmed by the irresistible force of passion and the confidence of agents unhindered by doubts or questions. So on the one hand, there is the tableau of the submissive knight on his knees before the yielding or fainting woman—their submission necessarily involuntary and possibly even coerced—and on the other, two consenting parties to the consummation, who desire nothing else than to be doing what they are doing. Regardless of the degree of passivity or agency in the moment of passion, the coupling ought to comprise a knight owing loyalty to his fellowship and a lady under no obligations to another man, yet often this is not the case, and then the contradiction grows more severe. The degree of the anomaly is proportionate to the silence it provokes. All Malory vouchsafes of Guinevere's liaison with Lancelot is that ancient love and modern love are not the same (Malory 1969: 2.426). Meanwhile Lancelot offers to prove by arms to the satisfaction of Arthur, in spite of all the well-attested accusations to the contrary, "that my lady, Queen Guenever, is a true lady unto your lord as any as is living" (2.477).

Daniel Heller-Roazen sums up this state of affairs in romance, where affirmation and denial, and innocence and sin keep changing places with each other, in the following equation: "x is possible if, and only if, x is impossible" (Heller-Roazen 2003: 108). The equation is installed in so much of the language of romance, despite the unquestionable certainties and doubtless absolutes it proclaims, that the figure of the double negative is often deployed to affirm a thing by denying its opposite ("I do not hate you"), a habit of thought and expression that infects even the language of Clara Reeve's interlocutors in her *Progress of Romance* when they say things like, "I have not heard you with insensitivity" (Reeve [1785] 1970: 1.104). When Johnson traveled to the Highlands of Scotland, the closest he ever got a real land of romance, he could speak in little else but litotes: when shown a tree of insignificant size in this largely treeless landscape, but informed proudly that there was a much larger one a few miles away, he recalls, "I was still less delighted to hear that another tree was not to be seen nearer" (Johnson 1971: 10).

\* \* \*

The islands of romance mentioned by Peter Heylyn, combined with the first voyages of European discovery, provide scurvy with its early modern fictions: Amerigo Vespucci is used by Thomas More; Vasco da Gama supplies Camoens and then Milton; Dampier is Lemuel Gulliver's cousin; while Antonio Pigafetta's journal of Magellan's navigation of the Pacific provides the hint for Bacon's *New Atlantis* (1627). In many of these utopias, the wretched state of the voyagers forms an extreme contrast with the amenity of the island and the happiness of the islanders. Their neediness contrasts with the health, beauty, and wisdom of the newest of new worlds, a paradise free from any countervailing negative principle, save for that embodied in the unlucky sailors, who are sick, starving, and unlovely. But no matter how the simplicity and abundance of the island is represented, some portion of its admirable economy is going to reveal imperfections analogous to those of its imperfect visitors. The Utopians have their Zapoletes, mercenaries living in the borderlands whom they exploit but would rather see extirpated; the Houyhnmns have their Yahoos, whose labor they command yet whose genocide they perpetually meditate; Henry Neville's *The Isle of Pines* contains a dissident sept of Phills, likewise headed for destruction. Inconsistencies like these declare the true illegitimate grounds of alliance between utopian fictions and romance: In both, a self-perpetuating system is held together not by symmetry but by compromised lines of descent, negatives, and reversible propositions. Paradisiacal innocence pursues wisdom whose implications are serpentine; self-denial holds hands with voluptuous pleasure; definitive action is marred by unpredictable consequences; and language is spoilt by equivocations and self-cancellations.

The *New Atlantis* hews closely to the program of scientific salvation that Hooke borrowed from Bacon and announced in his *Micrographia* (1665). The innocent fruit of knowledge, he promised, would remove the mortal taste of the primordial forbidden kind. Bacon's is a fiction of a fellowship of philosophers, inventors, and experimentalists who, having cleared away the rubbish from the doors of perception, now see divine, natural, and artificial knowledge in their true light. But the tale begins with a scorbutic emergency in which sin and scurvy are closely associated. The author and his crew are driven across the Pacific by a storm that leaves them without food or a position—peculiar, you might say. When they spot the verdant island of Bensalem and anchor in its harbor, they report to an official who meets them that they have seventeen sick with scurvy out of their complement of fifty-one, and those "in very ill Case; so that if they were not permitted to land, they ran danger of their lives" (Bacon 2014: 3). The crew is given citrus, "like an Orenge ... of colour between Orenge-tawney and Scarlett; which cast a most excellent Odour" (5). More of these oranges are forthcoming, "which (they said) were an assured Remedy for sickness taken at sea. Ther [*sic*] was given us also a box of small gray, or whitish Pills, which they wished our Sicke should take, one of the Pills, every night, before sleepe; which (they said) would hasten their Recovery" (7). Subject to this impressive antiscorbutic regime, the sick are soon better, "who thought themselves cast into some Divine Poole of Healing, they mended so kindely, and so fast" (8).

After this Bethesda-like miracle, they are told of another that happened in this island of Bensalem twenty years after the death and ascension of Christ: a pillar of light was seen with a wooden ark at its base, a heavenly covenant granting the fortunate inhabitants the knowledge of all things they now possess and celebrate. Shortly afterward, the redemption of the Bensalemites was proclaimed by the prophet Bartholomew. As a result, Bensalem now is "free from all Pollution, or foulenesse. It is the Virgin of the World" (31). The only ripple in the allegorical mirror of all that is "worthy to hold men's eyes" is, on the one hand, the arrival of people polluted and foul with scurvy, and on the other the travesty of scientific wisdom exhibited in the houses of deceits, where machines capable of mimicking the sensory extravagances of scorbutic prostration, reverie, and a depraved imagination are all to be seen. The sudden accommodation of the sick mariners, despite a prohibition against such incursions, finds an equivalent in the impostures and fictions generated by engines reproducing the ancient imperfections of postlapsarian sense organs. A sort of antiprostheses, these inventions are the opposite of the "helps for the sight" that make remote or small objects perfectly visible and colors prismatically pure; or helps for the ear that make sounds resonant; or helps for the voice that let it be projected whence and whither the speaker chooses; or helps for the nose permitting mixtures of smells to be accurately distinguished. That is to say they are not like the machines invented by Hooke; rather they resemble the catoptrical inventions of the Dutch, in that they are designed not to correct the senses but to fool them, and to restore the sense of an original infirmity—Adam's poor eyesight for instance—instead of removing it. Since all inventions in Salomon's House serve a purpose, these must be supposed to stand as a warning, otherwise there is no good reason for their existence. Even so, it is a strange way for the Bensalemites to commemorate the means of their redemption. The narrative of the *New Atlantis* begins and ends, therefore, with two purges for the taint of scorbutic misrecognition, one that is natural and embodied in the recovered sailors, the other artificial and precariously exemplary, lodged in the houses of deceits.

If the rapid admission of disease into Bensalem is a surprise, even more surprising is the instant application of antiscorbutics. It indicates that this commonwealth of health and innocence, pent in its virgin island and embraced by a doctrine of exceptionalism for almost two millennia, nevertheless shares a history with the fallen world that is remembered and can in fact be detected in everything these people make and do. Doubtless, the remedy dates back three thousand years to the period when Bensalemite navies crossed the world, driving an extensive trade in the Mediterranean two vast oceans away; but that it should be preserved fresh and ready to use now the island has drawn so completely into itself indicates that scurvy is as present to the memories of the inhabitants as the fantastic delusions in the houses of deceits. The innocence of the island is defined at all points by its opposite. A character in Timberlake Wertenbaker's play *Our Country's Good* says, "You have to be careful of words that begin with 'in' . . . innocent ought to be a beautiful word, but it isn't, it's full of sorrow. Anguish" (Wertenbaker 2015: 41).

The paradoxes of scientific salvation in the *New Atlantis* provide, nevertheless, opportunities for corroboration that otherwise would not have been available. Having defined corroboration as the coincidence of an imagined and an experienced thing—Coleridge's mariner dreaming of drinking and then waking to find his body soaking up rain—we can see that scurvy promotes this sort of doubling every time the invention of a "help" functions as a prop or supplement for an imperfect faculty, making what was wished for appear, and every time the imperfect faculty is reproduced in the distortions of a device of imposture. The contrast between the physical and mental prostration of the sailors and the luxury of their stay in the hospital—"a picture of our Salvation in Heaven ... a place where we found nothing but Consolations" (10)—is less subtle, but it serves to show that even pathological fantasies may collaborate with knowledge instead of obliterating it. When the Bensalemites imagine how the story of their island will be received in Europe as a dream, the hypothesis is exactly the same as Bacon's when he chose to advance the cause of truth in science by means of a fiction of a nonexistent island. It means that "x" and "non-x," the possible and the impossible, can find grounds or at least intuitions of compatibility. Its promise is heard in the litotes of the senior official who says to his visitors, "Something I may tell you, which I think you will not be unwilling to hear" (157).

Although like the *New Atlantis*, it deals with a utopia of experimental scientists in a ship under stress of weather with a sick crew that has been driven beyond the limits of the known world, Cavendish's *The Blazing World* presents things differently in. The single survivor of the vessel, soon to be the Empress of this icy shore called Paradise, is aware that the pole of the earth is conjoint with the pole of another world (hence the redoubled polar cold that kills the mariners and the "huge pieces of Ice" drifting in the sea, not to mention the bi-polar bears that visit the ship). She is also aware that she has crossed the border between what Alice, falling down her rabbit hole, calls "the antipathies." The first topics she wishes the scientists of this New World to handle concern the contrary mixtures of such things as snow, icebergs, and amphibians.

The bird-men explain that snow is made of a mixture of water and fire; the fish-men tell her that ice is caused by a salt vapor exhaling from the ocean, condensing into water and then congealing into ice; the worm-men report that they do not know what happens to the saline element when the ice returns to water; the fish-men and the worm-men assure her that there are creatures of a mixed nature, "partly Flesh, and partly Fish," whose systems of respiration differ and can accommodate various methods of breathing (Cavendish 1994: 139–47). These scientists are themselves chimerae, humans mixed with other species, reporting phenomena as mixed as themselves. They belong to a New World explicitly advertised as a fiction that comes about from fancying the juxtaposition of two antithetic worlds: "a Fiction ... made as an Appendix to my *Observations upon Experimental Philosophy* ... having some Sympathy and Coherence with each other ... joined together as Two several Worlds, at their

Two Poles" (124). So everything in and about these two worlds—one of reason and the other of imagination—is immersed in mixture, from the original conceit of twinned globes to the narrative of the actual business of discovering and explaining what is in them, which itself is modeled on a mixture of three genres of fiction: "romancical," philosophical, and fantastical.

In many places, Cavendish's disagreements with Bacon and his disciples refer to coalitions of motion and matter, causing the hybrid manifestations of light, heat, cold, and air; for she has blended Hobbes's materialism with Locke's skepticism about the capacities of the human sensorium. So she tells the bear-men to break their telescopes and microscopes, for such machines are "mere deluders" and will never let the eye pierce beyond the surface of things. However, she relents and allows them to be kept not for knowledge but for entertainment, cherished by the bear-men as "our only delight and dear to us as our lives" (142). The machines of knowledge are redefined then as toys, flattering and soothing the senses instead of correcting them—not quite imposture, but certainly not truth. The concession is not anomalous in a world where the limits of reason are shrewdly assessed, and the usefulness of fiction, if only as a recreation of the mind, is justified in the preface of the work itself. There is no ulterior motive in all of this, for if the end of reason is evidence fit for narrow human capacities, the end of fancy is only fiction. And as fiction affirms nothing, it cannot lie.

Addison will attempt some years later the same vindication of aesthetic pleasure that Cavendish is allowing here, especially in the conversation between the Empress and the worm-men, who teach her that color is not an accident lacking being, a mere delusion like the products of optical glasses, for "there is no body without colour, nor no colour without body, for colour, figure, place, magnitude and body, are all but one thing, without any separation or abstraction from each other" (Cavendish 1994: 151). So she corrects the awkward bear-men when they try to tell her that blackness is privation of light: "The Empress replied, that if all colours were made by reflection of light, and that black was as much a colour as any other colour; then certainly they contradicted themselves in saying that black was made by want of reflection" (143). Everywhere in her romance, Cavendish sets out the basis of corroboration, emphasizing the importance of mixture and hybridity to pleasures, which, as Hythloday reports of More's Utopians, are never to be despised, especially those "that be received by the eares, the eyes, and the nose" (More 1898: 106). So the union of imagination and sensation that makes its way sidelong into Bacon's fiction of virginal purity and innocence, is here emphasized and plainly enjoyed.

In Henry Neville's *The Isle of Pines*, the presentation of a utopian polity set in the South Seas begins like others, with a scorbutic ship, but after that, issues suppressed or only glanced at in the Bacon's and More's utopias, but much to the fore in Cavendish's (e.g., fiction, mixture and hybridity) are very salient and infest not only the unresolved disorders of the island but also the layered narrative structure that conveys it. The first George Pine reports a voyage in which sickness and death overtake some of the crew, followed by a storm that wrecks the five survivors, four women and one man, on an island. Clothes and tools they get from the wreck, fresh food is abundant, and soon all the women are pregnant by Pine. Between them, they bear him forty-seven offspring, who soon multiply into 565, and then into 1789, forming four tribes or families. Three are named after surnames of the first mothers— English, Sparkes, and Trevor—and the fourth is called Phill, after the given name (Philippa) of the black slave from whom it sprang.

His grandson records that Pine's son, his father, was so outraged by the sexual licence that abounded in the island by his time, particularly among the men, that he went to war to subdue the worst offenders; after a decisive victory, he imposed law upon the rest by means of six commandments in defense of religion, marriage, property, and the state. Evidently the law governing sexual probity was not much regarded, nor for that matter the authority of religion or the safety of the state and property, for the grandson, William Pine, calls upon his visitors, Henry van Sloetten and his company, to help quell an insurrection led by Henry Phill that arises from Phill's rape of the wife of a senior Trevor. After restoring peace, the visitors leave the island and return to Holland, where van Sloetten expects that some of his readers will refuse to believe his story, accounting it nothing but a fiction. Nevertheless he recommends the island to public notice as a settlement eminently worthy of expansion and development.

There is a set of contradictions lodged in the narrative that make it credible in spite of itself. The reason for the initial lifting of the sanctions against adultery, bigamy, and incest was for the continuity of the population, or so it is strongly urged by William, who says that his father was forced to institute the law against adultery because the descendants of the original Pine "did for wantonness ... what my Grand-father was forced to do for necessity" (Neville 2011: 22). But that is not what George the patriarch says, rather that "idleness and fulness of every thing begot in me a desire of enjoying the Women ... custome taking away shame (there being none but us) we did it ... as our Lusts gave us liberty" (19). The only difference between lust and wantonness in the beginning and what takes place later is the number of people involved. Otherwise, the standing laws against adultery, polygamy, and incest apply to all of them and always did, so that the sins of the many ("whoredoms, incests and adulteries" [22]) are no less odious than those of the few. We don't hear much from Henry Phill, but doubtless this is the theme of his defiance of authority—an authority, what's more, too weak to enforce its own sanctions without help from outside.

Everything that is justified in the Isle of Pines by appeal to the rule of necessity, property, religion, and authority is in fact subject to contingency and sin. Of the four women, George was least inclined to sleep with Philippa, the black slave, but when she crept into his bed in the dark to beguile him, he knew who it was, "yet being willing to try the difference, satisfied my self with her" (20). The patriarchal line that descends through the men who call themselves Pine is braided with the matrilineal tribes of the four mothers, a genealogy mightily confounded with the signature of illegitimacy, being composed of the natural children who can be called only by their mothers' names. That sign is most deeply imprinted on the Phills, who lack even a surname for their grandmother and are all of mixed blood in various proportions. Thus William Pine is the son of Henry Sparkes, who also called himself Pine, usurping a name as an enterprising person might usurp the surname King. This is something Henry Phill clearly means to do, precipitating a political crisis by an act of sexual anarchy that, far from an assault on the principles of the commonwealth, has been responsible for the founding of each tribe on the island.

A series of observations is made on this state of affairs, starting with the grandson who explains why this is not a utopia: "As it is impossible, but that in multitudes disorders will grow, the stronger seeking to oppress the weaker; no tye of Religion being strong enough to chain up the depraved nature of mankinde, even so amongst them mischiefs began to rise, and they soon fell from those good orders prescribed them by my Grandfather" (22). The consolation is at hand, however, for "as the Seed being cast into stinking Dung produceth good and wholesome Corn for the sustentation of mans life, so bad manners produceth good and wholesome laws for the preservation of Humane Society" (23). By accident and indirection, then, you find direction out and, as Mandeville is soon going to explain in his *Fable of the Bees* (1724), "Good springs up and pullulates from evil." However, when he asks van Sloetten to intervene in the war against the Phills, William Pine explains that "where the Hedge of Government is once broken down, the most vile bear the greatest rule," indicating that for every discovery of an alleged principle of order there is a contingency or a vile action ready to ruin it, and that it is the anarchic force of illegitimacy, not dynastic succession, that rules the roost

(27). Without a wholesale extirpation, the Phills will rise again and perhaps succeed in seizing the Pinehood of the island—a parallel emergency abridged by Gulliver's Houyhnhnms, who identify the same bastard seed of insurrection in their guest and order him to swim elsewhere.

The illegitimacy of succession is used by Neville to emphasize the length to which contingency and sin, stemming from a scorbutic landfall, can carry a state whose rule of law is not reinforced by sovereign power. It imitates the "swing" of Sir Kenelm Digby's bean plant, when he says of its germination, "It follows presently its own swing; and in that little natural body, we may read the fate which hangs over Political ones, when the inferior members ... have gotten the power into their hands: for, then every one of them following their impetuous inclinations, the whole is brought into confusion ... unless a superiour Architect ... come[s] to draw light and order out of that darkness and confusion" (Digby 1669: 21). The natural or bastard condition of the population of the Isle of Pines introduces this identical confusion because the architecture of the law is not strong enough to contain the pressures of natural as opposed to social reproduction, resulting not in a community or even a coherent tribe, but what Hobbes calls a "crowd" or multitude, of which he said: "There will be nothing about which the whole crowd, as a person distinct from every individual, can rightly say, this is mine more than another's. Nor is there any action which should be attributed to the crowd as their action" (Hobbes 1998: 75–76). Such unresolved disorder provides an analogy not only with the life force of a bean and the moral disorder of romance but also with the activity of imagination in a scorbutic reverie or with the inchoate situation in which scurvy supervenes when knowledge of where, when, and who one is all disappears; or with devices for juggling and deceit when they are not (as always they are in Salomon's House) strictly subordinate to the pursuit of knowledge; or with the chaotic relativity of dimensions and kinds that turns Gulliver at length into a centaur; or with the experience of that "Bastard of many Parents," scurvy itself, which partakes of all other maladies in order to leave its victims subject to the incomparable spirals of their own sensations. Through the distorting lens of the actions and testimonies of the three Pines, each compounded by the naiveties of van Sloetten's narrative and the short commentary of Abraham Keek, it is possible to read *The Isle of Pines* as the unwitting political testament of the illegitimate energy of the Empire of Love, which is opposed not by innocence but by arbitrary laws backed by weapons. It is Lancelot and Guinevere, Tristram and Iseult, all over again.

As John Scheckter points out, there is a problem of textual as well as political succession in *The Isle of Pines*, a "full-scale estrangement from known structures of knowledge, as conventions of location and duration break down" (Neville 2011: 34). Samuel Johnson gets a whiff of this disorientation when he reads novels with characters in them such as Tom Jones and Robert Lovelace, figures so equivocally conceived that the colors of right and wrong are confounded in everything they do. They are the creatures of authors who, pleased with this sort of confusion, "instead of helping to settle their boundaries, mix them with so much art, that no common mind is able to disunite them" (Johnson 1791: 1.17). Such characters are the illegitimate offspring (literally in Tom's case) of an indiscriminate taste and what he calls "promiscuous description" (16), a kind of literary adultery. Now the effect of such license in *The Isle of Pines* is akin to what we shall observe in *The Travels of Hildebrand Bowman* (1778) when the career of the Dutch captain John van Trump is found to run so closely parallel to that of the real (as opposed to the fictional) Tobias Furneaux, or when the disease of nyctalopia is defined as the opposite of what in fact it is.

The original Pine says in his testimony that they sailed into the Indian Ocean, where some of the crew died of sickness before the ship, cruising near the island of St. Laurence, was overtaken by a great storm that swept them "out of our own knowledge" and plunged them into darkness until he and the four women were wrecked on the island (Neville 2011: 17). Introducing his own account of how they came across this remarkable document, van Sloetten writes that his ship came into the Indian Ocean near the island of St. Laurence when a violent storm blew up and lasted for a fortnight, during which time many of his men fell sick, and some died, "the Weather being so dark all the while, and the Sea so rough, that we knew not in what place we were," until suddenly they came across the Isle of Pines (14–15). Whether Pine's narrative is the counterfeit of this or the other way round, it serves to show that the disorientation caused by the loss of position in space and time includes the uncertainty of authorship and weakness of judgment. Van Sloetten sees his task to proclaim the benefits of a commonwealth whose ruin began in exactly the same way as his own narrative: a scorbutic and ultimately a sinful confusion in which evidence of facts tries for a doctrinaire eminence, only to end up as fiction.

Erasmus Darwin lists all the varieties of reverie under the heading "Insensibility of the mind to the stimuli of external objects" (Darwin 1801: 3.361). The opposite pathology is "Increased Action of the Organs of Sense" and belongs to his fifth genus, Irritation. Yet in scurvy, the two are often compounded. This is the theme of a little known subtopian fantasy that grew out of an event which took place in Grass Cove in Queen Charlotte Sound, on the northern coast of New Zealand's South Island, in December 1773. In

November of that year, the *Adventure* (sharing a name with Gulliver's last ship) Tobias Furneaux commander parted company with the *Resolution* during a storm and was never reunited with her. The vessel anchored at Ship Cove, where the crew had stayed six months earlier trying to recover from a severe outbreak of scurvy. This time they experienced bad weather, suffered problems with spoiled food, and were heading for more problems with scurvy on their way to the Cape, which accounts for Furneaux's decision to send the cutter to a nearby bay called Grass Cove just before they left in order "to gather wild greens for the Ship's Company" (Cook 1961: 745). While they were there, the crew of the cutter ran foul of a large body of Maori that attacked, killed, and commenced eating them. Fanny Burney's brother James discovered the carnage and fetched away two hands, a head, a pair of trousers, a frock "& 6 shoes—no 2 of them being fellows," a dismal and disturbing repetition of Crusoe's catalogue of ownerless gear on the shores of his island (752). While they were rowing away, he thought he could hear someone calling from the shore, but got no answer when he shouted back, "& indeed I think it some comfort to reflect that in all probability every man of them must have been killd on the Spot" (ibid.; Fig. 27).

In 1778, William Strahan and Thomas Cadell, publishers of Hawkesworth's collection of the British voyages of the 1760s and of Cook's second and third voyages, brought out an anonymous narrative called *The Travels of Hildebrand Bowman*, "Written by Himself, Who went on shore in the *Adventure*'s large Cutter at Queen Charlotte's Sound New Zealand the fatal 17th of December 1773; and escaped being cut off, and devoured, with the rest of the Boat's crew, by happening to be a-shooting in the woods" (Anon. 1778: titlepage). The alleged eleventh man in the cutter identifies himself as a midshipman aboard the *Adventure* but so far no one has been credited confidently with the authorship of what is a clever blend of Condillac's *Treatise on the Sensations* (1754) with a travesty of stadial theory, for the story begins with his witnessing the slaughter and eating of his colleagues and then follows his escape via various communities, each exhibiting a different sensory specialization until he ends up back in London.

The first and most primitive people Bowman comes across are the Taupinierans, who see very well in the dark and go fishing by swimming underwater and taking their prey in their teeth, like Swift's Yahoos. The Olfactarians are next, who smell things more keenly than we do and who hunt like dogs with their noses. Then Bowman meets the Auditantes who cannot bear harsh sounds, and afterward the Bonhommicans who have the sixth sense of conscience or moral sense (214), ending up with the Luxo-voluptans whose taste, or palate, is highly refined and devoted to aesthetic and erotic refinements:

Fig. 27. Grass Cove, Marlborough Sound, New Zealand. Courtesy of Mark Adams.

"We also have [the sense] of Touch or Feeling in as exquisite a degree as human nature is capable of supporting, without turning pleasure into pain; especially in the commerce between the sexes" (267). Bowman's travels take him from the most primitive sensations to the most sophisticated, from the sharp eyesight that allows the Taupinierans to catch fish in the dark to the luxuries of Mirovolante, the capital of Luxo-volupto, which are clearly recognizable as belonging to modern London, where the great actor Garimond (Garrick) acts in plays by Avonswan, and everyone who can afford it is given over to adultery, scandal, gaming, and masked balls (282). On the way to this dissolute culmination, Bowman has paused in Bonhommica, situated in the land of the aboriginal Auditantes, a community of Elizabethan simplicity and candor apparently settled on the coast of North America; and it is to Bonhommica that he decides to return at the end of his adventures, having by now discovered the Great Southern Continent, which he thinks ought to be called, after him, Bowmania or Hildebrandia (397).

Although the author of *Hildebrand Bowman* has a pretty clear idea of the route of the *Adventure* and the details of the Grass Cove massacre, he departs (deliberately it seems) from some details, of which the most important from my point of view concern scurvy. He agrees that the cutter was sent to gather scurvy grass and wild greens, but nowhere does he acknowledge the difficulties with spoilt food and scurvy experienced by the *Adventure*. In fact, he makes a point of praising the shrewdness of the English navigators of the Pacific for lowering the dependence on salt meat by victualing their ships with a variety of foods and by paying close attention to cleanliness and ventilation. The Dutch captain who carries him back to Bonhommica, John van Trump, is presented as different from Furneaux, for although he has had a boat's crew cut off and eaten by cannibals, his men are scorbutic, and they are hurrying home "before the scurvy … quite disabled them" (379). In fact, the fictional van Trump and the historical Furneaux are exactly alike; so when van Trump asks Bowman about the British antiscorbutic system, "I gave him a fully account of our provisions and management.... I gave the surgeon instructions how the stores were to be managed," he is claiming an expertise no one on the *Adventure* possessed. His success is reported as the argument to the ninth chapter: "The Scurvy abates on board the Harlem Frigate, from the use of Malt and other things Moraveres [the Bonhommican skipper] spared them" (397).

At the same time Bowman, makes a strange mistake about nightblindness, often an accompaniment to scurvy caused by the lack of vitamin A, when he says apropos the Taupinierans (who like owls can see at night but are blind in daylight): "I have been told by a learned Physician since my return … that there is a disorder in the eyes (but a very rare one) called nycta lopia, which exactly resembles the sight of this species of people" (63). Although he gets the symptom entirely the wrong way round, Bowman is as fully alive as Trotter to the hyperactivity of the senses that seems always to accompany scurvy; and he is closely attentive to the cultural consequences fostered by each, with keen eyes and noses low in the scale of civility, and ears at the top along with the moral sense, while touch and taste lie somewhere between. The mistake about nyctalopia, which in fact makes the eye blind in poor light, seems to be linked to the overlapping of Bonhommica and Luxo-volupto as virtuous and corrupt versions of Britain, and to the curious mirroring of van Trump and Furneaux as the negative and positive exemplars of ship management, the one running a scorbutic vessel the other a clean one. Every "x" is accompanied by a "non-x", every affirmation by its negation. This twinning of positive and negative qualities is encouraged, it seems, by the presence of scurvy and the magnified sense-impressions that attend it.

When an alignment is made between scorbutic susceptibility and a utopian experimental or cultural model, fiction has a strange part to play, for while it offers to provide a Baconian mirror of the truth, it is also symptomatic of the opposite. Bacon's houses of deceits, Cavendish's frank acknowledgement of having invented a world that does not exist, the curiously improbable repetition of George Pine's voyage in van Sloetten's in *The Isle of Pines*, and Bowman's manifest falsification of the facts of the voyage of the *Adventure* are all examples of the real case being misrepresented. Yet the correlation with known facts is not hidden but transposed, so that the falsehood conspires with a truth being told in the wrong way or by the wrong person, allowing room for some sort of veracity to have a place. Thus van Trump is dogged by the same scorbutic emergencies that afflicted the actual voyage of the *Adventure*, now fictionalized as a ship whose antiscorbutic regime is a model for all others; the fish-men discourse objectively of amphibian creatures in *The Blazing World* as if they were a singular species. The doublings don't frame the truth, they mimic its alternation with its opposite, the "non-x" with the "x."

The author of *Hildebrand Bowman* was making an addition to a minor genre of voyage literature initiated after Anson's expedition. Fascinated by the accounts of the postscorbutic delights enjoyed by his men on Juan Fernandez and Tinian, Gabriel Francois Coyer wrote *A Supplement to Anson's Voyage round the World* (1752), in which he enlarged fantastically on the pleasures of the island. When the British sailors land, they are conducted by a slender Francophone connoisseur ("the Comptroller of Fashions") to a province at the center of Juan Fernandez called Frivoland, where they find an extensive variety of curiosities and bibelots, all superficially exquisite and utterly useless. The fruits are made of painted papier mâché, the trees are finely shaped but as brittle as glass, the horses are too frail to bear a rider, and the melodies of the birds are so wonderfully refined they cannot be heard. Trying to imagine what kind of trade might be driven between this distant land and the metropole, Coyer suggests "delicious Romances ... Operas fraught with melting Love" (Coyer 1752: 19). Coyer has in his eye no doubt the luscious descriptions Richard Walter provided of the paradise of Juan Fernandez and the mysterious splendors of the deserted island of Tinian, contributing a satirical treatment of the theme of ecstatic pleasure that is common in this branch of fiction. Diderot wrote his *Supplement au voyage de Bougainville* (1796) to make gentle mockery of Commerson's and Bougainville's enthusiasm for the sexual liberty of the Tahitians, and of the eagerness with which a European audience greeted Bougainville's picture of a Polynesian Venus rising from the sea to stand in resplendent nakedness before men long starved of all the comforts of life. Commerson's own letter describing the pleasures of this "isle of Cythera" as utopian appeared in an anonymous collection with the confusing title, *Supplement au voyage de M. de Bougainville, ou Journal d'un voyage autour du monde fait par MM Banks et Solander* (1772). So supplements, usually fictional embroideries of real voyages, could either reinforce or undercut the extravagant memories of scorbutic sailors. Horace Walpole is making supplemental fun of John Byron's misrecognitions of Patagonian giants in his mock-utopia called, *An Account of the Giants lately discovered* (1766). But Rousseau was doing the opposite with his *Julie, ou la nouvelle Heloise* (1761) when he took (like Coyer) Anson's voyage as a framework. The disappointed lover St. Preux joins the expedition to the South Seas, and when he returns, he explains to Julie that Juan Fernandez is "an asylum to innocence and persecuted love" and that Tinian, even more deserted and more delightful, has tempted him to lead there a life of charming solitude. When he enters her garden, he remembers (rather like Rousseau himself in his reveries) these desert islands: "I thought myself in the most wild and solitary place in nature, and I ... cried out in an involuntary fit of enthusiasm, O Tinian! O Juan Fernandez! Eloisa, the world's end is at your threshold" (Rousseau 1803: 2.132–33).

Bernardin de St. Pierre's romance *Paul and Virginia* ([1788] 1819), based on his visit to Mauritius in 1768, might be termed self-supplemental to the extent that it transforms scenes that at first caused the author intense disgust into a paradise where an idyll of young creole love is destroyed by metropolitan meddling. It is also a celebration of the innocence of bastardy: a story with a tendency quite the reverse of Neville's, but exhibiting strong affinities with Rousseau's reveries of being gloriously cast away on a desert island. "You are a natural child," Margaret tells her son, "you have no legitimate father ... you have no relation in the world but me" (Bernardin de St. Pierre [1788] 1819: 93). Paul's skin is dark, he has no second name, but the biological or natural energy that turns Henry Phill into a rapist and a rebel is focused by him lovingly and exclusively on Virginia de la Tour, daughter of a woman disinherited by her rich and ancient family for marrying beneath her rank. They grow up together in an isolated valley of Mauritius amidst trees, flowers, fruits, and birds that charm every sense. Paul's partner, like Adam's, is inseparable from the beauty of this natural abundance. He tells her, "Something of you, I know not how, remains for me in the air where you have passed, in the grass where you have been seated. When I come near you, you delight all my senses.... If I only touch you with my finger, my whole frame trembles with pleasure" (68–69). They come to maturity like plants, sensing that infant love is growing into pubescent passion, but not knowing what it is. Their love is inseparable, therefore, from the sensations of what Humboldt called plant geography, such as the exquisitely painful sympathy he himself felt for the

Andean waxwood palm, or the softer sensations that Bernardin de St. Pierre enjoyed amidst the wonders of a strawberry flower. It is when she is bathing in a pool Paul built for her that Virginia first feels the full force of her passion for him and her indistinct desire to consummate it, as if she were one of Erasmus Darwin's flowers in *The Botanic Garden*:

> She saw, reflected through the water upon her naked arms and bosom, the two cocoa-trees which were planted at her birth and that of her brother, and which interwove about her head their green branches and young fruit. She thought of Paul's friendship, sweeter than the odours, purer than the waters of the fountain, stronger than the intertwining palm-trees, and she sighed. Reflecting on the hour of the night, and the profound solitude, her imagination again grew disordered. Suddenly she flew affrighted from those dangerous shades, and those waters which she fancied hotter than the torrid sun-beam. (74)

> [Elle entrevoit dans l'eau, sur les bras nus et sur son sein, les reflèts des deux palmiers plantés à la naissance de son frère et à la sienne, qui entrelacaient au-dessus de sa tête leurs rameaux verts at leurs jeunes cocos. Elle pense à l'amitie de Paul, plus douce que les parfums, plus pure que l'eau des fontaines, plus forte que les palmiers unis, et elle soupire. Elle songe à la nuit, à la solitude, et un feu dévorant la saisit. Aussitot elle sort effrayée de ces dangereux ombrages, et de ces eaux plus brûlantes que les soleils de la zone torride]. (idem 1907: 61)

Here is a version of Ovid's Salmacis in her pool, or even of the young Yahoo girl in Gulliver's fourth book, both immersed in water and desiring a man nearby, but Bernardin de St. Pierre has purged the scene of any trace of sin, appetite, or comedy. Even the raw anarchy of Digby's bean plant, transferred so powerfully to Neville's island of bastards, is brought to heel by the girl's gentleness and fright. The serpent in this garden is the agent of Virginia's great-aunt, M. de la Bourdonnais, who arranges for her transfer to Paris, where she is to be inducted into a social rank from which poor Paul, a natural child in all senses, is entirely excluded. He is disqualified from a natural union by the laws of social succession which, as it turns out, are going to disqualify Virginia, too. The disappointment of the two lovers is spun out through the rest of the story as the perverse intentions of her great-aunt bear their bitter fruit, interfering with the biological swing of natural love first by separation and finally by Virginia's death at sea, causing Paul to die of grief soon afterward. By the time Mrs. Pexton visited the tomb of the lovers in 1817, the fiction had become history, and the paradisiacal valley a tourist attraction; but it was still faithful to a love rooted in plant life: "They are interred under 2 large trees planted at the time they were born" (Pexton 1817: n.p.).

Innocent love thriving amidst the vegetation of an isolated valley on a distant island is a fictional realization of Rousseau's and St. Preux's reveries, and like them, it lays a strong botanical emphasis on the unique flora of the place. Instead of thinking like Sir Joseph Banks and Carl Linnaeus of acclimatizing plants from the New World, or believing like Spenser that stock needed to be restored in metropolitan seed-banks such as his Garden of Adonis, Bernardin de St. Pierre shows his human plants renewing themselves at the periphery of the globe. However, as Diderot explained in his fictional enlargement of Bougainville's journal, the utopian promise of sexual love liberated from metropolitan restraints and hypocrisies that had fascinated French readers was itself a metropolitan fantasy, a sort of alibi for the promiscuity of the salon romances and for sexual adventures further afield. Commerson, Bernardin de St. Pierre, Kerguelen, and Pierre Poivre seem to have endorsed a cult of free love on Mauritius, closely linked to their botanical interests at the garden of Pamplemousse. Bernardin de St. Pierre had an affair with Poivre's wife, and Kerguelen learned from Commerson to carry a female companion on voyages of discovery, resulting in the hideous scenes of scorbutic jealousy on the second of the two voyages he made to his twice-discovered island in the Indian Ocean.

However, the real source of negative energy fueling this lyric account of arboreal innocence is to be found in Bernardin de St. Pierre's first landfall at Mauritius, when he was sick with scurvy on a ship with seven dead and eighty prostrate by the time it made Port Louis. Weary in his mind and "disgusted with everything," he said the island looked more like a colliery than a paradise. He described encounters with the flora and fauna very remote from the sensuous pleasures of the young lovers:

> Not a single flower adorns our meadows.... No plant bears flowers of a pleasant smell, nor is any shrub in the island to be compared to our white thorn. The *liannes* have not the fragrancy of honeysuckle or ivy. Not one violet in all the woods. As to the trees, they have large whitish trunks, that are bare, except a little kind of nosegay of leaves of a dull green. (Bernardin de St. Pierre 1800: 66)

The grass is full of prickles, some of the trees stink horribly, there is very little birdsong, just the croak of parrots and the shrilling of monkeys; the shrubs are venomous and the fish poisonous. The bark of the mapou tree inflames his throat; there is a pigeon whose flesh throws anyone who eats it into convulsions and a butterfly whose dust will blind you. He concludes, "Every sentiment of humanity is here depraved, nay, I may say extinct" (107). One can only conclude that the romance of *Paul and Virginia* emerged as a memory that didn't so much soften the acerbities of his first sensations on Mauritius as transfer their strength into the opposite register, from "x" to "non-x." Like William Hodges, Bernardin de St. Pierre uses the shapes and colors of a repellent scene of nature for a palimpsest of verdurous warmth; and like the Ancient Mariner, he turns horror into beauty, and then into tragedy. As the melancholy of the forlorn Paul deepens, he ends up feeling the same disgust for the island, now made foul by a hurricane, that distinguished Bernardin de St. Pierre's first acquaintance with it: "Every thing repulses me" (Bernardin de St. Pierre [1788] 1819: 127).

These switches of value and traverses of passion, so typical of utopian fantasies and their supplemental additions, are extreme in proportion as the intensities of pleasure or aversion ignore the mixture of passions and qualities on which inevitably they are based. For this reason, the sharpness of the division between the positive and the negative passions in *Paul and Virginia* appeals very powerfully to the imaginations of the first navigators and settlers of Australia. Pausing in the midst of his scorbutic circuit of the continent, Flinders took a walk at Coupang through the estate of the widow of the man who was governor when Bligh landed there: "I could not prevent my ideas from dwelling upon the happiness that a man whose desires were moderate might enjoy in this delightful retreat with the beloved of his heart; for here the summer sun could not scorch, nor was there any dread of winters cold." Almost immediately, Flinders shakes himself awake and predicts how his fantasy would conclude, "In the end I should fall a sacrifice to surrounding circumstances and become that mere inactive animal, or rather vegetable—a native of Timor" (Flinders 2015: 2.337). The only way the narrator of *Paul and Virginia* can deal with such violent antinomies is by transcending the senses and their contradictions. At the end of the story, the narrator makes a Cartesian sacrifice of empirical knowledge for a truth more bodiless: "My soul intuitively sees, tastes, hears, touches, what before she could only be made sensible of through the medium of our weak organs" (Bernardin de St. Pierre [1788] 1819: 162). As for Flinders, he returns to his original passion, one that will cost the lives of nine of his men, "that genuine spirit of discovery which contains all danger and inconvenience when put in competition with its gratification … that ethereal fire with which the souls of Columbus and Cook were wont to burn" (Flinders 2015: 2.311–12).

In his *Hygeia*, Beddoes selects a single example of fiction infected by the debility of exhausted nerves: Swift's *Gulliver's Travels* (1726), a narrative sitting as squarely at the center of the history and literature of scurvy as *The Rime of the Ancient Mariner*. This is what he says about it and its author:

> In his Gulliver, where his genius bursts out with such transcendent splendour, it still shines as if surrounded by a halo of malignant dissatisfaction. There is something in different parts of that work, from which the heart recoils.... Cut off from the participation of pleasure, he might serve, envy, and hate his species, till advancing years, perpetually irritating his secret sufferings, plunged him in the madness of misanthropy. Strong evidence ... is to be found in the delight he takes in images, "physically impure." (Beddoes 1802: 3.190)

Evidently, Beddoes believes that Swift's sexual capacity was disabled by excess, whether culminating in sheer exhaustion or venereal disease he doesn't say, but he sums it up in suitable romantic terms: "He entered the lists of love with all the flourish of a dauntless champion, but when he had proceeded to a certain point, threw down his arms, and turned his back like a recreant" (191). The "certain point" he indicates is Swift's own retreat from sexual intimacy when it reached the stage of apparent inevitability—with Esther Vanhomrigh and Esther Johnson, for instance—a démarche emphatically dramatized in Gulliver's hysterical refusal of sexual intercourse with the Yahoo female in the fourth book and his revulsion when his wife embraces him.

Beddoes reinforces the connections between Swift's misanthropy and the sinking state of national health in a letter where he imagines how the common level of credibility would be determined under a national regimen of perfect health:

> If there should ever be a state of things in which men employed their knowledge and powers for the advantage of their health, and faithful descriptions of such valetudinarians as abound among us, should fall into the hands of the people of those days, they will have as much difficulty in believing in their reality, as we have of the creatures so abominably distorted from humanity in *Gulliver's Travels*. (Stock 1811: lxiii)

He may be alluding to the Struldbrugs of the third book, but it is more likely he refers to the Yahoos of the fourth as the limit case of what bad diet, putrefying innards, stench, repulsive physique, and general malignancy is capable of producing by way of metamorphosis from the human into the beast. What he means to assert is not just an analogy but a causal connection between the tendency of nervous disease and the nonhuman images formed by a corrupted imagination, as if the attempt of the misanthrope to picture the vileness of the human species were in fact a self-portrait, the image of a negative Narcissus whose miserable self-love is shown to be as grotesque as it is incredible to anyone enjoying a normal state of body and mind.

So I shall briefly examine the fourth book of *Gulliver's Travels*, bearing in mind Trotter's analysis of the infectious potential of romance and Beddoes's objections to Swift's maritime utopia, namely that it originates in the bitterness and frustration incident to a valetudinarian rejection of normal sexuality. Like the other narratives I discuss in this chapter, Gulliver's fourth book begins with a scorbutic ship entering the South Seas. Now commander of the *Adventure*, Gulliver loses some of his men to calenture, although he doesn't specify their symptoms. After their replacements foment a mutiny and imprison their captain, more men die of a sickness that is now inevitably scurvy. Eight months at sea with a couple of short stops on its way to an island off the southern coast of Australia, the *Adventure* has sailed well within sphere of acute nutritional deficit, where Gulliver's "Victuals and Drink" could not by any stretch of the imagination have included greens or lemon juice. When marooned by the mutineers on the island of the horses, his sojourn begins with two scenes of misrecognition, each the reverse of the other: creatures later described as examples of "a perfect human figure" are mistaken for animals, and horses are mistaken for metamorphosed humans. Gulliver refers this predicament to a reader who "will easily believe that I did not much like my present Situation" (Swift 1995: 212, 209). It is one that introduces in its most drastic and radical form the sense of foreignness that permeates journal descriptions of these regions (Dampier's description of the coast of Australia for example, or Bernardin de St. Pierre's of Mauritius), as well as the "reverse gaze" convention of spy literature, oriental travelers, and it-narratives, where narrators are assumed to be entirely alienated from what they see and describe (Ballaster 2005: 149). But between Gulliver's entry into the strange world of the island and his emergence from it in an Australian cove, where the Portuguese sailors find him in a coat made of Yahoo skins, flat on his face behind a stone, and trying as hard as he can to be invisible, the reversal of the gaze is complete, with the familiar now as strange as what was truly strange before. They laugh at his speech, which is like the neighing of a horse, and when they talk, Gulliver says, "It appeared to me as monstrous as if a Dog or a Cow should speak, or a Yahoo in Houyhnhnm-Land" (430). It is clear that somewhere along the way to this colossal failure of communication on the shore of a New World, Gulliver has been metamorphosed into something like a horse, bereft of any sense of the necessary mutual antagonism of "x" and "non-x." The talent he maintained more or less throughout the first three books for negotiating the difference between what is empirically certain and yet conceptually impossible, and for representing the affective frictions precipitated by such miracles, has now been entirely abdicated. It has been replaced by a set of binary oppositions, of which horse and human define the unbridgeable positive and negative poles.

Three real-life reveries, two of them scorbutic, beg comparison with Gulliver's: Rousseau's seventh walk, Bernardin de St. Pierre's landing at Mauritius, and the zoologist Francois Peron's sudden love of seashells. All of Rousseau's walks are premised on his isolation from a world that has now become alien to him, his only reliable companions being plants, whose "sweet smells, bright colours and ... elegant shapes" vie for his attention (Rousseau 2004: 109). He feels himself isolated not only by the extent of his absorption in these delicious sensations but also by the degree of their disinterestedness. He is not fond of herbs and flowers for their medicinal value, what learned men call "the study of properties." He looks at them instead as charming accidents, living and dying to themselves, with no providential relevance to human life. This is Rousseau's refuge, plant life that is aesthetically satisfying without any shred of human utility to justify it. The thicker the congregation of plants, the more perfect his illusion that he has found his desert island, "a sanctuary unknown to the whole universe" (118). This discovery is all a joke because he finds he has pitched his idyll next door to a stocking factory, but the tight association of the noninstrumental intimacy of greenery and flowers with his forgetfulness of himself as a civil person is very similar to Gulliver's identifying as a horse. Even their gestures, excluding the foreign world of humans, are alike: flat on their faces, pressing themselves into earth so that the alien beings of whom once they were one might never see them again (117).

Bernardin de St. Pierre's arrival in Mauritius is exactly the reverse of Rousseau's penetration of the dense and welcoming vegetation of the Alps. He finds all nature there offensive to the senses: "The poison is in the island itself" (Bernardin de St. Pierre 1800: 85). But he has arrived with scurvy, so he brings half the poison with him. In this respect, the parallel with Gulliver is close, whose first sensation on being cast away on his island is disgust, caused by the smell of shit shed by Yahoos from the tree under which he is sheltering. Bernardin de St. Pierre's discovery that he can hold no fellowship with the plants and birds of Mauritius is his scorbutic contribution to natural history; and Gulliver's judgment of his own kind, strongly reminiscent of Dampier's contemptuous description of Aborigines, is no less sweeping an addition to the anthropology of the Southern Ocean: "I never beheld in all my Travels so disagreeable an Animal, or one against which I naturally conceived so strong an Antipathy" (334). Their smell, the color of their skin, the grossness of their sexual parts and proclivities, the rottenness of their food, the malignity etched so vividly in their faces are all revolting, providing sensational examples of creatures utterly foreign to a standard that is yet fully to be embodied in the rational and stoical horses.

Peron's narrative oscillates between notices of the terrible condition of the crew of the *Geographe* and his new enthusiasm for shells. Heading for Tasmania with humidity at 97 degrees and their food rotten, they lose Sautier, a botanist, to scurvy; then Hubert, a gunner, followed by Pougens and Courrager, two sailors. Men are drinking their own urine, and twenty-five of them are unable to move; meanwhile, he is adding furiously to his collection of 100,000 specimens, chiefly mollusks, which he has arranged "according to a constant and regular plan embrac[ing] all the details of the exterior organisation of the animal, explain[ing] all its characters in an absolute manner [that] will, in consequence, survive all the revolutions of methods and systems" (Peron 1810: 163, iv). So human chaos is confronted by lucid systematic order in the sphere of seashells.

Four elements are in play in each of these accounts, and disgust defines each of them positively or negatively: these are plants, humans, animals, and shells. Gulliver and Rousseau are disgusted by humans in general; Bernardin de St. Pierre is disgusted by the trees and flowers of a particular island, nor is he overly impressed by its human population; Peron is disgusted by the scurvied ship and impatient with his commander Baudin as well as the animals which used to be the legitimate object of his expertise. Gulliver consoles himself with the company of animals, Rousseau with plants, and Peron with shells, while Bernardin de St. Pierre presently lacks any alternative to Mauritian fauna and flora, although eventually he will come to love what sickened him at first. Of the four, Bernardin de St. Pierre is in the optimum position because his disgust doesn't oblige him to espouse an alternative endowed with what his previous affiliations have been found to lack—simplicity, charm, fragrance, color, reason, or beauty. In fact, he shows that Rousseau's love of plants, pursued as recompense for human depravity, is hazardous if disgust is what you are trying avoid. Like Shakespeare of the *Sonnets*, Bernardin de St. Pierre knows that festering lilies smell far worse than weeds; and like the Ancient Mariner, he knows, too, or at least he doesn't not know, that slime-born serpents moving on the ooze of a rotten ocean can in a moment change into sleek and exquisite creatures of light. Rousseau has room to laugh at his own metamorphosis, cast away on the edge of a stocking factory, but Peron and Gulliver have none.

Fig. 28. David Jones, *Gulliver and the Female Yahoo*. © Trustees of the Estate of
David Jones.

This is the source of the injustice of Peron's account insofar as it concerns
Baudin; and it is the problem with Gulliver's fourth book. Both voyagers exhibit the exceptionalism of the landed scorbutic, who is so amazed and disgusted by every face but his own that he chooses another species for company.
The confidence of Peron's embrace of shells, in which, as a zoologist, he can
claim little prior expertise, many a time prompts him to desert his colleagues
and to ignore his rendezvous with them, as if he had found far finer company.
His behavior compares with the decisiveness of Gulliver's rapid adoption of
a finer species than his own, even though he has no secure footing in its norms
and can never hope to be its specimen.

If Gulliver's metamorphosis were complete, and he were a horse writing
of humans to horses, or even conjectured horses, then the reversal of familiarity into foreignness and vice versa would make the sort of sense that can be
found in Apuleius's *The Golden Ass*, David Garnett's *Lady into Fox* (1922), or
the fable told by La Fontaine of the cat turned lady who still went mousing
(La Fontaine 1865). In these examples, the metamorphosis has taken place
and the interest of the story lies in what follows. But Gulliver is stranded between the species he admires and the one he detests, making feeble attempts
at mimicry and disguise, such as clothes made of Yahoo skin and trotting
when he walks. It isn't until he finds himself giving an account to his master
of the cruel treatment of horses in England, only to find he is a liar here and
a madman there, that he understands he has no community in either place.
The scene with the female Yahoo (Fig. 28) emphasizes Gulliver's difficulty,
for it replays the story of Salmacis and Hermaphroditus in Ovid where the
diffident male who refuses the naked female who accosts him in a pool is
transformed for his want of gallantry into a mixture of man and woman,
neither one thing nor the other, or what Cavendish would call a horse-man
(Fig. 15). Gulliver's status is likewise that of a hybrid whose appeals to either

party—the human reader or the master horse—are vain because he will not acknowledge the mixture. He wants to be all horse and nothing of a Yahoo.

The two occasions when his insistence on his singular integrity collapses illustrate what Beddoes takes to be the pathological direction taken by the sexual proclivities of Gulliver and Swift: the encounter with the Yahoo female and his meeting with his wife when he returns to Redriff. In both instances, Gulliver is forced to acknowledge his humanity: of the incident in the river, he confesses, "Now I could no longer deny, that I was a real Yahoo, in every Limb and Feature, since the Females had a natural Propensity to me" (242). And at the domestic reunion, he has to remember what he has done with the female now clasping him to her bosom: "When I began to consider, that by copulating with one of the Yahoo-Species, I had become a Parent of more; it struck me with the utmost Shame, Confusion and Horror" (261). An equilibrium that would have been maintained had Gulliver's judgments been more provisional has now migrated from his sense of things to the reader's sense of Gulliver. Instead of the hero exploring his contrary impressions of the Lilliputians and the Brobdingnagians, he himself strikes the reader as a bastard or a centaur, and never more so than when he is insisting on the purity of his language (saying only the thing which is) or his bloodline (incapable of mating out of kind). How far Beddoes is justified in equating Swift's valetudinarian tendencies with Gulliver's, it is impossible to say, but within the symptomology of scurvy and reverie, it is possible to see how much of Gulliver's sense of "outness" has been lost.

Exactly what Beddoes's idea of a valetudinarian might mean to Trotter becomes clearer when he mentions the two significant alterations to the balance of the mind caused by scurvy. The first is a gloominess that grows extreme and very isolating, "He flies into hiding places from his duty, broods over his own feelings in solitude, and indulges the most gloomy ideas of his safety, as if hypochondriacal;" and then he "not only forgets all [his] old attachments but shews the utmost signs of dislike to those who had been most dear." The second is the infinite care taken by the victim to deliver at length and in great detail the exact degree and pressure of his sensations, "as earnestly conducted, as we sometimes observe hypochondriacs in relating their feelings, from any ruffle of temper occasioned by ... slight causes" (Trotter 1792: 44–45; 1804: 3.364). The valetudinarian, that is to say, is singular in the same way as a scorbutic seaman, a resemblance that extends to Rousseau on his imagined desert island, Gulliver in the land of the Houyhnhnms, and Peron on a shelly beach: each cast away, estranged from all that was familiar, repelled by how strange it now seems. Peron's love of seashells, like Rousseau's love of plants, takes the form of a purposeless and interminable accumulation, resembling the exhaustive lists of symptomatic particulars and peculiar sensations compiled by scorbutic hypochondriacs, all given over to contingencies while pretending to assemble the building blocks of some grand taxonomy.

Gulliver's demonstration has to do almost entirely with the details of his disgust, which he wants to instance, again and again, as proof of his exemption from the filth and corruption of his own kind. Gulliver is not destined to recover (like Bramble and Bernardin de St. Pierre) a sense of familiarity with what belongs to him, he can only stress how gingerly and precisely he must go when he is in its vicinity: "During the first Year I could not endure my Wife or Children in my Presence, the very Smell of them was intolerable: much less could I suffer them to eat in the same Room" (434). Under the authority of an absolute dualism, Gulliver fully realizes and inhabits the hypothesis behind Descartes's *Meditations*: "What if what I do know, I didn't?" But the intended "amendment" he mentions as the purpose of his narrative has nothing to do with truth, any more than the disgust he feels for his spouse has anything to do with her depravity. The whole point of the exercise is aesthetic, designed to mitigate the sensory insult of the smell and the sight of a human: and the sense organ he wishes to neutralize above all others is his nose, and secondarily, his eye. All he wants from his own species is that they not come into his presence. Unlike the Ancient Mariner, for example, Gulliver cannot find his way back from disgust to beauty.

How this affects his language bears on the failure of the ellipse in Gulliver's case, owing to the absence in his sensations of any twin impulse of repulsion and attraction. He really wants to be operating in a circle with a single center, but when he is thrown into an alien curve, as he is whenever he begins to talk of England and Europe, there is no order or end to his accumulation of horrid examples. Like Dampier and all the other visitors to Australia's shore, Gulliver enumerates what is not there, but for him it is an absence to be celebrated, not mourned. His negatives define the extent of the amenity of the Houyhnhnm utopia, just as Peron's shells provide the structure of a grand and immortal system of calcareous nature. Here is the beginning of one of Gulliver's many lists of foreign iniquities, none of which has any reason to stop: He could go on piling up negatives forever: "Here were no Gibers, Censurers, Backbiters, Pickpockets, Highwaymen, House-breakers, Attorneys, Bawds, Buffoons, Gamesters, Politicisans, Wits Spleneticks, tedious Talkers, Controvertists, Ravishers" (Swift 1995: 250). His attachment to negative statements of a positive case is of course typical of the rhetoric of scorbutic passion, but it infiltrates his prose in a most peculiar way once he is back in England trying to pretend that he can come to terms with his aversions.

There are two examples in his last chapter where litotes gets out of hand because it is being used to enforce what ought to be a clear affirmative. Here is the first: "I am not a little pleased that this Work of mine can possibly meet with no Censurers" (263). Why does he not say, "I am pleased that there is no possibility of meeting censurers of my work"? His attempt to expel all possibility of censure by means of negatives actually weakens the claim, leaving the chance of such a meeting up in the air. Here is the second example: "I dwell the longer upon this Subject from the Desire I have to make the Society of an English Yahoo by any Means not insupportable" (266). Gulliver believes he is making an urgent and unequivocal statement, but somehow he cannot say, "I desire to make supportable the society of a Yahoo by any means." He finds it equally impossible to say, "I find the company of English Yahoos insupportable." So what he offers in this farewell to the reader is a modification of the truth according to Houyhnhnms, saying the thing which is, by blending it with his own habit of heaping up negative instances of his ideal commonwealth ("Here were no Gibers, Censurers, Backbiters," and so on). He chooses to say, very awkwardly it must be confessed, the thing which is not not, or at any rate, not yet. Language itself seems to sidle up and crepitate against sensations that Gulliver had been determined to suppress, installing a "non-x" wherever our centaur-narrator simply wants an "x." Language, or Swift, or both, are faithful to the double sense of loss and pleasure that drives Ishmael to say that the man who perishes of calenture will sink "no more to rise forever." But Gulliver, whose professed interest in some sort of amelioration of humans is entirely disingenuous, cannot sustain that kind of elliptical energy.

In a fable ascribed to John Gay, Ay and No are about to fight a duel when Ay says to No, "Why then should Kinsfolks quarrel thus? / For, Two of You make One of Us" (Gay 1974: 2.379). This is a treaty often ratified and sometimes broken in the literature and fiction of scurvy. When it works, the possibility of corroboration grows stronger, that is to say when the passion associated with the lack of something meets and mingles with the passion associated with the supply of it. This can only occur when the link between an imagined satisfaction and the reality of it is sustained, sometimes in good faith and sometimes by subterfuge. What is certain is that a reverie totally sealed off from "outness" and the world at large, such as Gulliver's is, has no prospect of enjoying the only consummation scurvy affords.

* * *

In view of the shaping power scurvy has upon the imaginations of those involved in the first discoveries of the Australian coastline and in the early colonization of the hinterland, it would be apt to conclude this chapter, with a look at some of the fiction it produced. In her review of a book called *The Bondwoman's Narrative* (2002), a story told in the first person of the life of a female slave who eventually escapes to freedom, Hilary Mantel judged the blend of seeming fact and outright plagiarism of nineteenth-century fiction in the memoir to have been necessary for the author's psychological survival: "Long before she was free in fact, she had escaped in imagination. She had extracted herself from degrading circumstances and inserted herself into others, more flattering, as a persecuted heroine in a romance" (Mantel 2004: 430). A similar argument is broached in Timberlake Wertenbaker's play, *Our Country's Good*, a play based on the first performance of a play in Australia: George Farquhar's *The Recruiting Officer*, in 1789, performed by a largely convict cast a year after the First Fleet arrived. Dabby Bryant, a female convict, looks at the company at the end of the dress rehearsal and says, "I saw the whole play, and we all knew our lines, and Mary, you looked so beautiful, and after that I saw Devon and they were shouting bravo, bravo Dabby, hurray, you've escaped, you've sailed thousands and thousands of miles on the open sea, and you've come back to your Devon, bravo Dabby, bravo" (Wertenbaker 2015: 87). Acting in the play is like a dream or a reverie, or like using *Bleak House* (1853) to report a rainy day in Washington DC, for it can restore everything that is missing from the gross realities of exile and servitude: hope, action, speech, love, community, and applause. The importance of such a fiction does not lie at all in any supplementary fidelity to facts, it disputes them at every point: it turns female into male (Mary as Sylvia becomes a man), the cruel sergeant into the criminal, the criminal into a character with options, the tyrant into a lover, bastards into heroes, evil into good. It covers up anguish with something like innocence. These metamorphoses appealed to the convict players. *Richard III* and *The Poor Soldier* were performed in Norfolk Island in 1793, and during 1796, Sydney saw *Jane Shore*, *The Wapping Landlady*, and *The Revenge* (Russell 2011: v.).

Perhaps the most important word in dictionaries of flash language is "fakement" meaning booty, forgery, or deceit. The verb has more extensive meanings: rob, wound, shatter: "fake your slangs" means break your shackles (Vaux 1819: 2.171–72). Put the two together and the insurgent meaning would be something like, "a fiction designed as a breakout from the mind." The convicts at Port Jackson who walked north to find China, or who set off in small boats for Timor, were performing fakements by refusing to internalize the shocking dreariness of their situation. Forgery exerted the same fascination on Thomas Watling and Joseph Lycett: it was a release from the terrible burden of unrelieved originality. Fanny Davis, a convict who played Alicia in *Jane Shore*, had been transported for a crime committed while impersonating a man (Russell 2011: 10). Henry Savery, author of one of the first colonial novels (*Quintus Servinton* [1832]), and Mr. Micawber fashioned between them reverse fakements, the last of Servinton's many forgeries being a story of his return to England, a free man; while Micawber (possibly happier in his lie) informs his wife he is bound for "a distant country expressly in order that he may be fully understood and appreciated for the first time" (Scheckter 1998: 88, 25).

The model of criminal fiction for criminals, as opposed to the phony providentialism of the *Newgate Calendar* (1824) and the routine falsehoods retailed at the gallows and then printed as Last Dying Words, was the memoir of Bampfylde Moore Carew, scion of a reputable family in the west of England who joined the fraternity of itinerant gypsy-beggars and made a handsome living by inventing characters, localities, genealogies, and friends that a gullible public found entirely probable. Accounts of his life appeared in 1745 and 1748, linking him to the Stuart cause in the rebellion of 1745. Usually his fiction would include a story of a shipwreck or similar accident (a herd of cows drowned in a flood, for example) accounting for his present destitution, which strangers are only too willing relieve. Each scene is a performance in which the beggar defies the penetration of his audience, so that the parallels between itinerant begging and traveling theater are very close, both relying on techniques of amusement, taking the word in its most extensive sense. Carew is twice transported to the American colonies, and at least twice dresses as a woman to levy his tax of charity on the public. That this is an impost on moral shortfall rather than mere gullibility is a point made by the narrator when he says, "It will be no disgrace to our hero if among [civil hypocrites] he appears polished as the best, and puts on a fresh disguise as often as it suits his conveniency" (Carew 1813: 125). No matter how genially Carew enters into his schemes, he is never deluded into thinking he has anything in common with those he deceives. Like James Hardy Vaux, he appends a dictionary of thieves' cant to his memoir, as if to emphasize the extent of the cultural division between mendicants and the public, one that Henry Fielding thought honored the former more than the latter (see Lynch 1998: 83–84). However, it leaves him strangely hollow, no more than the sum of his successive roles or fictions, like a knight in romance who is devoted to adventures for adventures' sake, with no prospect beyond.

A successful fakement requires first of all a hero in despair, destitute of material and psychological props, operating with a blank in the mind as complete as the "vast and unknown country" into which he has been spilled. "'What have you been?' That question is never asked in the colony" (O'Reilly [1880] 1975: 8). Before an opportunity arises of cultivating the inversions and reversals so typical of convict slang, there is just noise, like the roar heard by John Boyle O'Reilly vibrating from the scorbutic hold of a transport, making "every hatch-mouth a vent of hell," demons yelling "things even more repulsive than [their] physical appearance" (168, 190). John Mitchel recalled that convict etiquette required everyone "to cram as much brutal obscenity and stupid blasphemy into their common speech as it will hold" (Mitchel 1868: 140). Such eldritch figures, once landed, provoked even in him, who was predisposed to pity, feelings of disgust quite as unequivocal as Gulliver's: "I gaze on them with horror, as unclean and inhuman monsters" (295).

Mitchel's fakement during the voyage out on the scurvy-stricken *Neptune* was to adopt the role of the Ancient Mariner, wantoning with his thirst, "patiently eschewing the black ship liquid and lime-juice, and lustfully eye-ing the wealth of sweet water that, 'kerchiefed in a comely cloud,' comes this way sailing like a stately ship of Tarshish." When it obliges, he stands naked in the rain to let his body drink; and when it doesn't, he quotes, "The very deep did rot, Oh Christ! / That this should be" while thinking "of the mariner who had to bite his arm and suck the blood before he could sing out" (Mitchel 1864: 86–87). It is in the same vein as O'Reilly's transformation of his hero Moondyne into Wyville, leader of the penological reform movement in Lon-don, with generous plans for remodeling the vicious regime of Millbank. Both are loose fantasies indulged in the corners and quiet moments of the convict system, which itself is as total, absolute, and self-sufficient as any sys-tem of love or chivalry—"almost too complete," as Mr. Meekin says in Mar-cus Clarke's *For the Term of His Natural Life* (Clarke 1899: 353).

In order to oppose such a system, a fakement must exploit the condition to which the convict has been reduced. This requires that noise not be si-lenced or avoided but that it be transformed—for instance, into the kind of instrumental nonsense that keeps communication secret—or as a reorga-nization of meanings and hierarchies menacing the symmetry of the system. Peter Carey's Ned Kelly recognizes it in the conversation of Joe Byrne and Aaron Sherrit, "They had a queer and private way of conversing they said THAT PLACE & THAT COVE & THAT THING and only they knew what it meant" (Carey 2000: 222). A character on a Parliamentary committee in Moondyne says, "I begin to realize the meaning of the Antipodes: their common ways are our extraordinary ones—and they don't seem to have any uncommon ones" (O'Reilly [1880] 1975: 141). That is to say, in a world where nothing is normal, the possibilities of naturalizing the unusual provide fakement with its best opportunities. This is to be understood through the medium of another cant word, "cross." "Going on the cross" is to defy the law by "cross-work" or "cross doings"; cross-cattle have been stolen by cross-stockmen, now cross-coves, and rebranded: cross-marked. Someone playing the part of a gentleman is a cross-swell.

In Rolf Boldrewood's *Robbery Under Arms* (1888), a gang of bushrangers find their way to fakements by going cross-ways, following their elegant captain, Starlight, into adventures that require a high degree of sang froid and violence, and an equal degree of dramatic skill: "as good as a play," "a fancy-dress ball with real characters," "like playing at 'Robinson Crusoe' only there's no sea ... an Australian Decameron without the naughty stories." These are the interludes where they are either holed up in their happy valley, or consorting with real swells at the Turon races. Their final fakement is doomed, "He somehow didn't expect the fakement to turn out well," but as always, Starlight makes the best of it: "There's no help for it, Dick. We must play our parts gallantly, as demons of this lower world, or get hissed off the stage" (Boldrewood 1951: 476, 596, 397). The alternative would be the shameful condition of those three convict shepherds Mitchel mistook for Yahoos, who have not only no country but no flair, no bush panache.

In his Jerilderie Letter, Ned Kelly begins to state this alternative, but so vast is the inventory of reasons for it, its antithetic possibility is lost in vituperation: ("[a Man] who has no alternative only to put up with the brutal and cowardly conduct of a parcel of big ugly fat-necked wombat headed big bellied magpie legged narrow hipped splay footed sons of Irish Bailiffs or english landlords which is better known as Officers of Justice or Victorian Police who some calls honest gentlemen but I would like to know what business an honest man would have in the police" [Kelly n.d.: 43]). Kelly would like to know why Irishmen would think it an eligible choice "to serve under a flag and a nation that has destroyed massacreed and murdered their fore fathers" (45). Mitchel has the same complaint, and so does Swift, namely that Britain beggared Ireland then mocked the Irish as stupid and incoherent tatterdemalions. Mitchel sees that if some attitude of resistance doesn't intervene, the terminus a quo and the terminus ad quem of Irish misery in the convict system is exactly the same: "Starving wretches were transported for stealing vegetables by night" and then starving wretches were flogged for stealing vegetables by night (Mitchel 1864: xvii). The leading architect of the convict system, George Arthur understood its symmetry as an infernal vortex, "a continual circulation of convicts ... a natural and unceasing process of classification" (RSHC 1837: Appendix 1, 55). The nursery of any idea resistant to the system lies not in any nostalgia for a lost home but in the exasperation born of brooding misery: as Darwin might say, the volitional disease of reverie turning into active irritation. This is how it starts: "There is somewhat dreamlike indeed in this life I am leading. My utter loneliness in this populous ship amidst the strange grandeur of the ocean, and for so many days ... [makes] all my life ... the seeing of the eye only" (Mitchel 1864: 93).

Like Molyneux's newly sighted blind man in Berkeley's account of it, Mitchel is reduced to "only the empty amusement of seeing, without any other benefit arising from it" (Berkeley 1972: 53). The eye needs corroboration from another source. It is not until he experiences the savage delight of eating oranges in Pernambuco that his vitality finds a focus in exquisite sensation.

*For the Term of His Natural Life* is the most bitterly extensive of all the convict fictions, drawing largely on the period of George Arthur's command at the Tasmanian penal colony. The main part of the action takes place in the immediate aftermath of the move from Macquarie Harbour to Port Arthur in 1833, a period chronicled by James Backhouse in his journal—a source liberally cited by Clarke. The hero is introduced to the reader with a curse of bastardy coming out of the mouth of the man he thought was his father. "Silence, bastard … this impostor, who so long has falsely borne my name … shall pack" (Clarke 1899: 3). Arrested soon afterward and wrongly sentenced to transportation, Richard Devine becomes Rufus Dawes the convict, silenced by his father but also by the shame it would bring on his mother were he to speak of his plight; and pack he does, all the way to Australia. On his descent through the successive circles of the systematic convict hell, Dawes loses everything: name, identity, voice, and hope. We often see him overwhelmed, either by the evil noise around him or the cruelty of his guards, falling into a painful reverie. "His faculties of hearing and thinking—both at their highest pitch—seemed to break down … no longer stimulated by outward sounds, his senses seemed to fail him" (54). Sentenced to solitary confinement on Grummet Rock, he suffers a "lethargy of body and brain" that seals him off from all sensations, leaving him haunted by phantoms from the past in a situation that is itself no better than dilirium (109). The birth of Dawes's one great corroborative idea is preceded by another reverie. He is plunged into it when he sees that Sarah Island has been deserted: "The shock of this discovery almost deprived him of reason…. He struck himself to see if he was not dreaming. He refused to believe his eyesight…. He felt as might have felt that wanderer in the enchanted mountains, who, returning in the morning to look for his companions, found them turned to stone" (122). But then he sees a column of blue smoke that like the shaft of light in the New Atlantis he interprets as a covenant or a blessing.

Although scurvy is silently included in the many scenes of starvation in the novel, it is never mentioned as anything but lethargy and sickness; nevertheless, we assume that these trances in which the senses are alternately roused and blocked are in fact scorbutic. Backhouse's account of how scurvy developed in Tasmania is closely echoed in Clarke's narrative. The standard prison diet is the familiar fare of salt meat and flour, with a little tea and sugar. The chain gang is fed porridge and flour with salt meat every other day. On Grummet Island, Dawes has nothing but a small portion of flour for each day. Vegetables are grown only for the officers; and fresh meat is reserved for them and for the dogs at Eaglehawk Neck. So all the conditions for the scorbutic decline of the prisoners recorded by Backhouse and the local surgeons are in place in place for Dawes. Gabbett, the monstrous psychopath whose career shadows Dawes's, has several times escaped and survived on human flesh, the only fresh food he will have eaten in Tasmania. The great fakement of the tale involves Dawes in two activities that declare their paradoxical descent from his unpleasant father. The elder Devine was a shipbuilder and a victualer, making a fortune out of "measly pork and maggoty biscuit." His son is about to embark on the building of a boat and the supplying of fresh goat meat for his new community.

At the source of the blue woodsmoke, he finds Julia Vickers, ailing wife of the commandant, Sylvia her daughter, and Lieutenant Maurice Frere, all three marooned by mutineers who seized the boat in which they were sailing for the new settlement. Everyone is hungry, but Dawes is starving. He falls on his knees and croaks the single word, "Food." Frere roughly denies him any, but Sylvia hands him damper, saying, "Here, poor prisoner, eat!" (145). It is a scene with a long life in convict literature, and here it inaugurates (as it does in *Great Expectations* [1861] and *Jack Maggs* [1997]) a fakement within a romance. The focus is Sylvia, the quaint and beautiful child surrounded by the squalor and cruelty of a penal colony, whose imagination has been fed with literature, principally *Paul and Virginia*, which she is twice found reading, first in English and then in French. Generally, she exhibits the stately confidence of romance heroines, especially when confronted by the awkward attentions of Frere. "I won't kiss you," she tells him, "*Kiss* you indeed! My goodness gracious!" (103). "There are persons," she informs him later, "who have no Affinity for each other. I have no Affinity for you. I can't help it, can I?" (145). Like Virginia, she is fond of tableaux, "I am the Queen of the Island … and you are my obedient subjects. Now then, the Queen goes to the Seashore surrounded by her Nymphs. Pray, Sir Eglamour, is the boat ready?" (169–70). Of course, she has an affinity for Dawes and doesn't demur when he kisses her in the overflow of his gratitude for her addressing him as "Good Mr Dawes." The bond between them, originating in the pantomime of courtliness, is strengthened as the boatbuilding gets under way, with Dawes definitely in charge. He gets the idea of a coracle from something Sylvia says, and then he imagines how to bring it about. With Frere working as his assistant, he catches the goats he needs for skins, builds the framework of saplings, sheathes and caulks it, and it swims. A by-product of this splendid corroboration of imagination by the advent of a real thing is a plentiful supply of fresh meat. Dawes is Robinson Crusoe and Paul and Sir Eglamour all at once.

Instead of being whisked off to Europe, Sylvia is always going to be living within a few miles of Dawes, but in circumstances so radically different from his she may as well be on the other side of the world. This division is the result of her amnesia, for having fallen into a fever at the outset of the voyage in the coracle, she can remember nothing of the island sojourn and, with the death of her mother, Dawes's heroism has now no witness except Frere, who steals all the glory for himself, plunging Dawes into a second silence from which he will never emerge. Shortly afterwards, Frere marries a grateful Sylvia. Thus Dawes's story has been twice stolen from him, first of all in effect by his mother and now again by his tormentor. Thereafter, an evil combination of chance and relentless malice ensures that no one will ever know it. His captivity lasts until the final pages, when he makes his escape in the same boat as Sylvia, only to be overtaken by a hurricane that kills them both. Bernardin de St. Pierre's drowned Virginia is found clasping Paul's picture; the corpse of Sylvia lies on the remains of the shattered mainmast fast in the embrace of her dead lover, seen only by the sun as the wreckage drifts out to sea. The fiction enclosing Dawes's single fakement has been robbed of energy, just as the fakement is robbed of any consequence. There is no paradox, reversal, or transposition capable of wresting the story out of Frere's hands. In his rage, Dawes later tells the truth, but the logic of the place finds no difficulty in setting that aside as a lie. No one but his author hears or sees him; and that last look at the wreckage through the eye of the sun seems as false a transcendence of the senses as that proclaimed by the narrator at the end of *Paul and Virginia*. It puts authorship into doubtful territory, as if paradoxicalisation has migrated from the action to the narration, with "x" as Clarke and "-x" as Frere, his semblable.

Dawes and his author are both pessimists, believing that fakements are very seldom corroborated by the realities of the case. Dawes briefly exults in the fertility of his own brain, and then the triumph is over. Most of *Robinson Crusoe* consists of the repetition of this happy union of idea and fact. He is able to raft ashore his booty from the ship because, "As I imagined it, so it was" (Defoe 1983: 51); and the end of his story, he traps the mutineers because, "I imagin'd it to be as it really was" (264). But Dawes is forced to understand in the end that "he had, by his own act, given himself again to bondage" (Clarke 1899: 173). If Defoe's novel is about the recovery of the civil contract by the improvement of nature into property, then those of Bernardin de St. Pierre and Rousseau are about its opposite: the restoration of the irenic order of plants and its lenitive effect upon the mind. Clarke finds neither alternative satisfactory. As the vile Frere points out, the prison system is fully supported by nature: "Port Arthur couldn't have been better if it had been made on purpose... and all up the coast from Tenby to St Helen's there isn't a scrap for human beings to make a meal on. The West Coast is worse" (142). The ill accommodation afforded by the land is brought to the perfection of misery by the lash, chained labor, and the management of nutritional deficits. Bernardin de St. Pierre has experienced an inhospitable land and a cruel society, but he has not chosen to represent them simultaneously. Clarke exhibits a world of simultaneous evil, both natural and social: wild streams are poisoned by their own vegetation and a system of justice drives guiltless wretches into despair. Sylvia's innocence is nothing but a fantastic counterpart to this tangle of evil—really no more than an extended reverie—and even that faint opposition attenuates when, after marriage, all her romantic certitude turns to diffidence and uncertainty.

Peter Carey's Ned Kelly is a little like Rufus Dawes in terms of narrative resources, not endowed with many and certainly not keen to inherit them. He is under no illusions about how few props for the imagination came out of Ireland aboard the convict ships, only the Banshee, the herald of death. On the *Rolla*, the *Tellicherry*, the *Rodney*, and the *Phoebe Dunbar* "the BANSHEE sat herself on the bow and combed her hair all the way from Cork to Botany Bay.... It were clear St Brigit had lost her power to bring the milk down from the cow's horn ... but the Banshee were thriving like blackberry in the new climate" (Carey 2000: 99–100). Kelly, therefore, is deeply troubled by cultural imports that conflict with frontier notions of masculinity, such as his father's cross-dressing, "his broad red beard his strong arms his freckled skin all his manly features buttoned up inside that cursed dress" (19). The wearing of dresses as a gesture of rural resistance is explained in subsequent fakements where Steve Hart, "the horrid thing that had previously worn a dress" (222), plays a part. Hart talks of his heroes Robert Emmett, Thomas Meagher, and Smith O'Brien but, as Kelly acidly notes, "He never seen them men but he were like a girl living in Romances and Histories always thinking of a braver better time" (223). Wearing blackface and a dress, Steve enters the hut like "the Dame in the pantomime" explaining to everyone that he is a son of Sieve, appropriately accoutred: "Its what is done in Ireland ... its what is done by the rebels" (309). Although in his own eyes he is a White Boy or a Ribbon Man, to Mary he seems like a Molly, as indeed Ned's father seemed to Ned when he wore his "cursed dress." Ned's plan to make armor from the quarter-inch steel plate stripped from the moldboards of ploughs is to shift the key of defiance decisively from cross-dressing to chivalry. Egged on by Thomas Curnow's reading of the rousing speech from

*Henry V* ("We would not die in that man's company / That fears his fellow-ship to die with us"), Ned accepts the doomed fakement of the fight at Glen-rowan, believing perhaps the promise of Shakespeare's line, "This story shall the good man teach his son." But it is one formed out of a strangely awkward alliance of history-in-the-making and romance that Curnow appreciates more fully than Kelly and to which he has, possibly, a greater title as author than Carey, for all his dubious motives in faking it. Authorship begins to look possibly like dirty work.

Carey's earlier convict novel probes more deeply the relation of Australian fakements to metropolitan life by coiling the fiction of *Jack Maggs* around Dickens's *Great Expectations*. Although set at the time when Dickens was still writing his novel, *Jack Maggs* acts as its supplement, for it is by no means at one with the sentiments or motives of Dickens, which are treated as shabby and opportunist by Carey's narrative voice. The narrative shell of Carey's *True History of the Kelly Gang* (2000) resembles that of *Jack Maggs* inasmuch as authorship is characterized as a rogue activity by a voice not at all times easily distinguished from the author's own. Oates stands in as close a relation to Carey as he does to Dickens, likewise the character Curnow transforms history into the fakement to which Carey puts his name. The same terrible ambiguity links Maurice Frere to Marcus Clarke. How then do we read the paragraph beginning, "The death of children had always had a profound ef-fect on him ... for Tobias had been a poor child ... and he was ... famously earnest in defence of the child victims of mill and factory owners" (Carey 1997: 130)? I don't think we are supposed to doubt the profound effect, or the strength, of the appeal abused and dead children make to everyone's heart and conscience. The only explanation for the callous tone ("famously earnest") in which this remark is made must lie in the sequel to the scene that is common to these three convict fictions, where a child gives a starved and chained criminal some food: Sylvia handing Dawes the damper; Pip giving Magwitch a pie intended for Christmas dinner; Henry offering Maggs the pig's trotter he wanted at that moment more desperately than anything else in the world.

The profound effect on the adult is what is important, because it is from that surge of passionate thankfulness—not just for the food but for the rec-ognition of an abandoned person's humanity—that three parallel fakements grow, as do the three novels recording them. We have seen that Dawes's fake-ment is the weakest; but the other two are fueled by sentiments initially pow-erful and direct, and then over time seasoned with nostalgia and self-pity, which corrupt the original passion and blur its outcome, or at least that is what Carey seems to suggest. Magwitch is so deeply moved by the action of a small boy who commits the first theft of his life to fill the belly of a starving man that he is determined to construct a romance in which the boy will assume the part of hero. Young Henry Phipps gives famished Jack Maggs a pig's trotter, even holds it so he can chew it because his hands are held close by the chain, and Jack builds a similar fiction in which Henry is quite content to play his part of a spoilt young gentleman as long as there is plenty of money in it. Both convicts break their parole to return to England in order to view their handiwork and to bring it to its consummation, which in both cases is meant to include a sentimental reunion between the generous old varmint and the grateful youth.

An inevitable consequence is that both varmints put themselves in danger of their lives for breaching their parole; but another is that the stories they meant to tell by means of their benefactions have become confounded with others, equally dangerous. No story tangential to these two fakement-fictions entirely belongs to the person of whom it is told, or to the person who tells it. They are like one of Watling's or Lycett's paintings. Just as Compeyson tracks Magwitch, so the sinister Partridge has an interest in Maggs, and each is fleshing out a tale intended to bring the quarry to the gallows. Meanwhile, Oates's various plans for his novel *The Death of Maggs* is fraught with new dangers for the hero every time the author fiddles with his plot, or unearths by mesmerism a new cache of information from the vessel that never leaks, namely Maggs's cerebrum. In fact, deaths proliferate as each story collides with another—Compeyson, Magwitch, and Partridge dead, and Oates frequently close to it.

Maggs makes elaborate plans to protect his own secrets from hostile readers, writing them in mirror script in invisible ink, an innovation in penmanship analogous to cant in speech and not unlike the scriptural code used by John Rix in *For the Term of His Natural Life* that is cracked by Maurice Frere. However, the status of the possession of a secret is badly compromised by the mesmerism, and also by Oates's empty promise made to Maggs, "All your secrets shall be returned to you" (232). A returned secret is a little like renovated innocence in the *New Atlantis*. Magwitch has many fewer resources than Maggs, relying for immediate help on Pip and Herbert, but for all his long-term needs, he has been dependent upon Jaggers, the repository of all secrets in *Great Expectations* and for whom there is no equivalent in Jack Maggs except Tobias Oates, at once author and archive.

Jaggers controls entirely the story of Pip and Estella that is told within the fictional world of *Great Expectations*; he has controled it ever since Estella was adopted by Miss Havisham; and of course, that secret was the offspring of another concerning her father and her mother in which Jaggers himself played a legal part, brilliantly manipulating a hypothesis that forced a jury to find a woman almost certainly guilty of murder, innocent. So none of these secrets was founded on truth or on respect for the law; they were authorized by the power generated jointly among those who defy it and those who manipulate it, just like the fakements of the Australian convicts into which these lies have been wreathed. In Clarke's novel, the narrative voice controls a synthesis of three imperfect stories—Dawes's, Rex's, and Frere's—that constitutes the true account, of which none of the three is aware. That is why the eye of the sun looks awfully like the eye of God in the final scene and why there is nothing like Jaggers's or Oates's collections of secrets to complicate the dynamic of the action, which becomes increasingly repetitive.

Maggs feels the force of this dynamic when he says of his fakement, quoting Ned Kelly, "It was not what I planned, but such is life" (183). He knows how powerfully the counter-fictions of his adversary are conceived, but even more how brilliantly they are performed, such as the astonishing pantomimic grotesqueries of Sir Spencer Spence, the plague doctor with red lips who places the house in Great Queen Street under quarantine while brandishing two terrifying mock-surgical tools. Maggs's exit from this pressure on the story he was trying to tell, and indeed from the story itself, is executed in what must be assumed a deliberately spare and perfunctory manner when he goes back to his two children in Australia, "the real Jack Maggs" on the Portsmouth Mail with Mercy Larkin, a women who, like her partner, will never appear in the fiction Tobias Oates is going to publish.

Dickens's *Great Expectations* has two alternative endings stemming from the moment when Magwitch imagined, or was imagined imagining, that if he made it rich, he could forge a gentleman out of the little boy who gave him food on the marshes of the Thames estuary. Pip will never marry Estella because she won't have him; he will marry Estella because they are destined for one another. The third ending is Jaggers's choice, knowing as he does the advantages of fiction over true history, advantages that have been as horribly abused as the children for whom that abuse of fiction was partly calculated, and it runs as follows: Pip cannot marry Estella because he knows the secret of her birth, which is almost as bad as the secret of his own great expectations, and he can never tell her that he knows it. In the scene where Jaggers confirms Pip's suspicions about her origins, there is a pantomime as rich in mock-forensics as Sir Spencer Spence's was in music hall medicine. When he puts a case, like the case in defense of his servant Molly, Estella's mother, Jaggers demands a jury follow where it leads, or to set up their own instead: "You set up the hypothesis," he warns them, "You must accept all the consequences of that hypothesis" (Dickens 2008: 360). But now, after prefacing each secret in Estella's story with the proviso that keeps it speculative ("Put the case ... Put the case ... Put the case" [377–78]), he says to Pip that no consequences at all should flow from these hypotheses. They must end in silence and inaction because no one, least of all Estella, would benefit from having her secret told. " 'Add the case that you had loved her, Pip, and had made her the subject of those "poor dreams" ... then I tell you that you had better chop off that bandaged left hand of yours with your bandaged right hand, and then pass the chopper on to Wemmick there, to cut *that* off, too.' ... Wemmick ... gravely touched his lips with his forefinger. I did the same" (379). It is a terrible travesty of the hypothetic work Johnson and Gallagher expect readers to perform in reading novels, "entertaining various hypotheses [until it becomes clear that] the reality of the story itself is a kind of suppositional speculation" (Gallagher 2006: 1.346). Instead of supposing that what you don't know, you do, the Cartesian alternative is active here and elsewhere in cognate genres, supposing instead that what you do know, you don't. Then hypothesis fulfills the same function as litotes when Wemmick talks to Pip of Magwitch as "a certain person not altogether of uncolonial pursuits, and not unpossessed of portable property," one who has been in a distant place "not quite irrespective of government expense" (Dickens 2008: 337). As extravagant instances of legal nicety, this is all very comic, but somewhere in the background, you can hear the whisper of Flinders's grass; see the endless list of nothings that greeted Dampier's eye; appreciate the expedience of the alliance forged between Gay's Ay and No; and detect the outlines of the utopian geography of the not impossible. The language of utopia, Michel de Certeau has argued, is glossolalia, the linguistic equivalent of trompe l'oeil, fooling the ear with the mockery of articulate speech by " 'saying' without saying something." Glossolalia emerges from sheer noise in the same way that fakements grow out of reveries, moving from "cannot say" to "can say" by way of "can say nothing" (de Certeau 1996: 29–31). It is a very limited achievement and hedged about with negatives, but it offers a faint sound of what otherwise would never have been heard at all.

I will end with a non-Australian scorbutic convict fiction whose moment of corroboration is wonderfully clear, George Orwell's *1984* (1949). It always comes as a surprise in reading this dystopian nightmare that the Party has carefully planned Winston and Julia's affair to take place in the seedy, bug-ridden bedroom above Charrington's junkshop. There the dissidents make what they think is private love to the accompaniment of the working-class woman singing beneath their window of the *chagrin d'amour* while she pegs out her washing. Inside this carefully prepared setting that looks so very like a happy accident and so authentically proletarian, Charrington tempts Winston with the bric-a-brac of a past of which everyone is ignorant because it has been remade and erased so many times. But the temptation is cleverly done because each object feeds Winston's desire to have some empirical support for a history he knows only as nostalgia. Around the little he can remember of his mother and the blank which represents his father, he has assembled sensations and images designed to counteract the worst moments of his loneliness in time and space, which he figures as an underwater forest patrolled by a monster who is himself. He dreams of his mother and his sister sinking beneath the sea, "looking up at him through the green water, hundreds of fathoms down and still sinking" (Orwell 1981: 29), their eyes filled with a deep and complex sorrow. When Charrington offers him a piece of coral set in a hemisphere of heavy lead glass, it is as if he holds the mystery of the undersea in his hand. "The inexhaustibly interesting thing was not the fragment of coral but the interior of the glass itself.... He had the feeling that he could get inside it, and that in fact he was inside it, along with the mahogany bed and the gateleg table, and the clock and the steel engraving and paperweight itself" (120). The glass composes the odds and ends of Charrington's shop into "a sort of nostalgia, a sort of ancestral memory" (81). This coral ornament is the first thing that is smashed when the thought police invade the room, putting an explosive end to the fragile fantasy it sustained.

Of course such nostalgia has nothing ancestral about it, being entirely a fiction of the State. After his arrest and a savage beating, Winston's imagination evacuates the marine landscape, coming up through the depths to surface in the interrogation room (192). During this period he is not fed—he sees a fellow-prisoner dying of starvation—and by the time of his release he has lost 25 kilos. The ulcer on his leg has opened up, and as he pulls out Winston's loose teeth, O'Brien describes his victim's evidently scorbutic condition: " 'Look at the condition you are in ... look at that running sore on your leg. Do you know that you stink like a goat? ... Your hair is coming out in handfuls.... Nine, ten, eleven teeth left ... and the few you have left are dropping out of your head.' ... He seized one of Winston's remaining front teeth [and] wrenched the loose tooth out by the roots. 'You are rotting away,' he said, 'you are falling to pieces. What are you? A bag of filth?' " (219). At this point, Winston does what he has been doing regularly ever since he was arrested, and what he will do until he dies: he bursts into tears.

The physical symptoms of scurvy and the utterly illusory condition of nostalgia are tied together by two ingenious additions to Charrington's stage set. One is a fragment of a nursery rhyme recalling the tones and rhythms of London's church bells: "Oranges and lemons, say the bells of St Clement's, / You owe me three farthings, say the bells of St Martin's, / When will you pay

me, say the bells of Old Bailey, / When I grow rich, say the bells of Shoreditch."
The other is the song sung by the large woman hanging out the clothes. The
immemorial cure of scurvy, the juice of an orange or a lemon, stands in opposition to the emptiness of everything in the nostalgic tableau superintended
by Charrington because its truth can be attested by the organs of the bodily
senses. Winston's first tears are like Odysseus's at the court of Alcinous insofar as they represent the ill of emptiness calling out to be filled. The nursery
rhyme was inserted into the scenario for the same reason as the old woman.
She sings not of objects of longing, but the sweet pointlessness of the feeling of longing itself: " 'Twas only an 'opeless fancy" (114). Produced by a machine in a subsection of the music department of the State, the song's words,
dreams, smiles, and tears have no referent except an ersatz passion. Oranges
and lemons were intended be of a piece with the whole nostalgic tableau arranged by O'Brien and his subordinates, he even supplies the lines Winston
can't remember, and to this extent, pain is no different from yearning for a
home in the past insofar as the State knows exactly what both are like. Not
even Winston's subsequent agony under torture is his own, for O'Brien can
describe it to him while inflicting it.

When Winston proposes his seditious erotic passion as resistance to the
State, his fakement, he strikes the paradoxical poses of a Lucifer or a Faust:
"I hate purity," he tells his corrupt nymph, knowing that no emotion is pure
because "everything was mixed up with fear and hatred" (105). So the erotic
cult of corruption is, he believes, a political act. The State responds with an
avalanche of factory-made paradoxes, of which doublethink is the formal
mode of thought and Newspeak the linguistic vehicle. Although the one
seems to accommodate the equality of rival ideas ("War is Peace," "three is
four," and so on) while the other is awash with surplus negatives such as "un-
light" for dark or "undark" for light, it is clear that the purpose of both systems is stark uniformity with the ideology of a totalitarian state, not the
volatility of conjoint opposites. Could the word "innocent" ever function in
Newspeak? Winston's embrace of corruption is no different, for with one
exception, nothing he can do or imagine is beyond the capacity of the State
to anticipate. The exception lies in his memory of the reality of lemons, so
sharply different from the homogeneous sweetness of Julia's chocolate. "I've
seen oranges," says Julia, "They're a kind of round yellow fruit with a thick
skin." Winston can do more than just recall the sight of lemons, "They were
so sour that it set your teeth on edge even to smell them" (120). Somewhere
in his brain "the empty amusement of seeing, without any other benefit" is
complicated by synesthesia, a complexity the State could never compass. The

memory of a real food persists, the kind of food that puts an end to lost teeth, falling hair, and a stinking, ulcerous body. Thus Adam's dream, where corroboration anticipates the act of sin, but also provides the means of salvation:

> Each Tree
> Load'n with fairest Fruit, that hung to the Eye
> Tempting, stirrd in me a sudden appetite
> To pluck and eate; whereat I wak'd, and found
> Before mine Eyes all real, as the dream
> Had lively shadowd. (Milton 1958: 8, ll. 306–11)

# Coda
### James May and Fiona Harrison

The physical symptoms of extreme vitamin C deficiency, scurvy, have been described in numerous ships' logs, diaries, and medical texts, stretching back for hundreds of years. These include *hemorrhage*, and the characteristic broken blood vessels under the skin; *hyperkeratosis*, or changes in hair such as thinning, alopecia, and corkscrew hairs on the limbs; *hematologic abnormalities* including anemia and the poor healing of wounds, or opening of old wounds; and *hypochondriasis*, or multiple physical complaints, both imagined and real. Using modern techniques in neuroscience research, we are now beginning to unravel the highly complex roles of vitamin C in the brain, which may have changed the behavior of those experiencing long periods of nutritional deficiencies. The most important roles for vitamin C, also known as ascorbate and ascorbic acid, are in the synthesis of neurotransmitters, the chemical messengers of the brain, and for protection of neurons (brain cells) against damage by a constant barrage of free radicals. Using genetically engineered mice to model vitamin C deficiency, we can study how altered brain vitamin C levels affect neurotransmitter levels and, more importantly, how they affect behavior. The most notable changes in the mice mirror those observed in humans: extreme swings in activity level that recover quickly following the reintroduction of the vitamin to the diet. The brains of scorbutic animals also show changes in neurotransmitter levels and damage to cells.

Scurvy itself is relatively rare, although not unheard of, in modern Western populations. Hypovitaminosis for vitamin C—chronic low levels of ascorbic acid that do not present as scurvy, also known as occult scurvy or, in the eighteenth century, scorbutic diathesis—are actually common in a number of populations, from college students to the elderly [1–5]. This finding has prompted a number of recent studies in how such deficiencies may affect disease processes such as cancers, Alzheimer's disease, diabetes, and cardiovascular ailments. This research has been significantly advanced by the development of a number of genetically engineered mouse models in which vitamin C synthesis and transport can be manipulated [6–8], allowing researchers better to understand the functions of vitamin C in normal and pathological states. These models have been particularly useful in the investigation of behavioral abnormalities and biochemical changes in the brain, data that relate not only to diseases that are common today but also to the symptoms of sailors and explorers exposed to scurvy in the past.

## Transport and Requirements

The majority of mammals are able to synthesize their own vitamin C in the liver. However, humans and other primates, plus a few notable other exceptions such as the guinea pig, have lost this ability due to an inactive form of the enzyme *gulonolactone oxidase*, which is responsible for catalyzing the final stage in the synthesis process. We are, therefore, reliant on dietary intake in order to maintain our stores of the vitamin. The human body is estimated to hold a pool of approximately 1500 mg on average, although maximal or optimal levels are not known. For maintenance of this pool, it is recommended to ingest between 60 and 120 mg per day, with required intake depending on factors such as body size, gender, pregnancy, smoking, alcohol intake, and certain diseases. As little as 10 mg per day may be enough to ward off the more serious symptoms associated with scurvy, although probably not indefinitely. In fact, it takes higher doses of 200–400 mg to fully saturate the cells in the blood stream [9], from where the vitamin can be circulated around the body to each organ, including the brain. Of all of the organs in the body, the brain maintains by far the highest concentrations of vitamin C, two to five times greater than spleen, liver, kidneys, heart, and lungs. This is taken as an indicator of its importance to neuronal tissues and its success in harnessing the bulk of available supplies. Of special interest is the differential compartmentalization of vitamin C stores within the brain. The more metabolically active areas, crucial for complex cognitive processing and behavioral control, have the highest levels of vitamin C and maintain them more aggressively in the face of deficiency. This may reflect both increased need for the vitamin in these highly active areas as well as the differential distribution of cell types, including density of the distribution of cell bodies.

For reference, one large, fresh orange, of approximately 180 g can provide as much as 100 mg of vitamin C. Time and storage conditions, however, can rapidly contribute to the degeneration of the vitamin within food substances and decrease the content of this critical nutrient in food stuffs. Over time, with insufficient intake levels, the body's stores gradually drop, with peripheral organs being depleted first, followed by the vitamin stores in the brain. Ultimately, despite the best preservation tactics employed within the body, scurvy can develop. In individuals who already had poor intake before the extended period of deficiency, eventual decline and sickness occur much more quickly.

Vitamin C transport and storage are carefully controlled in the body by specific transporters and recycling systems. There are two Sodium-dependent Vitamin C Transporters (SVCT1 and SVCT2). SVCT1 is required to carry vitamin C from the food into the blood stream, and also to reabsorb vitamin C into the blood stream within the kidney to offset its excretion in urine. SVCT2 transports the circulating vitamin C into the cells of the organs where it is needed. In particular, SVCT2 is found in the choroid plexus, which forms part of the barrier between the circulatory system and the brain parenchyma (the specialized cells required for brain function). This transporter is also on neurons themselves. At each of these locations, the transporter works to transfer vitamin C against a concentration gradient, leading as noted earlier to levels in the brain that are far higher than those found anywhere else in the body. This system, combined with additional chemical recycling systems, helps maintain the high levels in the brain, even during times of depletion. Nevertheless, under conditions of prolonged insufficiency, even brain vitamin C levels begin to decline, and the changes that occur as a result may include behavioral abnormalities, such as extreme lethargy, sudden laughter, or copious weeping.

## FUNCTIONS OF VITAMIN C

### Antioxidant

The reason that vitamin C is so important, and why scurvy is so devastating, is the large range of functions it organizes in the brain. The first is its agency as a major antioxidant. Generation of energy for normal cellular functions creates highly reactive molecules termed "free radicals." These radicals can react with and damage proteins, fats, and even DNA itself—all crucial to the operation of the cell—leading ultimately to cell death. To prevent this damage, the body produces a wide variety of antioxidants and obtains more from the diet. Under normal circumstances, a balance is maintained between damage caused by free radicals and the repair of the cell's constituents. Vitamin C is a crucial antioxidant that often provides the first line of defense against oxidative damage. Not only does it combat the free radicals, it also supports other antioxidants such as vitamin E and glutathione, and even helps to regenerate them. When the free radicals outweigh the antioxidant potential, and damage exceeds the rate of repair, our systems are said to be under "oxidative stress." Of all our organs, the brain is most vulnerable to oxidative damage because it is so metabolically active and because production of new neurons (neurogenesis) is extremely limited in adults, especially when compared to other organs that can regenerate cells in high numbers (e.g., the skin and liver).

## Neurotransmitter Synthesis

Neurotransmitters are the chemical messengers of the brain. They are vital molecules that are released from one neuron into the synapse (the functional connection between two neurons), where they are detected by another neuron and thus pass a signal from one cell to another. The formation and strength of synaptic connections are the basis of learning, memory, and cognitive function. Disruption of the different neurotransmitter systems is associated with almost all disease states having a psychological or psychiatric symptom set: Alzheimer's disease, autism, ADHD, schizophrenia, bipolar disorder, and Parkinson's disease to name just a few. Vitamin C is involved in the synthesis pathway of at least three of these neurotransmitters. They are dopamine, norepinephrine, and serotonin. Its role is mainly to keep particular enzymes in an active form, which it does via very specific antioxidant reactions, typically acting on transition metals that are part of the enzyme complex, such as iron ($Fe^{3+}$) or magnesium ($Mg^{2+}$). Put very generally, dopamine is strongly involved in neuronal activity and perhaps even more so in reward-motivation and addiction behaviors. So altered dopamine levels affect the psychological reward system associated with risk-taking. In the eighteenth century, it was thought that victims of scurvy were liable to gamble excessively and certainly their imaginations were known to work overtime in picturing the things they craved so badly. There is a recent theory implicating dopamine in schizophrenia, whose positive symptoms include hallucinations and delusions, and whose negative ones include avolition (lack of desire or interest in action). The implication of dopamine in these types of imagined experiences is an interesting one. Vitamin C, with other antioxidant vitamins such as vitamin E, has been tried as an adjuvant therapy in a number of trials in schizophrenic patients, suggesting that lower vitamin C levels may be common in psychiatric disease [10, 11]. Such a result could be due to increased demands for antioxidants in sufferers [12] or lack of appetite or ability to follow good dietary guidelines. Either way, the association is intriguing and implies that those prone to some forms of psychiatric illness may require greater vitamin C and thus be more prone to scurvy, but also that its deficiency may exacerbate their symptoms. Norepinephrine has a key role in the ability to maintain mental ability and concentration. Serotonin has roles in social and emotional function related to depression and happiness. Thus it can be seen that any major disruption to these systems are likely have a dramatic impact on behavior.

There is no direct evidence that scurvy can cause psychoses, although it is associated with hypochondriasis, including lassitude, depression, and irritability. Unfortunately, these sorts of behaviors are also quite challenging to study in animal models in ways that are directly relatable to humans. However, we do have evidence of affected neurotransmitter systems from animal models under altered vitamin C states [13, 14]. It is conceivable that in cases of preexisting susceptibility, whether the result of genetic makeup or environmental factors, chronic and severe vitamin C deficiency would accentuate psychological changes. Nevertheless, in the situations described throughout this book, it is never certain that the individuals portrayed were suffering from a single nutritional deficiency. Symptoms listed as scorbutic were often owing to low levels of vitamins A, B, and D, as well as C, causing illnesses such as nightblindness, pellagra, beriberi, and weakness in the bones. This is a further complication in the story of brain health.

## Other Roles

Glutamate is the major excitatory neurotransmitter. That means it is critical for neuronal activation; but if it is not cleared quickly from the synapse, it will soon become excitotoxic, triggering neuronal damage and cell death. The major part of glutamate clearance is by astrocytes, specialized support cells that surround the synapse. As glutamate is cleared via a specialized transporter (GLT-1), vitamin C is released from the astrocyte [15], offering acute protection for the neurons. This direct connection between vitamin C and normal function of a neurotransmitter system is yet another example of the critical role that vitamin C has in normal brain function and provides a further area susceptible to damage and dysfunction under low vitamin C conditions.

For the most part, the other major roles of vitamin C are also related to its role as a cofactor in enzymatic reactions. This applies to the synthesis of collagen, the protein whose loss is so dramatically visible in scurvy. The breakdown of collagen matrices is responsible for the putrid gums, loose teeth, reopened wounds, and disappearing cartilage so often mentioned in narratives of scurvy. Vitamin C is also used in the synthesis of carnitine, which is important in the production of energy from fats within cells, as well as several peptide hormones (small, water-soluble proteins capable of inducing changes in cellular function). It is needed for the proper absorption of iron, as it must be available to convert ferric iron ($Fe^{3+}$) to the ferrous iron ($Fe^{2+}$) that is absorbed. It is this reaction that explains why those with hemochromatosis (a disorder of excess iron) would require greater vitamin C intake levels, and it accounts for symptoms of anemia presenting in scurvy. More recent research has shown that vitamin C is required for the action of specific enzymes (e.g., ten-eleven-translocation [TET] enzymes) that are involved in the modification of DNA and that this may be of particular importance in dopaminergic systems [16]. Very small changes to DNA, such as the addition or removal of small molecules at particular sites (methylation and demethylation), can alter its shape or conformation. This can limit the ability of other molecules to interact with the protein at the correct active sites and ultimately modify how efficiently genes are able to produce functional proteins, impacting cellular function in the brain as well as throughout the body. Also affecting DNA is a transcription factor termed hypoxia-inducible factor $1\alpha$ (HIF-$1\alpha$). Activation of a specific prolyl hydroxylase enzyme by a vitamin C leads to degradation of HIF-$1\alpha$ [17, 18]. HIF-$1\alpha$ functions as an "oxygen sensor" to regulate a variety of metabolic pathways, especially in cancer. Decreased vitamin C causing higher levels of HIF-$1\alpha$ could thus activate mechanisms needed to sustain growing cancer cells or unwanted vessel proliferation (such as in diabetes).

## BRAIN AND BEHAVIORAL CHANGES IN ANIMAL STUDIES

As described above, in order to test the effects of extreme deficiency in animals, it is necessary to find a model that does not synthesize its own vitamin C. Although guinea pigs have naturally lost function of *gulonolactone oxidase*, they are not widely used in behavioral studies. In contrast, the use of mice in them is widespread, and the tasks used are well defined and validated. Our own work using the *gulolactone oxidase* knockout mouse model, which was genetically engineered to have a nonfunctional copy of the gene, has confirmed that the most reliable reproducible behavioral changes observed in mice kept under vitamin C–deficient conditions are an extreme decrease in voluntary locomotor behavior, and physical or muscular weakness [14]. These symptoms directly mirror the behaviors often reported in humans. In fact, James Lind reports in his *Treatise on the Scurvy* (1753) that "the first indication of the approach of this disease, is generally a change of colour in the face ... with a *listlessness to action, or an aversion to any sort of exercise* .... Mean while, the person eats and drinks heartily, and seems in perfect health; except that his countenance and *lazy inactive disposition* may portend an approaching scurvy." Further, he notes that this "former aversion to motion degenerates soon into an *universal lassitude, with a stiffness and feebleness of their knees upon using exercise; with which they are apt to be much fatigued*" [19]. Newer work from our lab, and others, has shown that severe vitamin C deficiency is almost always associated with oxidative stress, including damage to proteins and lipids (fats, such as those that make up cell membranes) [14, 20]. There is also evidence for disruption of neurotransmitter systems, particularly the dopaminergic system [14, 21]. These findings are not unexpected, given the known roles for vitamin C, and for dopamine in regulation of activity levels, but they provide excellent support for the notion that the physical and psychological changes reported in sufferers of scurvy were due to pathophysiological changes and not just the stark conditions of the vessels or environments in which they found themselves.

## Individual Differences in Populations

There are several factors that have not previously been given much consideration, but could dramatically affect the extent to which each individual is affected by vitamin C deficiency, or at least the speed at which they succumb to scurvy. First, there are a number of small mutations, known as single nucleotide polymorphisms (SNPs) that can change the function of the vitamin C transporters—although this occurs to a much larger extent in SVCT1 than in SVCT2. This is owing to the fact that insufficient vitamin C provided to the brain by the SVCT2 is lethal to developing embryos and would therefore be selected out of most populations. Prevalence of some SNPs in the SVCT1 has been estimated to occur in 6%–17% of the African American population, with one mutation leading to a decrease of nearly 80% of the transport capability when tested in a model system in the laboratory [8]. The prevalence of these mutations varies in particular populations, but can affect how efficiently vitamin C can be taken up, or how well it can be conserved. Particular mutations have been associated with several disease states including cancers, but are perhaps more relevant to the dental issues reported with scurvy, such as periodontitis [22]. So it is highly plausible that particular individuals would require an even greater intake of vitamin C to obtain the same tissue levels as other members of their group and to avoid scurvy.

Second, men are also larger on average than women are and therefore require greater vitamin C intake. Their kidney function is also poorer than that of women. Compared to the younger boys that often also made up part of a ship's crew, and who often fared better than their older colleagues in terms of scurvy, a full-grown male would require a larger amount of vitamin C to meet his needs. That means adult males falling into scurvy are less able to conserve their critical vitamin C stores, if indeed any is still circulating in their blood. Alcoholism is often associated with vitamin C deficiency, and in many cases, this is due to poor diet, but alcohol is also thought to increase the need for vitamin C, and thus diminish total body stores further [23]. Many sailors who endured long voyages at sea had serious addictions to both nicotine and alcohol, and combined with the rough and active life, and lack of treatment for many minor and chronic illnesses, there were likely multiple additional drains on their limited vitamin C stores.

Third, there are a number of disease states that can increase oxidative stress in the body and thus would increase the need for additional vitamin C, including diabetes, high cholesterol, smoking, neurodegenerative diseases, and cancers. It is not currently known whether psychiatric illnesses could also create an excess drain on neural vitamin C stores, rather than just potentially intensifying under conditions of deficiency. Two disease states require further comment. First is hemochromatosis. This is a disorder that causes a person to absorb and then store too much iron. The iron can build up over time and become toxic. As described above, vitamin C is required for the absorption of iron, but increased systemic iron uses up more vitamin C. Hemochromatosis is, in fact, the most common genetic disorder in Caucasians and is caused by changes in the Hemochromotosis (HFE) gene, which monitors iron levels and governs levels of absorption, storage, and release. One particular mutation (C282Y) is thought to affect around 11% of the Irish population [24]. Although most cases of hemochromatosis are caused by a genetic mutation, other cases have also been linked to liver disease and alcoholism. Haptoglobin (Hp) is protein made by the liver that binds and helps to remove cell-free hemoglobin from the bloodstream, thus preventing its now readily available iron from generating damaging free radicals. There are several polymorphisms of Hp that determine its antioxidant capability. When it is in short supply, then vitamin C is used up instead, so Hp-genotype is likely another direct modifier of vitamin C intake requirements [25].

The second disease in which vitamin C might well play a role is diabetes, and diabetic retinopathy in particular. Diabetic retinopathy is a leading cause of blindness in Western countries [26]. In adult-onset or type 2 diabetes, visual loss is often due to macular edema, which results from the leakage of plasma from vessels in the primary visual area. Swelling of this area causes visual distortion that can become permanent. Vitamin C is known both to prevent death of cells that line the vessels in this area [27, 28] and to tighten the barrier formed by living cells against the leakage of plasma [29]. Since persons with poor glucose control of their diabetes have plasma and tissue vitamin C concentrations as low as 40% of normal [30–32], it is possible that repleting this vitamin C could decrease vascular leakage in the eye and thus prevent visual loss.

## CONCLUSION

There is clearly still much that can be learned about the plight of the scorbutic patient through the use of modern neuroscience techniques. It is often difficult to tease out the effects of scurvy from other deficiencies, and the environment and situations that are described in historical texts. In a well-controlled scientific experiment, those same problems do not exist. Perhaps even more important, is how vitamin C deficiency, which is so easily avoided, can impact human health today, from prenatal development, through to normal and pathological aging.

## REFERENCES

1. Johnston, C. S., Solomon, R. E., Corte, C., *Vitamin C status of a campus population: College students get a C minus.* J Am Coll Health, 1998. **46**(5): pp. 209–13.

2. Johnston, C. S., Thompson, L. L, *Vitamin C status of an outpatient population.* J Am Coll Nutr, 1998. **17**(4): pp. 366–70.

3. Raynaud-Simon, A., Cohen-Bittan, J., Gouronnec, A., Pautas, E., Senet, P., Verny, M., Boddaert, J., *Scurvy in hospitalized elderly patients.* J Nutr Health Aging, 2010. **14**(6): pp. 407–10.

4. Mosdol, A., Erens, B., Brunner, E. J., *Estimated prevalence and predictors of vitamin C deficiency within UK's low-income population.* J Public Health, 2008. **30**(4): pp. 456–60.

5. Schleicher, R. L., Carroll, M. D., Ford, E. S., Lacher, D. A., *Serum vitamin C and the prevalence of vitamin C deficiency in the United States: 2003–2004 National Health and Nutrition Examination Survey (NHANES).* Am J Clin Nutr, 2009. **90**(5): pp. 1252–63.

6. Maeda, N., Hagihara, H., Nakata, Y., Hiller, S., Wilder, J., Reddick, R., *Aortic wall damage in mice unable to synthesize ascorbic acid.* Proc Natl Acad Sci USA, 2000. **97**(2): pp. 841–6.

7. Sotiriou, S., Gispert, S., Cheng, J., Wang, Y., Chen, A., Hoogstraten-Miller, S., Miller, G. F., Kwon, O., Levine, M., Guttentag, S. H., Nussbaum, R. L., *Ascorbic-acid transporter Slc23a1 is essential for vitamin C transport into the brain and for perinatal survival.* Nat Med, 2002. **8**(5): pp. 514–7.

8. Corpe, C. P., Tu, H., Eck, P., Wang, J., Faulhaber-Walter, R., Schnermann, J., Margolis, S., Padayatty, S., Sun, H., Wang, Y., Nussbaum, R. L., Espey, M. G., Levine, M., *Vitamin C transporter Slc23a1 links renal reabsorption, vitamin C tissue*

*accumulation, and perinatal survival in mice.* J Clin Invest, 2010. **120**(4): pp. 1069–83.

9. Levine, M., Wang, Y., Padayatty, S. J., Morrow, J., *A new recommended dietary allowance of vitamin C for healthy young women.* Proc Natl Acad Sci USA, 2001. **98**(17): pp. 9842–6.

10. Dakhale, G. N., Khanzode, S. D., Khanzode, S. S., Saoji, A., *Supplementation of vitamin C with atypical antipsychotics reduces oxidative stress and improves the outcome of schizophrenia.* Psychopharmacology (Berl), 2005. **182**(4): pp. 494–8.

11. Sandyk, R., Kanofsky, J. D., *Vitamin C in the treatment of schizophrenia.* Int J Neurosci, 1993. **68**(1–2): pp. 67–71.

12. Suboticanec, K., Folnegovic-Smalc, V., Korbar, M., Mestrovic, B., Buzina, R., *Vitamin C status in chronic schizophrenia.* Biol Psychiatry, 1990. **28**(11): pp. 959–66.

13. Bornstein, S. R., Yoshida-Hiroi, M., Sotiriou, S., Levine, M., Hartwig, H. G., Nussbaum, R. L., Eisenhofer, G., *Impaired adrenal catecholamine system function in mice with deficiency of the ascorbic acid transporter (SVCT2).* Faseb J, 2003. **17**(13): pp. 1928–30.

14. Ward, M. S., Lamb, J., May, J. M., Harrison, F. E, *Behavioral and mono-amine changes following severe vitamin C deficiency.* J Neurochem, 2013. **124**(3): pp. 363–75.

15. Wilson, J. X., Peters, C. E., Sitar, S. M., Daoust, P., Gelb, A.W., *Glutamate stimulates ascorbate transport by astrocytes.* Brain Res, 2000. **858**(1): pp. 61–6.

16. He, X. B., Kim, M., Kim, S. Y., Yi, S. H., Rhee, Y. H., Kim, T., Lee, E. H., Park, C. H., Dixit, S., Harrison, F. E., Lee, S. H., *Vitamin C facilitates dopamine neuron differentiation in fetal midbrain through TET1- and JMJD3-dependent epigenetic control manner.* Stem Cells, 2015. **33**(4): pp. 1320–32.

17. Yin, R., Mao, S. Q., Zhao, B., Chong, Z., Yang, Y., Zhao, C., Zhang, D., Huang, H., Gao, J., Li, Z., Jiao, Y., Li, C., Liu, S., Wu, D., Gu, W., Yang, Y. G., Xu, G. L., Wang, H., *Ascorbic acid enhances Tet-mediated 5-methylcytosine oxidation and promotes DNA demethylation in mammals.* J Am Chem Soc, 2013. **135**(28): pp. 10396–403.

18. Minor, E. A., Court, B. L., Young, J. I., Wang, G., *Ascorbate induces ten-eleven translocation (Tet) methylcytosine dioxygenase-mediated generation of 5-hydroxy-methylcytosine.* J Biol Chem, 2013. **288**(19): pp. 13669–74.

19. Lind, J., *A treatise on the scurvy in three parts: containing an inquiry into the nature, causes, and cures of that disease: together with a critical and chronological view of what has been published on the subject.* Special ed. The Classics of Medicine Library. 1980, Birmingham, AL: Classics of Medicine Library. xiv.

20. Harrison, F. E., Meredith, M. E., Dawes, S. M., Saskowski, J. L., May, J. M., *Low ascorbic acid and increased oxidative stress in gulo(-/-) mice during development.* Brain Res, 2010. **1349**: pp. 143–52.

21. Chen, Y., Curran, C. P., Nebert, D. W., Patel, K. V., Williams, M. T., Vorhees, C. V., *Effect of vitamin C deficiency during postnatal development on adult behavior:*

*functional phenotype of Gulo(-/-) knockout mice.* Genes Brain Behav, 2012. **11**(3): pp. 269–77.

22.   de Jong, T. M., Jochens, A., Jockel-Schneider, Y., Harks, I., Dommisch, H., Graetz, C., Flachsbart, F., Staufenbiel, I., Eberhard, J., Folwaczny, M., Noack, B., Meyle, J., Eickholz, P., Gieger, C., Grallert, H., Lieb, W., Franke, A., Nebel, A., Schreiber, S., Doerfer, C., Jepsen, S., Bruckmann, C., van der Velden, U., Loos, B. G., Schaefer, A.S ., *SLC23A1 polymorphism rs6596473 in the vitamin C transporter SVCT1 is associated with aggressive periodontitis.* J Clin Periodontol, 2014. **41**(6): pp. 531–40.

23.   Zloch, Z., Ginter, E., *Moderate alcohol consumption and vitamin C status in the guinea-pig and the rat.* Physiol Res, 1995. **44**(3): pp. 173–8.

24.   Delanghe, J. R., De Buyzere, M. L., Speeckaert, M. M., Langlois, M. R., *Genetic aspects of scurvy and the European famine of 1845–1848.* Nutrients, 2013. **5**(9): pp. 3582–8.

25.   Delanghe, J. R., Langlois, M. R., De Buyzere, M. L., Na, N., Ouyang, J., Speeckaert, M. M., Torck, M. A., *Vitamin C deficiency: more than just a nutritional disorder.* Genes Nutr, 2011. **6**(4): pp. 341–6.

26.   *Centers for Disease Control, National Diabetes Statistics Report.* 2014. p. 6.

27.   Hink, U., Li, H., Mollnau, H., Oelze, M., Matheis, E., Hartmann, M., Skatchkov, M., Thaiss, F., Stahl, R. A., Warnholtz, A., Meinertz, T., Griendling, K., Harrison, D. G., Forstermann, U., Munzel, T., *Mechanisms underlying endothelial dysfunction in diabetes mellitus.* Circ Res, 2001. **88**(2): pp. E14–22.

28.   May, J. M., Jayagopal, A., Qu, Z. C., Parker, W. H., *Ascorbic acid prevents high glucose-induced apoptosis in human brain pericytes.* Biochem Biophys Res Commun, 2014. **452**(1): pp. 112–7.

29.   Meredith, M. E., Qu, Z. C., May, J. M., *Ascorbate reverses high glucose- and RAGE-induced leak of the endothelial permeability barrier.* Biochem Biophys Res Commun, 2014. **445**(1): pp. 30–5.

30.   Maxwell, S. R., Thomason, H., Sandler, D., Leguen, C., Baxter, M. A., Thorpe, G. H., Jones, A. F., Barnett, A. H., *Antioxidant status in patients with uncomplicated insulin-dependent and non-insulin-dependent diabetes mellitus.* Eur J Clin Invest, 1997. **27**(6): pp. 484–90.

31.   Stankova, L., Riddle, M., Larned, J., Burry, K., Menashe, D., Hart, J., Bigley, R., *Plasma ascorbate concentrations and blood cell dehydroascorbate transport in patients with diabetes mellitus.* Metabolism, 1984. **33**(4): pp. 347–53.

32.   Yue, D. K., McLennan, S., McGill, M., Fisher, E., Heffernan, S., Capogreco, C., Turtle, J. R., *Abnormalities of ascorbic acid metabolism and diabetic control: differences between diabetic patients and diabetic rats.* Diabetes Res Clin Pract, 1990. **9**(3): pp. 239–44.

# BIBLIOGRAPHY

*Abstract of the Evidence delivered before a Select Committee of the House of Commons (1790–91) concerning the Slave Trade.* 1792. Bury St. Edmonds.

Addington, Anthony. 1753. *An Essay on the Sea-Scurvy*. Reading, UK: C. Micklewright.

Addison, Joseph, and Richard Steele. 1907. *The Spectator*. Edited by G. Gregory Smith. 8 vols. bound as 4. London: J. M. Dent & Co.

Agamben, Giorgio. 1999. *Remnants of Auschwitz: The Witness and the Archive*. Translated by Daniel Heller-Roazen. New York: Zone.

Alexander, Caroline. 1919. *The Endurance: Shackleton's Legendary Antarctic Expedition*. New York: Knopf.

Amundsen, Roald. (1912) 1976. *The South Pole*. Translated by A. G. Chater. London: John Murray; repr. London: C. Hurst & Co.

Anon. 1824. *Aureus, or The Life and Opinions of a Sovereign*. London.

Anon. 1962. "Scott and Scurvy." *Canadian Medical Journal* 87: 32–33.

Anon. 1778. *The Travels of Hildebrand Bowman, written by himself*. London: W. Strahan and T. Cadell.

Annual Returns of Disease and Death. 1838. *House of Assembly Journals* (Hobart). HAJ/5/1860/105.

Armstrong, Alexander. 1858. *On Naval Hygiene and Scurvy*. London: John Churchill.

Backhouse, James. 1843. *A Narrative of a Visit to the Australian Colonies*. London: Hamilton, Adams and Co.

Bacon, Francis. 2014. *New Atlantis*. Edited by G. C. Moore Smith. Cambridge: Cambridge University Press.

Ballaster, Ros. 2005. *Fabulous Orients: Fictions of the East in England 1662–1785*. Oxford: Oxford University Press.

Banks, Joseph. 1962. *The Endeavour Journal of Joseph Banks 1768–71*. 2 vols. Edited by J. C. Beaglehole. Sydney: Public Library of New South Wales and Angus and Robertson.

———. 2007. *The Scientific Correspondence of Sir Joseph Banks 1765–1820*. 6 vols. Edited by Neil Chambers. London: Pickering and Chatto.

Barrère, Pierre. 1753. *Observations anatomiques tirees des overtures d'un grand nombre de cadavers*. Perpignan.

Bateson, Charles. 1959. *The Convict Ships 1787–1868*. Glasgow: Brown, Son & Ferguson.

Baudin, Nicolas. 2004. *The Journal of Post-Captain Nicolas Baudin*. Translated by Christine Cornell. Adelaide: Libraries Board of Australia.

Beale, Thomas. 1839. *A Natural History of the Sperm Whale*. London: John van Voorst.

Beddoes, Thomas. 1793. *Observations on the Nature and Cure of Calculus, Sea Scurvy, Etc.* London: John Murray.

———. 1796. *Considerations on the Medicinal Use of Factitious Airs.* Part 1. 2nd ed. Bristol: J. Johnson.

———. 1802. *Hygeia: Essays Moral and Medical on the Causes Affecting the Personal State of our Middling and Affluent Classes.* 3 vols. Bristol: J. Mills.

Beer, Gillian. 1996. *Open Fields: Science in Cultural Encounter.* Oxford: Oxford University Press.

Bentham, Jeremy 1803. *A Plea for the Constitution, Shewing the Enormities Committed . . . in and by the Design, Foundation and Government of the Penal Colony of New South Wales.* London: Mawman and Hatchard.

Berkeley, George. 1734. *The Analysis; or, A Discourse Addressed to an Infidel Mathematician.* London: J. Tonson.

———. 1972. *Essay towards a New Theory of Vision.* London: Dent.

Bernardin de St. Pierre, Jacques-Henri. 1800. *A Voyage to the Isle of France.* London: J. Cundee.

———. 1804. *Etudes de la nature.* 5 vols. Paris: Imprimerie de Crapelet.

———. (1788) 1819. *Paul and Virginia.* Translated by Helen Maria Williams. London: John Sharpe.

———. (1788) 1907. *Paul et Virginie.* London: J. M. Dent & Co.

Betagh, William. 1728. *A Voyage round the World.* London: T. Combes.

Bewell, Alan. 1999. *Romanticism and Colonial Disease.* Baltimore: Johns Hopkins University Press.

Blane, Gilbert. 1799. *Observations on the Diseases of Seamen.* 3rd ed. London: Murray and Highly.

Bligh, William. 1937. *The Log of the Bounty.* Edited by Owen Rutter. 2 vols. London: Golden Cockerel.

———. 2011. *Account of the Rebellion of the New South Wales Corps.* Edited by John Currey. Malvern, Australia: Colony Press for the Banks Society.

Boldrewood, Rolf. 1951. *Robbery under Arms.* Oxford: Geoffrey Cumberlege at Oxford University Press.

Bollet, Alfred J. 2004. *Plagues and Poxes.* New York: Demos.

Bonnemains, Jacqueline, Elliott Forsyth, and Bernard Smith, eds. 1988. *Baudin in Australian Waters.* Melbourne: Melbourne University Press and the Australian Academy of Humanities.

Boswell, James. 1980. *The Life of Samuel Johnson.* Edited by R. W. Chapman. Oxford: Oxford University Press.

Bougainville, Louis-Antoine de. 2002. *Pacific Journal of Louis-Antoine de Bougainville 1767–68.* Translated and edited John Dunmore. London: Hakluyt Society.

Boulaire, Alain. 1997. *Kerguelen, le Phenix des Mers Australe.* Paris: France-empire.

Bown, Stephen. 2003. *Scurvy: How a Surgeon, a Mariner and a Gentleman Solved the Greatest Medical Mystery of the Age of Sail*. Camberwell, Victoria: Penguin Books Australia.

Boyle, Robert. 1670. *Experiments and Considerations touching Colours*. London: Henry Herringman.

———. 1671. *The Excellence of Theology compar'd with Natural Philosophy*. London: T. N. for Henry Herringman.

———. (1672) 1999. *Of the Great Efficacy of Effluviums*. In *The Works of Robert Boyle*, 14 vols., edited by Michael Hunter and Edward B. Davis, 7.257–71. London: Pickering and Chatto.

———. (1673) 1999. *Of the Strange Subtlety of Effluviums*. In *Works* (1672) 1999, 7.233–54.

———. 1687. *The Martyrdom of Theodora and Didymus*. London: H. Clark.

———. 1690. *An Essay on the Effects of Languid and Unheeded Motion*. London: Samuel Smith.

———. 1996. *A Free Enquiry into Vulgarly Received Notions of Nature*. Edited by Edward Davis and Michael Hunters. Cambridge: Cambridge University Press.

———. 1999. "Relations about the Bottom of the Sea." In *Works*, 7.413–17.

Boym, Svetlana. 2001. *The Future of Nostalgia*. New York: Basic Books.

Bradley, William. 1969. Journal of Lt. William Bradley (1786–92). Facsimile. Sydney: Trustees of the Public Library of New South Wales with Ure Smith Pty Ltd.

Brossard, Maurice Raymond de. 1970. *Kerguelen le Decouvrier et ses iles*. 2 vols. Paris: Editions France Empire.

Brown, William. 1788. "A Letter from William Brown, Physician at Kolyvan in Siberia ... giving an Account of the Scurvy which prevailed in Russia." In *Medical Commentaries for the Year 1787 collected by Andrew Duncan*, vol. 2. Edinburgh: C. Elliott, T. Kay & Co.

Bulkeley, John, and John Cummins. 1927. *A Voyage to the South Seas in His Majesty's Ship the* Wager, *1740–41*. London: R. Walker.

Burnett, John. 1831. Letter, Port Arthur. CSOI-484–10750. 10 November.

Burton, Richard F., trans. 1881. *Camoens: His Life and His Lusiads*. 2 vols. London: Bernard Quaritch.

Byron, George Gordon, Lord. 1824. *The Two Foscari*. London: Benbow.

Byron, John. 1768. *An Account of the Great Distresses on the Coasts of Patagonia*. London: S. Baker and G. Leigh.

———. 1773. "An Account of a Voyage round the World 1764–66." In Hawkesworth 1773, 1–139.

Cameron, Charles. 1829. *Copy of a Letter to the Commissioners for Victualling his Majesty's Ships*. London.

Camoens, Luis Vaz de. 1881. *Camoens: His Life and His Lusiads*. 2 vols. Translated by Richard F. Burton. London: Bernard Quaritch.

————. 1940. *The Lusiads*. Translated by Richard Fanshawe. Edited by Jeremiah D. M. Ford. Cambridge, MA: Harvard University Press.

————. 1997. *The Lusiads*. Translated by Landeg White. Oxford: Oxford University Press.

Campbell, Alexander. 1747. *Sequel to Bulkeley's and Cummins's Voyage*. London: A. Campbell.

Carew, Bampfylde Moore. 1813. *The Surprising Adventures of Bampfylde Moore Carew, King of the Beggars*. London: A. K. Newman.

Carey, Peter. 1997. *Jack Maggs*. London: Faber and Faber.

————. 2000. *The True History of the Kelly Gang*. St. Lucia: University of Queensland Press.

Carpenter, Kenneth. 1981. *Pellagra*. Stroudsburg, PA: Hutchinson Ross.

————. 1986. *The History of Scurvy and Vitamin C*. Cambridge: Cambridge University Press.

Carter, Harold. 1988. *Sir Joseph Banks, 1743–1820*. London: British Museum.

Carter, Paul. 1987. *The Road to Botany Bay: An Exploration of Landscape and History*. Chicago: University of Chicago Press.

Carter, Paul, and Susan Hunt. 1999. *Terre Napoleon: Australia through French Eyes*. Sydney: Historic Houses Trust and Hordern House.

Carteret, Philip. 1773. "An Account of a Voyage round the World 1766–69." In Hawkesworth 1773, 303–448.

————. 1965. *A Voyage Round the World 1766–1769*. Edited by Helen Wallis. 2 vols. Cambridge: Cambridge University Press for the Hakluyt Society.

Casey, Gavin. 1834. Letter, Port Arthur. CSOI-706–15459. 15 April.

————. 1834. Surgeon's Return for Port Arthur. CSOI-735–15912. 9 December.

————. 1835. Letter from Port Arthur to James Scott, Colonial Surgeon. CSOI-735–15912. 1 January.

————. 1835. Letter, Port Arthur. CSOI-641–14418. 11 January.

Cavendish, Margaret. 1994. *The Description of a New World, called The Blazing World*. Edited by Kate Lilley. London: Penguin.

————. 2001. *Observations upon Experimental Philosophy*. Cambridge: Cambridge University Press.

Cervantes Saavedra, Miguel de. 1991. *Don Quixote*. Translated by Peter Motteux. 2 vols. New York: Everyman.

Chambers, Ephraim. 1786. "Calenture." In *Cyclopaedia*. London: J. F. and C. Rivington et al.

Chaplin, Joyce. 2012. "Earthsickness: Circumnavigation and the Terrestrial Human Body." *Bulletin of the History of Medicine* 86: 515–42.

————. 2012a. *Round about the World: From Magellan to Orbit*. New York: Simon and Schuster.

Charleton, Walter. 1654. *Physiologia: Epicuro-Gassendo-Charletonianae; or, A Fabrick of Science Natural upon the Hypothesis of Atoms*. London: Thomas Newcomb.

———. 1657. *The Immortality of the Human Soul*. London: Henry Herringman.

———. 1659. *The Natural History of Nutrition*. London: Thomas Herringman.

———. 1670. *Natural History of the Passions*. London: James Magnes.

———. 1672. *De Scorbuto*. London: William Wells and Robert Scot.

Charters, Erica. 2009. "Disease, Wilderness, Warfare and Imperial Relations: The Battle for Quebec, 1759–60." *War in History* 16 (1): 5–28.

Cherry-Garrard, Apsley. 1994. *The Worst Journey in the World*. London: Vintage Classic.

Chick, Harriet. 1953. "Early Investigations of Scurvy and the Antiscorbutic Vitamin." *Proceedings of the Nutrition Society* 12 (3): 210–19.

Clark, Ralph. 1981. *The Journal and Letters 1787–92*. Edited by Paul G. Fidlon and R. J. Ryan. Sydney: Australian Document Library.

Clarke, Marcus. 1899. *For the Term of His Natural Life*. London: Macmillan.

———. 1952. *For the Term of His Natural Life*. Oxford: Oxford University Press.

Clendinnen, Inga. 2005. *Dancing with Strangers: The True History of the Meeting of the British First Fleet and the Aboriginal Australians, 1788*. Edinburgh: Canongate.

Cobley, John. 1963. *Sydney Cove 1789–90*. Sydney: Angus and Robertson.

Coleridge, Samuel Taylor. (1796) 1970. *The Watchman*. Edited by Lewis Patton. Vol. 75, Bollingen Series. London: Routledge and Kegan Paul.

———. 1930. *Coleridge's Shakespearian Criticism*. 2 vols. Edited by Thomas Middleton Raysor. Cambridge, MA: Harvard University Press.

———. 1956. *Collected Letters*. Edited by Earl Leslie Griggs. 6 vols. Oxford: Clarendon Press.

———. 1993. *The Rime of the Ancient Mariner*. In *Coleridge's Ancient Mariner: An Experimental Edition of Text of Revisions 1798–1828*, edited by Martin Wallen. New York: Station Hill.

———. 1995. *Shorter Works and Fragments*. 2 vols. Edited by H. J. Jackson and J. R. de J. Jackson. Princeton, NJ: Princeton University Press; London: Routledge.

———. 2002. *Notebooks*. Edited by Kathleen Coburn. London: Routledge.

———. 2004. *Complete Poems*. Edited by William Keach. London: Penguin.

Collins, David. 1798, 1802. *An Account of the English Colony in New South Wales*. 2 vols. London: T. Cadell and W. Davies.

Colnett, James. 1795. *A Voyage . . . into the Pacific Ocean*. London: James Colnett.

Condillac, Etienne Bonnot de. 1930. *Treatise on the Sensations*. Translated by Geraldine Carr. London: Favil Press.

Congreve, William. 1923. *The Complete Works of William Congreve*. Edited by Montagu Summers. 4 vols. London: Nonesuch Press.

Cook, G. C. 2004. "Scurvy in the British Mercantile Marine in the Nineteenth Century." *Postgraduate Medical Journal* 80: 334–29.

Cook, James. 1776. "The Method taken for preserving the Health of the Crew of His Majesty's Ship the Resolution during her late Voyage round the World." In

*Philosophical Transactions of the Royal Society of London 1776*, vol. 66, part 2. London: J. Bowyer and J. Nichols.

———. 1777. *A Voyage towards the South Pole and Round the World 1772–75*. 2 vols. London: W. Strahan and T. Cadell.

———. 1955. *The Voyage of the* Endeavour *1768–71*. Edited by J. C. Beaglehole. London: Hakluyt Society.

———. 1961. *The Voyage of the* Resolution *and* Adventure *1772–75*. Edited by J. C. Beaglehole. London: Hakluyt Society.

———. 1967. *The Voyage of the* Resolution *and* Discovery *1776–80*. Edited by J. C. Beaglehole. 2 vols. London: Hakluyt Society.

Corbin, Alain. 1986. *The Foul and the Fragrant: Odor and the French Social Imagination*. Cambridge, MA: Harvard University Press.

Costello, Con. 1987. *The Story of the Convicts Transported from Ireland to Australia 1791–1853*. Cork: Mercier Press.

Cowley, Abraham. 1881. *The Complete Works*. Edited by Alexander Grosart. 2 vols. Edinburgh: Constable.

Cowper, William. 1854. *Poetical Works*. Edited by Robert Aris Willmott. London: George Routledge and Sons.

Coyer, Gabriel Francois. 1752. *A Supplement to Lord Anson's Voyage round the World*. London: A. Millar.

Crosfield, R. T. 1797. *Remarks on the Scurvy*. London: R. T. Crosfeild.

Crowley, John E. 2011. *Imperial Landscapes*. New Haven, CT: Yale University Press for the Mellon Center for British Art.

Cudworth, Ralph 1731. *A Treatise concerning Eternal and Immutable Morality*. London: James and John Knapton.

Cullen, William. 1800. *Nosology; or, A Systematic Arrangement of Diseases by Classes, Orders, Genera and Species*. Edinburgh: William Creech.

———. 1827. *Works of William Cullen*. Edited by John Thomson. 2 vols. Edinburgh: Blackwood.

Cuppage, Francis E. 1994. *James Cook and the Conquest of Scurvy*. Westport, CT: Greenwood.

Curran, J. O. 1847. "Observations on Scurvy as It Has Lately Appeared throughout Ireland." *Dublin Quarterly Journal of Medical Science* 4 (August and September): 83–134.

Currey, John. 2000. *David Collins: A Colonial Life*. Carlton, Victoria: Melbourne University Press.

Dames, Nicholas. 2001. *Amnesiac Selves: Nostalgia, Suffering and British Fiction 1810–70*. Oxford: Oxford University Press.

———. 2007. *The Physiology of the Novel: Reading, Neural Science, and the Form of the Novel*. Oxford: Oxford University Press.

Damousi, Joy. 1997. *Depraved and Disorderly: Female Convicts, Sexuality and Gender in Colonial Australia*. Cambridge: Cambridge University Press.

Dampier, William. 1729. *A Collection of Voyages*. 4 vols. London: J & J Knapton.

———. 1939. *A Voyage to New Holland*. Edited by James A. Williamson. London: Argonaut Press.

———. 1999. *A New Voyage round the World*. London: Hummingbird Press.

Darwin, Charles. 1904. *The Expression of Emotion in Man and Animals*. London: John Murray.

Darwin, Erasmus. 1794. *Zoonomia*. 2 vols. London: J. Johnson.

———. 1801. *Zoonomia*. 4 vols. London: J. Johnson.

Daston, Lorraine. 1991. "Marvelous Facts and Miraculous Evidence in Early Modern Europe." *Critical Inquiry* 18 (1): 93–124.

Daston, Lorraine, and Katharine Park. 1998. *Wonders and the Order of Nature, 1150–1750*. New York: Zone Books.

Davis, John. 1880. *Voyages and Works*. Edited by A. H. Markham. London: Hakluyt Society.

Davy, Humphrey. 1800. *Researches, Chemical and Philosophical, Chiefly Concerning Nitrous Oxide*. London: J. Johnson.

———. 1851. *Consolations in Travel*. London: John Murray.

de Certeau, Michel. 1996. "Vocal Utopias: Glossolalias." *Representations* 56 (Fall): 29–47.

Defoe, Daniel. 1927. *The Farther Adventures of Robinson Crusoe*. 3 vols. London: Basil Blackwell.

———. 1966. *Journal of the Plague Year*. Oxford: Oxford University Press.

———. 1983. *The Life and Surprizing Adventures of Robinson Crusoe*. Edited by J. Donald Crowley. Oxford: Oxford University Press.

Delanghe, Joris R., Marc L. De Buyzere, Marijn M. Speeckaert, and Michael R. Langlois. 2013. "Genetic Aspects of Scurvy in the European Famine of 1845–8." *Nutrients* 5: 3582–88.

Delepine, Gracie. 1998. *L'Amiral de Kerguelen et les mythes de son temps*. Paris: L'Harmattan.

Descartes, Rene. 1998. *Treatise on Light*. In *The World and Other Writings*, translated and edited by Stephen Gaukroger, 1–75. Cambridge: Cambridge University Press.

Dettelbach, Michael. 2005. "The Stimulation of Travel: Humboldt's Physiological Construction of the Tropics." In *Tropical Visions in an Age of Empire*, edited by Felix Driver and Luciana Martins, 43–58. Chicago: University of Chicago.

Dibdin, Charles. 1779. *The Mirror; or, Harlequin Every Where*. London: G. Kearsly.

Dickson, Walter. 1866. "Scurvy in the Mercantile Marine." In *Transactions of the Epidemiological Society*, 2–22. London: T. Richards.

Dickens, Charles. 2008. *Great Expectations*. Edited by Margaret Cardwell. Oxford: Oxford University Press.

Diderot, Denis. 1992. *Supplément au Voyage de Bougainville*. Translated by John Hope Mason and Robert Wokler. In *Diderot: Political Writings*, edited by John Hope Mason and Robert Wokler, 31–75. Cambridge: Cambridge University Press.

Digby, Sir Kenelm. 1658. *Two Treatises: The Nature of Bodies and the Immortality of Reasonable Souls*. London: John Williams.

———. 1669. *A Discourse concerning the Vegetation of Plants*. London: John Williams.

———. 1968. *Loose Fantasies*. Edited by Vittorio Gabrieli. Rome: Edizione di Storia e Letteratura.

Douglas, Bronwen. 2005. "'Cureous Figures': European Voyagers and Tatau/Tattoo in Polynesie, 1595–1800." In *Tattoo: Bodies, Art, and Exchange in the Pacific and the West*, edited by Nicholas Thomas, Anna Cole, and Bronwen Douglas, 33–52. Durham, NC: Duke University Press.

Douglass, Frederick. 2003. *My Bondage and My Freedom*. Edited by John David Smith. London: Penguin Books.

Douthwaite, Julia. 2002. *The Wild Girl, Natural Man and the Monster*. Chicago: University of Chicago Press.

Druett, Joan. 2011. *Tupaia: Captain Cook's Polynesian Navigator*. Santa Barbara, CA: Praeger.

Dryden, John. 1808. *The Works of John Dryden*. Edited by Walter Scott. 18 vols. London: William Miller.

———. 1811. *Virgil's Aeneas*. London: J. Walker and J. Harris.

Dumont D'Urville, Jules S. C. 1987. *An Account of Two Voyages to the South Seas*. 2 vols. Honolulu: University of Hawai'i Press.

Duncan, Andrew, ed. 1788. *Medical Commentaries for the Year 1787*. Vol. 2. Edinburgh: C. Elliot, T. Kay & Co.

Dunlop, John. 1816. *The History of Fiction*. 3 vols. Edinburgh: James Ballantyne.

Dunmore, John. ed. 1994. *The Journal of J. F. de Galaup de La Perouse*. 2 vols. London: Hakluyt Society.

Edgeworth, Richard Lovell, and Maria Edgeworth. 1802. *Essay on Irish Bulls*. London: J. Johnson.

*Encyclopédie ou Dictionnaire Raisonné des Sciences, des Arts et des métiers*. 1969. 5 vols. New York: Pergamon Press.

Evans, Edward R.G.R. 1926. *South with Scott*. London: William Collins.

Fabian, Johannes. 2000. *Out of Our Minds: Reason and Madness in the Exploration of Central Africa*. Berkeley: University of California Press.

Falconer, William. 1788, 1791. *A Dissertation on the Influence of the Passions upon Disorders of the Body*. London: C. Dilly and J. Phillips.

Fanning, Edmund. 1989. *Voyages and Discoveries in the South Seas, 1792–1832*. Mineola, NY: Dover Publications. First published Salem, MA: Marine Research Society, 1924.

Farquhar, George. 1721. *The Recruiting Officer*. In *The Comedies of George Farquhar*. 2 vols. 4th ed. London: J. Knapton.

Flinders, Matthew. 1814. *A Voyage to Terra Australis 1801–1803*. 2 vols. London: G. and W. Nicol.

———. 2015. *Australia Circumnavigated: The Voyage of Matthew Flinders in* HMS Investigator, *1801–1803*. Edited by Kenneth Morgan. 2 vols. London: Ashgate for the Hakluyt Society.

Forster, Georg. 1777. *A Voyage round the World in the Sloop* Resolution, *1772–75*. 2 vols. London: G. Robinson.

Forster, Johann Reinhold. 1982. *Resolution Journal*. Edited by Michael E. Hoare. 4 vols. London: Hakluyt Society.

———. 1996. *Observations Made during a Voyage round the World*. Edited by Nicholas Thomas, Harriet Guest, and Michael Dettelbach. Honolulu: University of Hawai'i Press.

Fowell, John. 1789. MS Letter, MLMSS 4895/1/19.

Foxhall, Katherine. 2012. *Health, Medicine and the Sea: Australian Voyages c.1815–1860*. Manchester: Manchester University Press.

Funnell, William. 1707. *A Voyage Round the World 1703–4*. London: W. Botham & James Knapton.

Furphy, Joseph. 1981. *Such is Life*. Edited by John Barnes. St. Lucia: University of Queensland Press.

Gallagher, Catherine. 2006. "The Rise of Fictionality." In *The Novel*, 2 vols., edited by Franco Moretti, 1.336–63. Princeton, NJ: Princeton University Press.

Gay, John. 1974. *Poetry and Prose*. Edited by Vincent A. Dearing. 2 vols. Oxford: Clarendon Press.

Gilpin, William. 1792. *Observations on the River Wye*. London: R. Blamire.

Gold, John. 1827. Correspondence, Mitchell Library MSS 846. Vol. 4, 79–80.

Golinski, Jan. 2005. *Making Natural Knowledge*. Chicago: University of Chicago Press.

———. 2011. "Humphry Davy: The Experimental Self." *Eighteenth-Century Studies* 45 (1): 15–28.

Goodman, Kevis. 2008. "Romantic Poetry and the Science of Nostalgia." In *Cambridge Companion to Romantic Poetry*, edited by James Chandler and Maureen N. McLane, 195–216. Cambridge: Cambridge University Press.

———. 2010. " 'Uncertain Disease': Nostalgia, Pathologies of Motion, Practices of Reading." *Studies in Romanticism* 49 (Summer): 197–227.

Grimaud, Jean. 1789. *Memoires sur la nutrition*. St. Petersburg: Imprimerie de L'Academie Imperiale des Sciences.

Guicciardini, Niccolo. 2009. *Isaac Newton on Mathematical Certainy and Method*. Cambridge, MA: MIT Press.

Guly, Henry. 2012. "The Understanding of Scurvy during the Heroic Age of Antarctic Exploration." In *Polar Record*, 1–7. Cambridge: Cambridge University Press.

Guthrie, Matthew. 1788. "A Letter from Matthew Guthrie, Physician at Petersburg … [on] the Effects of a Cold Climate on the Land Scurvy." In Duncan, 328–38.

Hampton, Timothy. 2009. *Fictions of Embassy*. Ithaca, NY: Cornell University Press.

Harris, James. 1841. *The Works of James Harris*. Oxford: J. Vincent.

Harrison, Fiona. 2013. "Behavioural and Neurochemical Effects of Scurvy in Gulo Knockout Mice." *Journal for Maritime Research* 15 (1): 107–14.

Harrison, Fiona, with James May. 2009. "Vitamin C Function in the Brain: Vital Role of the Ascorbate Transporter SVCT2." *Free Radical Biology & Medicine* 46: 719–30.

Harrison, Mark. 2010. *Medicine in an Age of Commerce and Empire: Britain and Its Tropical Colonies, 1660–1830*. Oxford: Oxford University Press.

———. 2013. "Scurvy on Sea and Land: Political Economy and Natural History." *Journal for Maritime Research* 15 (1): 7–25.

Hartley, David. 1810. *Observations on Man*. 2 vols. London: Wilkie and Robinson.

Harvey, Gideon. 1675. *The Disease of London: or, A New Discovery of the Scorvey*. London: T. James for W. Thackery.

Haywood, Eliza. 2000. *Love in Excess*. Edited by David Oakleaf. Ontario: Broadview.

Hawkesworth, John. 1773. *An Account of the Voyages and Discoveries in the Southern Hemisphere*. 3 vols. London: W. Strahan and T. Cadell.

Hawkins, Richard. 1906. "Voyage into the South Seas." In *Samuel Purchas, Hakluytus Posthumus: or, Purchas his Pilgrimes*, 20 vols., 17.57–204. Glasgow: James Maclehose.

Hazlitt, William. 1904. *Collected Works*. Edited by A. R. Waller and A. Glover. 12 vols. London: J. M. Dent.

Heller-Roazen, Daniel. 2003. *Fortune's Faces: The* Roman de la Rose *and the Poetics of Contingency*. Baltimore: Johns Hopkins University Press.

———. 2013. *Dark Tongues: The Art of Rogues and Riddlers*. New York: Zone Books.

Henderson, Andrew. 1833. Surgeon's Journal, Convict Ship *Royal Admiral*, TNA ADM 101/65/2.

———. 1835. Surgeon's Journal, Convict Ship *Aurora*, TNA ADM 101/6/7.

Heylyn, Peter. 1667. *Cosmographie*. London: Philip Chetwode.

Hickey, Alison. 1997. *Impure Conceits*. Stanford: Stanford University Press.

*Historical Records of Australia*. 1914. Series I, vols. xix, xx. 1788–96. Canberra: Library Committee of Commonwealth Parliament.

———. 1921–23. Series III, vol. 1. Tasmania 1821–25.

———. 1924. Series I, vol. xx.

———. 1927. Series III, vi.

Hobbes, Thomas. 1839. *Elements of Philosophy*. Vol. 1 in *The Works of Thomas Hobbes*, edited by Sir William Molesworth. 11 vols. London: John Bohn.

———. 1998. *On the Citizen*. Edited by Richard Tuck and Michael Silverthorne Holmes. Cambridge: Cambridge University Press.

———. 2004. *Leviathan*. Edited by Richard Tuck. Cambridge: Cambridge University Press.

Hofer, Johannes. (1688) 1934. *Medical Dissertation on Nostalgia*. Translated by Carolyn Kiser Anspach. *Bulletin of the History of Medicine*: 376–87.

Homer. 1956. *The Odyssey*. Edited by Allardyce Nicoll. Translated by George Chapman. Bollingen Series, 41. Princeton, NJ: Princeton University Press.

Hooke, Robert. 1969. "General Scheme." In *Posthumous Works* (1705), edited by Richard Waller, 3–70. New York: Johnson Reprint Corporation.

———. (1665) 2003. *Micrographia*. London: John Martyn and James Allestry; repr. New York: Dover Publications.

Hoorn, Jeanette. 1990. *The Lycett Album*. Canberra: National Library of Australia.

Huber, Therese. (1801) 1966. *Adventures on a Journey to New Holland*. Translated by Rodney Livingstone. Sydney: Lansdown Press.

Hughes, Robert. 1988. *The Fatal Shore*. New York: Vintage Books.

Hulme, Nathaniel. 1768. *De Natura Scorbuti; and A Proposal for Preventing the Scurvy in the British Navy*. London: T. Cadell.

Hume, David. 1978. *Treatise of Human Nature*. Edited by L. A. Selby-Bigge and P. H. Nidditch. Oxford: Clarendon Press.

Hunter, John. 1968. *An Historical Journal of Events at Sydney and at Sea 1787–92*. Edited by John Bach. Sydney: Royal Australian Historical Society and Angus and Robertson.

Hutcheson, Francis. 2002. *Essay on the Nature and Conduct of the Passions and Affections*. Edited by Aaron Garrett. Indianapolis, IN: Liberty Fund.

Hutchinson, Lucy. 2012. *Works*. Vol 1 (two parts). Edited by Reid Barbour and David Norbrook. Oxford: Oxford University Press.

Illbruck, Helmut. 2012. *Nostalgia: Origins and Ends of an Unenlightened Disease*. Evanston, IL: Northwestern University Press.

James, Henry. 1947. *The Art of the Novel*. New York: Charles Scribner.

———. 1975. *The Spoils of Poynton*. London: Penguin Modern Classics.

Jay, Mike. 2009. *The Atmosphere of Heaven*. New Haven, CT: Yale University Press.

———. 2012. *The Influencing Machine: James Tilly Matthews and the Air Loom*. London: Strange Attractor Press.

Johnson, Samuel. 1760. "Calenture." In *A Dictionary of the English Language*. London: J. Knapton et al.

———. 1791. *Rambler*. 4 vols. in 2. London: W. Locke.

———. 1971. *A Journey to the Western Islands of Scotland*. Edited by M. Lascelles. New Haven, CT: Yale University Press.

Jones, Henry, ed. *Philosophical Transactions of the Royal Society 1700–1720*. 1721. Vol. 5. London: G. Strahan.

Kames, Home, Henry, Lord. 2005. *Elements of Criticism* Edited by Peter Jones. 2 vols. Indianapolis, IN: Library Fund.

Kamphuis, Amy Ruth. 2007. "Medicine and the Convict Body." BA Thesis. University of Tasmania (Hobart).

Kant, Immanuel. 2006. *Anthropology from a Pragmatic Point of View*. Edited and translated by Robert B. Louden. Cambridge: Cambridge University Press.

Keevil, J. J. 1957. *Medicine and the Navy 1200–1900*. Vol. 1 of 4. Edinburgh: E. & S. Livingstone.

Kelly, Celsus. 1966. *La Austrialia del Espiritu Santo: The Journal of Fray Martin de Munilla O. F. M.* 2 vols. Cambridge: Cambridge University Press for the Hakluyt Society.

Kelly, Ned. N.d. *Jerilderie Letter.* www.jerilderie.local-e.nsw.gov.au.

Kerguelen-Trémarec, Yves-Joseph de. 1782. *Relation de deux voyages dans les mers Australes & des Indes.* Paris: Knapen et fils.

Lafayette, Marie-Madeleine de. 1994. *The Princess of Cleves.* Edited by John D. Lyons. New York: Norton.

La Fontaine, Jean de. 1865. *Fables.* Paris: Furne et Cie.

Lamb, Jonathan. 2001. *Preserving the Self in the South Seas.* Chicago: University of Chicago Press.

———. 2005. "Inchoate Possession: How Captain Kerguelen Claimed an Island." *Journal of Maritime Research* (January): 1–9. Online at www.jmr.nmm.ac.uk /lamb.

———. 2009. "Scientific Gusto versus Monsters in the Basement." *Eighteenth-Century Studies* 42 (2): 309–20.

La Perouse, J. F. de Galaup de. 1994. *Journal.* Edited by John Dunmore. 2 vols. London: Hakluyt Society.

Larkin, Philip. (1988) 2003. "Church Going." In *Collected Poems,* edited by Anthony Thwaite, 58–59. London: Faber and Faber.

Latour, Bruno. 2002. *Iconoclash: Beyond the Image Wars in Science, Religion, and Art.* Edited by Bruno Latour and Peter Weibel. Karlsruhe, Germany: SKM/ Centre for Art and Media; Cambridge, MA: MIT Press.

Lavery, Brian, ed. 1998. *Shipboard Life and Organisation 1713–1815.* Vol. 138. Navy Records Society Publications. Aldershot: Ashgate.

Lawrence, Christopher. 1979. *Natural Order; Historical Studies of Scientific Culture.* Beverly Hills, CA: Sage Publications.

Leclerc, Georges-Louis, Comte de Buffon, and Louis-Jean-Marie Daubenton. 1758. *Histoire Naturelle.* 15 vols. Paris: Imprimerie Royale.

———. 1791. *Natural History: General and Particular.* 9 vols. Edited by William Smellie. London: A. Strahan and T. Cadell.

Lennox, Charlotte. 1973. *The Female Quixote.* Edited by Margaret Dalziel. Oxford: Oxford University Press.

Lery, Jean de. 1992. *History of a Voyage to the Land of Brazil.* Translated by Janet Whatley. Berkeley: University of California Press.

Levi, Primo. 1996. *Survival in Auschwitz.* Translated by Stuart Woolf. New York: Touchstone.

Lezra, Jacques. 1997. *Unspeakable Subjects.* Stanford: Stanford University Press.

Lind, James. 1753. *A Treatise on the Scurvy.* Edinburgh: J. Millar. 2nd ed., 1757.

Lindsey, D. T., and A. M. Brown. 2002. "Color Naming and the Phototoxic Effects of Sunlight on the Eye." *Psychological Science* 13 (6): 506–12.

Lloyd, Christopher, and Jack L. S. Coulter. 1961, 1963. *Medicine and the Navy 1200–1900.* Vols. 3 and 4 of 4 vols. London: E. & S. Livingstone.

Locke, John. 1979. *An Essay Concerning Human Understanding*. Edited by Peter H. Nidditch. Oxford: Clarendon Press.

Longinus. 1739. *On the Sublime*. Translated by William Smith. London: J. Watts.

Lucretius. 2006. *On the Nature of Things*. Edited by G. P. Goold. Cambridge, MA: Harvard University Press.

Luhmann, Niklas. 1986. *Love as Passion*. Translated by Jeremy Gaines and Doris L. Jones. London: Polity Press.

Lycett, Joseph. 1824, 1825. *Views in Australia*. London: J. Souter.

Lynch, Deidre. 1998. *The Economy of Character*. Chicago: University of Chicago Press.

Macalister, R. A. Stewart. 1937. *The Secret Languages of Ireland*. Cambridge: Cambridge University Press.

MacBride, David. 1767. *Experimental Essays*. London: A. Millar and T. Cadell.

Macdonald, Janet. 2004. *Feeding Nelson's Navy*. London: Chatham Press.

MacFie, Peter, and Marissa Bonet. 1985. "Convict Health at Port Arthur and Tasman Peninsula 1830–77: The Relationship between Diet, Work, Medical Care and Health." PhD dissertation. University of Tasmania (Hobart). Tasmania National Parks and Wildlife Service.

Mack, James D. 1972. *Matthew Flinders*. Penrith: Discovery Press.

Mahon, Henry. 1841. Surgeon's Journal, Convict Ship *Barossa*. TNA ADM 101/7/8.

Malieu de Meyserey, Guillaume. 1754. *La Médicine d'Armée*. Paris, Cavelier & fils.

Malory, Sir Thomas. 1969. *Le Morte d'Arthur*. 2 vols. Edited by Janet Cowen. Harmondsworth: Penguin.

Mantel, Hilary. 2004. "The Shape of Absence." In *In Search of Hannah Crafts*, edited by Henry Louis Gates and Hollis Robbins, 422–30. Cambridge, MA: BasicCivitas Books.

Markham, Albert Hastings, ed. 1880. *The Voyages and Works of John Davis*. London: Hakluyt Society.

Marvell, Andrew. 2007. *The Poems of Andrew Marvell*. Edited by Nigel Smith. London: Pearson Longman.

Marx, Karl. 1978. *Capital I*. In *The Marx-Engels Reader*, 2nd ed., edited by Mark C. Tucker, 294–438. New York: W. W. Norton.

Maxwell-Stewart, Hamish, and James Bradley. 2006. "Crime and Health—An Introductory View." In *Effecting a Cure: Aspects of Health and Medicine in Launceston*, edited by Paul Richards, 7–25. Launceston, UK: Myola House of Publishing.

May, James. 2013. "Medical Aspects of the Development of Scurvy: Past and Present." *Journal for Maritime Research* 15 (1): 95–105.

Maynwaringe, Everard. 1666. *Morbus Polyrhizos et Polymorpheaeus: A Treatise of the Scurvy*. 2nd ed. London: S. Thompson.

McCalman, Iain. 2014. *The Reef, A Passionate History*: New York: Scientific American and Farrar, Straus, and Giroux.

McLeod, A. D. 1983. "Calenture—Missing at Sea?" *British Journal of Medical Psychology* 56: 347–50.

Mead, Richard. 1794. *A Discourse on the Scurvy*. In *An Historical Account of a New Method for extracting the foul Air out of Ships … a friend, by Samuel Sutton*, 99–134. London: J. Brindley.

Melville, Herman. 1847. *Omoo: A Narrative of the South Seas*. London: John Murray.

———. 1972. *Moby-Dick; or, The Whale*. Edited by Harold Beaver. Harmondsworth: Penguin.

Melville, Thomas. N.d. "Journal of a Voyage in the Speedy, Whaler." Mitchell Library NSW, ML MS Q 36 (FM4/2223).

Milton, John. 1958. *The Poetical Works of John Milton*. Edited by Helen Darbishire. Oxford: Oxford University Press.

Mitchel, John. 1864. *Jail Journal; or, Five Years in British Prisons*. Dublin: James Corrigan.

———. 1868. *Jail Journal; or, Five Years in British Prisons*. New York: P. M. Haverty.

Mitchell, Mathew. 1740–43. "Log of the *Gloucester*." TNA ADM 5/402.

Mitchell, Thomas L. 1835. *Three Expeditions into the Interior of Australia*. 2 vols. London: T. and W. Boone.

Mitchill, Samuel. 1795. *Remarks on the Gaseous Oxyd of Azote or Nitrogene*. New York.

———. 1815. "The Phenomenae of the human Mind which belong to the … State … intermediate between waking and sleeping." In *Remarkable Sermons of Rachel Baker*, 124–40. London: E. Cox.

Mitford, Nancy. 1947. *The Pursuit of Love*. Harmondsworth: Penguin.

Mollon, John. 2013. Personal communication concerning tritanopia (July 21).

Montaigne, Michel de. 1711. *Essays*. Translated by Charles Cotton. 3 vols. London: Daniel Brown, J. Nicholson, R. Wellington, B. Tooke, B. Barker, G. Strahan, R. Smith, and G. Harris.

More, Thomas. 1898. *Utopia*. Translated by Raphe Robynson. London: J. M. Dent.

Morrell, Benjamin. 1832. *A Narrative of Four Voyages in the South Sea, 1822–31*. New York: J. and J. Harper.

Morrison, James. 1935. *The Journal of James Morrison*. Edited by Owen Rutter. London: Golden Cokerel Press.

Nagle, Jacob. 1988. *Journal*. Edited by John C. Dann. New York: Weidenfeld & Nicolson.

Neville, Henry. 2011. *The Isle of Pines*. Edited by John Scheckter. Farnham: Ashgate.

Newton, Isaac. 2003. *Opticks*. New York: Prometheus Books. O'Reilly, John Boyle. (1880) 1975. *Moondyne*. Adelaide: Bigby.

Orwell, George. 1981. *1984*. Harmondsworth: Penguin Putnam.

Ovid. 2000. *Metamorphoses*. Translated by Arthur Golding. Edited by John Frederick Nims. Philadelphia: Paul Dry.

Oxley, John. 1820. *Journals of Two Expeditions into the Interior of New South Wales 1817–18*. London: John Murray.

Park, Mungo. 2000. *Travels in the Interior Districts of Africa*. Edited by Kate Ferguson Marsters. Durham, NC: Duke University Press.

Pearn, John, and Catherine O'Carrigan. 1983. *Australia's Quest for Colonial Health*. Brisbane: Department of Child Health.

Peron, Francois. 1810. *A Voyage of Discovery to the Southern Hemisphere, 1801–4*. Vol. 11 of *A Collection of Modern and Contemporary Voyages and Travels*. London: Richard Phillips.

———. 2014. "Memoir on the English Settlements in New Holland." In Jean Formasiere and John West, *French Designs on Colonial New South Wales*, 125–200. Adelaide: Friends of the State Library of South Australia.

Mrs. Pexton. 1817. "Account of a Voyage on Board the Ship Pilot, Captain W. Pexton, to Port Jackson." Sydney: National Maritime Museum Library.

Phillip, Arthur. 1790. *The Voyage of Governor Phillip to Botany Bay*. 3rd ed. London: John Stockdale.

Pope, Alexander. 1787. *Essay on Man*. In *The Works of Alexander Pope*, 6 vols., edited by William Warburton, 2.37–100. London: C. Bathurst et. al.

Poupart, Francois. 1721. "Strange Effects of the Scurvy at Paris." In Jones 1721, 170–73.

Price, John Washington. 2000. *The Minerva Journal, 1798–1800*. Edited by Pamela Jeanne Fulton. Melbourne: Melbourne University Press.

Pridmore, Saxby. 1983. "A Note on the History of Disease in Tasmania." In Pearn and O'Carrigan 1983, 272–78.

Priestley, Joseph. 1790. *Experiments on Different Kinds of Air*. 3 vols. Birmingham: Thomas Pearson.

Pringle, Sir John. n.d. "Medical Annotations." Royal College of Physicians in Edinburgh, PRJ/1–9.

Purchas, Samuel. 1905–7. *Hakluytus Posthumus; or, Purchas His Pilgrimes*. 20 vols. Glasgow: James MacLehose and Sons.

Quigley, Killian. Forthcoming. "Epidemic Constitutions: Early Australia's Irish Plague." *Eighteenth-Century Literature*.

de Quiros, Pedro Fernandez. 1904. *The Voyages of Pedro Fernandez de Quiros*. 2 vols. Edited by Sir Clements Markham. London: Hakluyt Society.

Radcliffe, Ann. 1989. *The Mysteries of Udolpho*. Edited by Bonamy Dobree. Oxford: Oxford University Press.

Ravenstein, E. G., ed. 1898. *Journal of the First Voyage of Vasco da Gama 1497–99*. London: Hakluyt Society.

Reece, Bob, ed. 1993. *Irish Convict Lives*. Sydney: Crossing Press.

Reeve, Clara. [1785] 1970. *The Progress of Romance*. 2 vols. Colchester: W. Keymer; repr. (2 vols. in one). New York: Garland.

Reid, Thomas. 1822. *Two Voyages to New South Wales and Van Diemen's Land*. London: Longman et al.

———. 1997. *An Inquiry into the Human Mind on the Principles of Common Sense*. Edited by Derek R. Brookes. University Park: Pennsylvania State University Press.

*Report from the Select Committee of the House of Commons on Transportation*. 1812. London: H. M. Stationery Office.

———. 1837. London: H. M. Stationery Office.

*Report of the Commissioners of His Majesty's Navy*. 1792. London: Navy Office.

*Report of the Committee of the Admiralty into the Causes of the Outbreak of Scurvy in the Recent Arctic Expedition*. 1877. London: H. M. Stationery Office.

*Report of the Select Committee of the House of Commons relative to the ... Convict Establishment*. 1799. London: R. Shaw.

Rodger, N. A. M. 2004. *The Command of the Ocean*. London: Allen Lane and the National Maritime Museum.

Roggeveen, Jacob. 1970. *The Journal of Jacob Roggeveen*. Edited and translated by Andrew Sharp. Oxford: Clarendon Press.

Rosenthal, Michael. "Artless Landscape." Forthcoming.

Ross, Lynette. 1995. "Death and Burial at Port Arthur, 1830–77." BA Thesis, University of Tasmania (Hobart).

Roth, Michael S. 1993. "Rethinking Nostalgia." In *Home and Its Dislocations in 19th Century France*, edited by Suzanne Nash. New York: SUNY Press.

Rousseau, Jean-Jacques. 1803. *Eloisa: or a Series of Letters*. 4 vols. London: Vernor & Hood; Longman & Rees.

———. 1953. *Confessions*. Translated by J. M. Cohen. London: Penguin.

———. 2004. *Reveries of the Solitary Walker*. Translated by Peter France. London: Penguin.

*Rules and Regulations for the Penal Settlement of Tasman's Peninsula*. 1868. Port Arthur, Tasmania: W. Fletcher.

Russell, Gillian. 2011. *The Playbill and Its People*. Canberra: National Library of Australia.

Russell, John. 1831. Letter, Port Arthur. CSOI-483–10748. 26 April.

Rymer, James. 1793. *An Essay upon the Scurvy*. London: J. Rymer.

Saumarez, Philip. 1739–43. *Centurion* Log. NMM ADM/L/C/ 301.

Savery, Henry. 1832. *Quintus Servinton: A Tale Founded upon Incidents of Real Occurrence*. 3 vols. Hobart Town: Henry Melville; London: Smith Elder.

Schaffer, Simon. 1998. "Regeneration." In *Science Incarnate*, edited by Christopher Lawrence and Steven Shapin. Chicago: University of Chicago Press.

Scheckter, John. 1998. *The Australian Novel, 1830–1980*. New York: Peter Lang.

Schmidgen, Wolfram. 2013. *Exquisite Mixture: The Virtues of Impurity in Early Modern England*. Philadelphia: University of Pennsylvania Press.

"Scorbut." 1765. *Encyclopedie*. Vol 3 of 5. Neufchastel: Samuel Faulche.

Scoresby Jr., William. 1820. "On the Colour of the Greenland Sea," *Edinburgh Philosophic Journal*. Vol. 2. Edinburgh: Constable.

Scott, Robert Falcon. 1907. *The Voyage of the "Discovery."* 2 vols. London: Smith, Elder.

———. 1947. *Scott's Last Expedition*. Edited by Leonard Huxley. 2 vols. London: John Murray.

Shapin, Steven. 1994. *A Social History of Truth*. Chicago: University of Chicago Press.

———. 2010. *Never Pure*. Baltimore: Johns Hopkins University Press.

Shaw, A. G. L. 1966. *Convicts & the Colonies: A Study of Penal Transportation from Great Britain & Ireland to Australia*. London: Faber and Faber.

Shelvocke, George. 1726. *A Voyage round the World by way of the Great South Sea*. J. Senex.

Smith, Bernard, and Alwyne Wheeler, eds. 1988. *The Art of the First Fleet, and Other Early Australian Drawings*. New Haven, CT: Yale University Press.

Sparrman, Anders. 1944. *A Voyage round the World*. London: Golden Cockerel.

———. 1956. Cited in Bernard Smith. "Coleridge's Ancient Mariner and Cook's Second Voyage." *Journal of the Warburg and Courtauld Institute* 19 (1–2): 138.

Spinoza, Baruch. 1993. *Ethics*. Translated by A. Boyle and G.H.R. Parkinson. London: Everyman.

Sprat, Thomas. 1667. *The History of the Royal-Society of London*. London: J. Martyn.

Stalnaker, Joanna. 2010. *The Unfinished Enlightenment*. Ithaca, NY: Cornell University Press.

Staniforth, Mark. 1996. "Diet, Disease and Death at Sea on the Voyage to Australia, 1837–39." *International Journal of Maritime History* 8 (2): 119–56.

Stanley, Thomas. 1687. *The History of Philosophy*. London: Thomas Bassett.

Starobinski, Jean. 1966. "The Idea of Nostalgia." *Diogenes* 54: 81–103.

Steller, George Wilhelm. 1988. *Journal of a Voyage with Bering, 1741–42*. Edited by O. W. Frost. Translated by Margritt Engel and O. W. Frost. Stanford, CA: Stanford University Press.

Sterne, Laurence. 1992. *The Life and Opinions of Tristram Shandy*. Edited by Ian Campbell Ross. Oxford: Oxford University Press.

Stevenson, Robert Louis. 1922. "The Paumotus: Atolls at a Distance." In *The Works of Robert Louis Stevenson*. Vailima edition. Vol. 16. London: William Heinemann.

Stock, John Edmonds. 1811. *Memoirs of Thomas Beddoes MD*. London: John Murray.

Sturt, Charles. 2002. *Journals of the Central Australian Expedition of 1844–1846*. Edited by Richard C. Davis. London: Hakluyt Society.

Sutton, Samuel. 1799. *An Historical Account of a New Method for extracting the foul Air out of Ships . . . a friend, by Samuel Sutton*. London: J. Brindley.

Swift, Jonathan. 1958. *The Poems of Jonathan Swift*. Edited by Harold Williams. 3 vols. Oxford: Clarendon Press.

———. 1995. *Gulliver's Travels*. Edited by Christopher Fox. Bedford: Bedford Books of St. Martin's Press.

Tench, Watkin. 1789. *A Narrative of the Expedition to Botany Bay*. London: J. Debrett.

———. 1793. *A Complete Account of the Settlement at Port Jackson in New South Wales*. London: G. Nicholl and J. Sewell.

Thomas, Pascoe. 1745. *Journal of a Voyage to South Seas in His Majesty's Ship the Centurion*. London: S. Birt, J. Newbery, J. Collyer.

Thomson, John. 1839. *The Life, Lectures, and Writings of William Cullen*. 2 vols. Edinburgh: William Blackwood.

Thurmond, John. 1774. *Harlequin Sheppard: A Night-Scene in Grotesque Character*. London: J. Roberts.

Trotter, Thomas. 1792. *Observations on the Scurvy*. London: T. Longman.

———. 1795. *Additional Observations on Scurvy*. In *Medical and Chemical Essays*, 3 vols., 1.5–118. London: J. S. Jordan.

———. 1804. *Medicina Nautica: An Essay on the Diseases of Seamen*. 3 vols. 2nd ed. London: Longman.

———. 1807. *A View of the Nervous Temperament*. London: Longman, Hurst, Rees & Orme.

———. 1812. *A View of the Nervous Temperament*. Newcastle: Edward Walker.

Turnbull, John. 1805. *A Voyage round the World*. 3 vols. London: Richard Phillips.

*The Universal Spectator* (25 August) 1744. "Dialogue between an officer of the Centurion and a friend." In Williams 1967, 241–43.

Vancouver, George. 1984. *A Voyage of Discovery in the North Pacific Ocean and round the World*. 4 vols. Edited by W. Kaye Lamb. London: Hakluyt Society.

van der Merwe, Pieter. 2006. " 'Icebergs' and Other Recent Discoveries in Paintings from Cook's Second Voyage by William Hodges." *Journal for Maritime Research* 8 (1): 34–45.

Vaux, James Hardy. 1819. *A New and Comprehensive Vocabulary of the Flash Language, Bound with Memoirs of James Hardy Vaux, Written by Himself*. 2 vols. London: W. Clowes.

Vickers, Neil. 2004. *Coleridge and the Doctors*. Oxford: Clarendon Press.

Wakefield, Edward Gibbon. 1829. *A Letter from Sydney*. London: Joseph Cross et al.

Wales, William. 1772–74. MS Journal transcript. Mitchell Library. CY Safe 1/84.

Walker, Robin, and Dave Roberts. 1988. *From Scarcity to Surfeit: A History of Food and Nutrition in New South Wales*. Kensington: New South Wales University Press.

Walpole, Horace. (1766) 1964. "An Account of the Giants lately discovered." In *Byron's Journal of his Circumnavigation 1764–1766*, edited by Robert E. Gallagher, 200–209. Cambridge: Cambridge University Press for the Hakluyt Society.

———. 1937–83. *Correspondence*. Edited by W. S. Lewis. 45 vols. New Haven, CT: Yale University Press.

Walter, Richard. 1838. *A Voyage Round the World in the Years 1740–44*. London: SPCK.

Warren, Christopher. 2013. "John Milton and the Epochs of International Law." *European Journal of International Law* 24 (2): 557–81.

Watling, Thomas. 1793. *Letters from an Exile in Botany-Bay to his Aunt in Dumfries*. Penrith: Ann Bell.

Watt, James. 1979. "Medical Aspects and Consequences of Cook's Voyages." In *Captain Cook and His Times*, edited by Robin Fisher and Hugh Johnston. Camberra: Australian National University Press.

———. 1998. "The Medical Bequest of Disaster at Sea." *Journal of the Royal College of Physicians* 32 (6): 572–79.

Watt, James, E. J. Freeman, and W. F. Bynum. 1981. *Starving Sailors*. Bristol: National Maritime Museum.

Weaver, Paul. 1993. "The Voyages of the Robert Small and Phoebe Dunbar to Fremantle in 1853." In *Irish Convict Lives*, edited by Bob Reece. Sydney: Crossing Press.

Weddell, James. 1827. *A Voyage towards the South Pole 1822–24*. London: Longman.

Wertenbaker, Timberlake. 2015. *Our Country's Good*. London: Bloomsbury Methuen Drama.

Westall, William. 1962. *Drawings by William Westall*. 1962. Edited by T. M. Perry and Donald H. Simpson. London: Royal Commonwealth Society.

White, John. 1790. *Journal of a Voyage to New South Wales*. London: J. Debrett.

Whitehead, Alfred North. 1967. *Science and the Modern World*. New York: Free Press.

Williams, Bernard. 1978. *The Project of Pure Enquiry*. London: Routledge.

Williams, Elizabeth A. 2012. " 'Digestive Force' and the Nature of Morbidity in Vitalist Medicine." In *Vital Matters: Eighteenth-Century Views of Conception, Life, and Death*, edited by Helen Deutsch and Mary Terrall. Toronto: University of Toronto Press.

Williams, Glyndwr. 1967. *Documents Relating to Anson's Voyage 1740–44*. Vol. 109. London: Navy Records Society.

———. 1999. *The Prize of all the Oceans*. London: Harper Collins.

———. 2009. *Arctic Labyrinth: The Quest for the Northwest Passage*. Berkeley: University of California Press.

———. 2013. *Naturalists at Sea: From Dampier to Darwin*. New Haven, CT: Yale University Press.

Willis, Thomas. 1681. *The Anatomy of the Brain*. In *The Remaining Medical Works*, translated by Samuel Pordage, 55–127. London: T. Dring et al.

———. 1683. *Two Discourses concerning the Soul of Brutes*. Translated by Samuel Pordage. London: Thomas Dring.

————. 1684. *A Tract of the Scurvy*. In *Dr. Willis's Practice of Physick*, translated by
  Samuel Pordage. London: T. Dring et al.
Wilson, Catherine. 2008. *Epicureanism at the Origins of Modernity*. Oxford:
  Oxford University Press.
Wilson, Edward. 1967. *Diary of the Discovery Expedition 1901–4*. Edited by Ann
  Savours. New York: Humanities Press.
Woodall, John. 1617. *The Surgeon's Mate*. London.
Wordsworth, William. 1984. *William Wordsworth: Major Works*. Edited by
  Stephen Gill. Oxford: Oxford University Press.
Worgan, George. 2009. *Sydney Cove Journal 20 January–11 July 1788*. Edited by
  John Curly. Malvern, NSW: Colony Press for the Banks Society.
Worrall, David. 2007. *Harlequin Empire*. London: Pickering and Chatto.

# INDEX